Frontiers in Magnetic Resonance

(*Volume 1*)

Electron Paramagnetic Resonance in Modern Carbon-Based Nanomaterials

Edited by

Dariya Savchenko

Department of Analysis of Functional Materials, Division of Optics, Institute of Physics CAS, Prague, Czech Republic
Department of Physics and Solid State Physics, National Technical University of Ukraine, "Igor Sikorsky Kyiv Polytechnic Institute", Kyiv, Ukraine

&

Abdel Hadi Kassiba

Institute of Molecules and Materials, UMR-CNRS, Le Mans University, Le Mans, France

Frontiers in Magnetic Resonance

Volume # 1

Electron Paramagnetic Resonance in Modern Carbon-Based Nanomaterials

Editors: Dariya Savchenko and Abdel Hadi Kassiba

ISSN (Online): 2589-708X

ISSN (Print): 2589-7071

ISBN (Online): 978-1-68108-693-4

ISBN (Print): 978-1-68108-694-1

Bentham Science Publishers Ltd.
Executive Suite Y - 2
PO Box 7917, Saif Zone
Sharjah, U.A.E.
Email: subscriptions@benthamscience.org

BENTHAM SCIENCE

CONTENTS

FOREWORD

Carbon based materials include nanographites, conducting carbon nanomaterials, carbon nanotubes, graphene oxides, nanodiamonds, hybrids like carbon nanotubes embedded into polymer composites or functionalized molecular groups. These compounds are being implemented in multiple architectures with versatile chemical bonding, organization and morphologies leading to the unique physical properties such as exceptional electrical, thermal, structural dependent dimensionalities, mechanical and tribological performances. For instance, pure diamond is an excellent electrical insulator while some graphite based materials are more or less good electrical conductors, depending on their composition and pre-treatment. Carbon and graphite foams are very good thermal insulators, even at very high temperatures. On the other hand, diamond is used for the heat sink in electronics due to its very high thermal conductivity. Mechanical properties of carbon materials also differ considerably, depending on the type of the material. Since carbon allotropes and hybrids materials may contain different chemical bonding, multi-functional compounds can be tailored for various applications in nanoelectronics, integrated optoelectronics, energy storage and conversion, sensors, biomedicine, *etc.*, being both already implemented in working devices and currently under development.

Intrinsic electronic features originating from doping or structural defects critically contribute to physical properties of carbon-based materials. These features may be exhaustively characterized by various electron magnetic resonance techniques including continuous wave (CW) or pulse electron paramagnetic resonance (EPR), electron nuclear double resonance (ENDOR) and other advanced electron magnetic resonance methods. Using a variety of complementary EPR techniques provides detailed insight into the local environment and the electronic peculiarity of defect structures in carbon-based systems. Moreover, extraordinary sensitivity of EPR techniques to the relaxation times (both spin-lattice and spin-spin) of paramagnetic species as well as the capability of selective control and detection of defects paves the way for the understanding of spin dynamics, which is extremely important in quantum computation or implementations of non-volatile memory devices. Thus, EPR techniques, based on different instrumental functionalities and methodologies, due to their specific window of time scales allow probing structural and electronic features of intrinsic and engineered spin systems in carbon based materials with the aim to open new challenges toward advanced and emerging technologies.

Dr. Alexander I. Shames
Laboratory of Magnetic Resonance,
Department of Physics
Faculty of Natural Sciences
Ben-Gurion University of the Negev
Be'er-Sheva
Israel

PREFACE

Volume 1 of *Frontiers in Magnetic Resonance* comprises 13 chapters on topics of high importance in the field of electron paramagnetic resonance study of carbon-containing nanomaterials. The topics and authors were selected from recently published papers in highly cited journals (Nat. Commun., Sci. Rep., J. Mater. Chem. C, Phys. Chem. Chem. Phys., Phys. Rev. B, Appl. Phys. Lett., J. Appl. Phys., Chem. Phys. Lett., Phys. Status Solidi B, Appl. Magn. Reson., *etc.*).

The first chapter by Prof. S.K. Misra will give the reader the fundamentals of EPR spectroscopy in regards to its application to the carbon-containing materials. The focus of chapter 2 by Dr. A. Barbon *et al.* is set to the resolution of the EPR signals attributable to different species, or structures, that are present in graphite and graphene-like materials. Chapter 3 by Prof. L. Forró *et al.*, presents the ESR characterization of spin dynamics of conducting carbon nanomaterials. In chapter 4, by Dr. V. Likodimos *et al.*, the EPR spectroscopy is exploited to investigate spin dynamics of DWCNTs and composites of oxidized MWCNTs embedded in an elastomeric poly(ether-ester) block copolymer. Chapter 5 by Dr. rer. nat. E. Erdem *et al.,* focuses on discussing EPR investigations on graphene oxide, reduced graphene oxide, and carbon nanotubes with different chemical functionalities. In Chapter 6, Dr. K. Marumoto reviews the ESR spectroscopy of semiconducting single-walled CNT thin films and their transistors. Chapter 7 by Dr. S. Ruthstein *et al.* describes the findings on the oxygenation processes of coal and graphene materials using in-situ EPR experiments at various atmospheric environments. In Chapter 8, Prof. A.B. Więckowski *et al.* describe the application of the two-temperature EPR measurement method to carbonaceous solids. In Chapter 9, Prof. P. Baranov *et al.* review the characterization of impurities in nanodiamonds by means of multifrequency CW and pulse EPR techniques. Chapter 10 by Dr.Sc. S. Andronenko *et al.* shows the application of multifrequency EPR to the study of SiCN nanoceramics. In Chapter 11, Prof. A. Kassiba *et al.* discuss the study of paramagnetic centers in SiC nanomaterials by means of CW and pulse EPR techniques. In Chapter 12, Dr. D. Savchenko reviews the size effects observed in EPR and ENDOR spectra of SiC nanoparticles and nanodiamonds. Finally, in Chapter 13, DrSc. E. Kalabukhova *et al.* review the EPR study of paramagnetic defects in amorphous $a\text{-}Si_{1-x}C_x$:H and $a\text{-}SiC_xN_y$ thin films.

We would like to express our gratitude to all the authors for their excellent contributions. We would also like to thank the entire team of Bentham Science Publishers, particularly Mr. Shehzad Naqvi (Senior Manager Publication) and Dr. Faryal Sami (Assistant Manager Publications), for their excellent efforts. We are confident that this volume will receive wide appreciation from students and researchers.

Dr. Dariya Savchenko
Department of Analysis of Functional Materials,
Division of Optics, Institute of Physics CAS, Prague
Czech Republic
Department of Physics and Solid State Physics,
National Technical University of Ukraine
"Igor Sikorsky Kyiv Polytechnic Institute", Kyiv
Ukraine

Prof. Abdel Hadi Kassiba
Institute of Molecules and Materials, UMR-CNRS,
Le Mans University, Le Mans,
France

List of Contributors

Abdel Hadi Kassiba Institute of Molecules and Materials, UMR-CNRS, Le Mans University, 72085 Le Mans, France

Andrzej B. Więckowski Institute of Physics, Faculty of Physics and Astronomy, University of Zielona Góra, Szafrana 4a, 65-516 Zielona Góra, Poland
Institute of Molecular Physics of the Polish Academy of Sciences, Smoluchowskiego 17, 60-179 Poznań, Poland

Andreas Pöppl Institute of Experimental Physics II, Faculty of Physics and Earth Sciences, Leipzig University, Leipzig, 04103, Germany

Angeliki Diamantopoulou Section of Solid State Physics, Department of Physics, National and Kapodistrian University of Athens, Panepistimiopolis, 15 784, Greece

Anna Szymczyk Institute of Physics, West Pomeranian University of Technology, Al. Piastow 48, 70-311 Szczecin, Poland

Antonio Barbon Department of Chemical Sciences, University of Padova, Padova, Italy

Bálint Náfrádi Laboratory of Physics of Complex Matter (LPMC), Ecole Polytechnique Fédérale de Lausanne, CH-1015 Lausanne, Switzerland

Chuyen V. Pham Laboratory for MEMS Applications, Department of Microsystems Engineering - IMTEK, University of Freiburg, Georges-Koehler-Allee 103, D79110 Freiburg, Germany

Dariya Savchenko Department of Analysis of Functional Materials, Division of Optics, Institute of Physics of the Czech Academy of Sciences, Prague, 182 00, Czech Republic
Department of Physics and Solid State Physics, National Technical University of Ukraine “Igor Sikorsky Kyiv Polytechnic Institute”, Kyiv, 03056, Ukraine

Emre Erdem Institute of Physical Chemistry, University of Freiburg, Albertstr. 21, 79104 Freiburg, Germany

Francesco Tampieri Department of Chemical Sciences, , University of Padova, Padova, Italy

George Mamin Department of Quantum electronics and radiospectroscopy, Kazan Federal University, Kazan, 420000, Russian Federation

Grzegorz P. Słowik Institute of Physics, Faculty of Physics and Astronomy, University of Zielona Góra, Szafrana 4a, 65-516 Zielona Góra, Poland

Grzegorz Zolnierkiewicz Institute of Physics, West Pomeranian University of Technology, Al. Piastow 48, 70-311 Szczecin, Poland

Kazuhiro Marumoto Division of Materials Science, University of Tsukuba, Tsukuba, Ibaraki 305-8573, Japan
Tsukuba Research Center for Interdisciplinary Materials Science (TIMS), University of Tsukuba, Tsukuba, Ibaraki 305-8571, Japan

Mohammad Choucair School of Chemistry, University of Sydney, Sydney, New South Wales 2006, Australia

Michael Krueger Carl-von-Ossietzky University Oldenburg, Institute of Physics, Carl-von-Ossietzky Str. 9-11, D-26129 Oldenburg, Germany

Nikolaos Guskos	Institute of Physics, West Pomeranian University of Technology, Al. Piastow 48, 70-311 Szczecin, Poland
Ortal Marciano	Bar-Ilan University, Faculty of Exact Sciences, Department of Chemistry, 5290002, Ramat Gan, Israel
Pavel Baranov	Microwave Spectroscopy of Crystals Laboratory, Ioffe Institute, RAS, St Petersburg, 194021, Russian Federation
Sergey I. Andronenko	Department of Physics, Concordia University, Montreal, Canada, H3G 1M8 Institute of Physics, Kazan Federal University, Kazan, 420008, Russian Federation
Sergej Repp	Institute of Physical Chemistry, University of Freiburg, Albertstr. 21, 79104 Freiburg, Germany
Sergei Orlinskii	Department of Quantum electronics and radiospectroscopy, Kazan Federal University, Kazan, 420000, Russian Federation
Sharon Ruthstein	Bar-Ilan University, Faculty of Exact Sciences, Department of Chemistry, 5290002, Ramat Gan, Israel
Spyridon Glenis	Section of Solid State Physics, Department of Physics, National and Kapodistrian University of Athens, Panepistimiopolis, 15 784, Greece
Sushil K. Misra	Department of Physics, Concordia University, Montreal, H3G 1M8, Canada
Vlassis Likodimos	Section of Solid State Physics, Department of Physics, National and Kapodistrian University of Athens, Panepistimiopolis, 15 784, Greece
Victor Soltamov	Microwave Spectroscopy of Crystals Laboratory, Ioffe Institute, RAS, St Petersburg, 194021, Russian Federation Department of Quantum electronics and radiospectroscopy, Kazan Federal University, Kazan, 420000, Russian Federation

CHAPTER 1

Fundamentals of Electron Paramagnetic Resonance in Modern Carbon-based Materials

Sushil K. Misra*

Department of Physics, Concordia University, Montreal, H3G 1M8, Canada

Abstract: The advantages of using multifrequency Electron Paramagnetic Resonance (EPR) in studying carbon-based materials are discussed. The details of designing continuous-wave EPR spectrometers operating at different frequencies are presented. Designs of CW and pulse Electron Nuclear Double Resonance (ENDOR) spectrometers, which are very important techniques for studying precisely hyperfine interactions and local environment of paramagnetic ions in carbon-based materials are included. Analysis of EPR spectra, spin Hamiltonians, EPR lineshapes, evaluation of spin-Hamiltonian parameters, and simulation of single-crystal and powder spectra are also explained. A short review of carbon-based materials studied by EPR is given.

Keywords: Carbon-based materials, Continuous Wave EPR, Davies ENDOR, Electron spin echo (ESE), Electron Spin Echo Envelope Modulation (ESEEM), Evaluation of spin Hamiltonian parameters, Electron Nuclear Double Resonance (ENDOR), EPR, EPR lineshape, EPR spectrometer, High-frequency spectrometers, Hyperfine interaction, Hyperfine Sublevel Correlation Spectroscopy (HYSCORE), Mims ENDOR, Pulse EPR, Pulse ENDOR, Simulation of EPR spectrum, Spin Hamiltonian, Zeeman effect.

ELECTRON ZEEMAN EFFECT: EPR RESONANCE CONDITION

In electron paramagnetic resonance (EPR), one observes the resonant absorption of microwave (mw) energy by an unpaired electron making a transition from a lower-energy state to a higher-energy state in the presence of an external magnetic field (The term EPR will be used throughout this chapter, although electron spin resonance (ESR) and electron magnetic resonance (EMR) are also used in the literature). These energy levels are due to the interaction of the electronic magnetic moment with the applied external magnetic field. An unpaired electron is equivalent to a small bar magnet due to its magnetic moment.

* **Corresponding author Sushil K. Misra:** Department of Physics, Concordia University, Montreal, H3G 1M8, Canada ; Tel: +01-514-482-3690; E-mail: sushil.misra@concordia.ca

If it is aligned with B_{ext}, then its energy is lower than when it is aligned opposite to the direction of B_{ext}, as shown in Fig. (**1**). This effect is called the Zeeman Effect.

The unpaired electron possesses the spin 1/2, so its lower and higher energy states are designated by the electronic magnetic quantum numbers $M_S = -1/2$ and $M_S = +1/2$, respectively. Then expressed for the two values of M_S the energies E of an unpaired electron in an external magnetic field, B_{ext}, are:

$$E = g\mu_B M_s B_{ext} = \pm\frac{1}{2} g\mu_B B_{ext} \quad \textbf{(1)}$$

where, the dimensionless constant, g, termed as g-factor, is expressed in terms of the gyromagnetic ratio, γ, as $\gamma = g\mu_B/\hbar$, where μ_B is the Bohr magneton (=9.274 × 10^{-24} J/T; 1 Tesla [T] = 10.000 Gauss) and $\hbar$ is the reduced Plank's constant (= $h/2\pi$, with h = 6.626 x 10^{-34} J×s being Planck's constant).

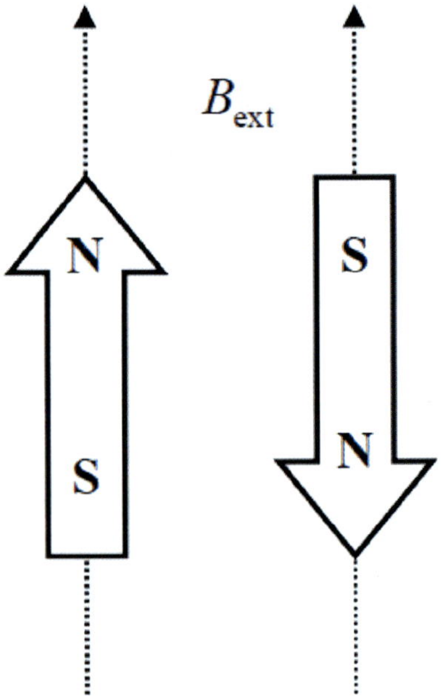

Fig. (1). Lower (left) and higher (right) possible orientations of an unpaired electron's magnetic moment in the external magnetic field, B_{ext} (Adapted from [1] with the Permission from John Wiley and Sons).

Under the action of an oscillating mw radiation of frequency υ_0 in Hertz, the resonance condition is satisfied when

$$B_{ext} = B_0, \quad \textbf{(2)}$$

where, $B_0 = h\upsilon_0/g\mu_B$, so that the energy level separation between the two energy levels represented by $M_S = \pm 1/2$, is equal to $h\upsilon_0$. In that case, resonant absorption of mw radiation will take place for $B_{ext} = B_0$. This is shown in Fig. (**2**).

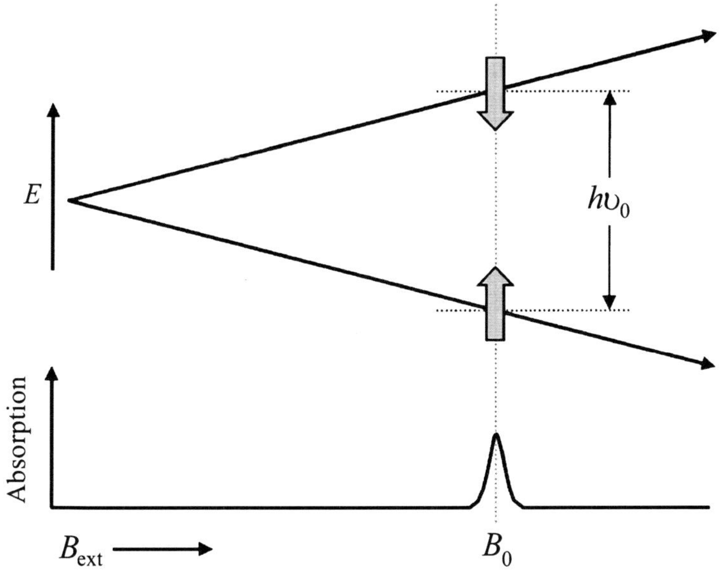

Fig. (2). A plot showing the splitting due to Zeeman effect by the external magnetic field B_{ext} (neglecting hyperfine splitting and higher order zero-field splitting terms) and resonant absorption of mw radiation as the magnetic field is swept; adapted from [1] with the Permission from John Wiley and Sons).

HYPERFINE SPLITTING

The energy of the unpaired electron is influenced by its local surrounding. For instance, if there is a nucleus possessing magnetic moment near an unpaired electron, the magnetic field seen by the unpaired electron at its site would change due to the interaction between the electron and nuclear spin, known as hyperfine interaction (HFI) [2]. In that case, the external magnetic field required to satisfy the resonance condition given by eq. (2) will shift from B_0 accordingly. Therefore, if the effective local magnetic effect, B_1, depending on the HFI constant, of the nucleus on the unpaired electron is in the direction of B_{ext}, the resonance will occur at $B_{ext} < B_0$, and if B_1 opposes B_{ext} at the location of the unpaired electron, the absorption will occur at $B_{ext} > B_0$, for a given υ_0 (more details on HFI terms in the spin Hamiltonian are given below).

EPR LINE SHAPES

There are three types of line shapes, *i.e.* EPR absorption line intensity versus the magnetic field, usually observed in EPR. They are (i) *Lorentzian*, (ii) *Gaussian*, and (iii) *Dysonian.* These line shapes are expressed as:

$$y = \frac{a}{\left(1 + b(x - x_0)^2\right)}; \quad \textbf{(3)}$$

$$y = a\exp\left(-b\left(x - x_0\right)^2\right), \quad \textbf{(4)}$$

and

$$y = \frac{\Delta x + \alpha\left(x - x_0\right)}{\left(x - x_0\right)^2 + \Delta x^2} + \frac{\Delta x + \alpha\left(x + x_0\right)}{\left(x + x_0\right)^2 + \Delta x^2} \quad \textbf{(5)}$$

respectively, where y is the line intensity; x is the magnetic field; x_0 is the resonance field; α is the fraction of the dispersion component added into the absorption signal for Dysonian line shape [3], and Δx is the linewidth. One obtains for the first-derivative absorption signal an expression, which is proportional to $Y'_{\mathrm{max}}(\Delta B_{\mathrm{pp}})^2$, where $2Y'_{\mathrm{max}}$ is the amplitude of the peak-to-peak derivative, whereas ΔB_{pp} is the peak-to-peak linewidth.

When there exists superposition of many components which differ from spin to spin, the line shape becomes Gaussian; it is known as *inhomogeneous broadening*. In contrast, if there exists no variation the line shape from one spin to another spin, a *Lorentzian* line shape is observed; it is known as *homogeneous broadening*.

A mixture of the absorptive and dispersive components of the susceptibility produces an asymmetric *Dysonian* line shape. This happens due to the non-uniform distribution of the mw electromagnetic field when the sample size is bigger than the skin depth. In addition to this, the movement of the paramagnetic centers also produces the asymmetry. The two terms in the Dysonian lineshape expression above (eq. (5)) take into account the clockwise and the anticlockwise circular polarizations of the mw radiation [3, 4]. The theory of Dysonian line shapes due to conduction electrons in metals is very complex.

An experimental EPR spectrum for anthracite reported by Tadyszak *et al.* [5] illustrates best the Dysonian lineshape when the sample becomes conducting as depicted in Fig. (**3**), showing a plot of the asymmetry parameter *A*/*B*, versus the reduced dimension for $\lambda = d/\delta = 3.5$, where d is the sample thickness and δ is the skin depth. (Here A and B are the magnitudes of the amplitudes of the maxima of the first-derivative lineshape with respect to the baseline before and after the center of the lineshape). It can be seen from Fig. (**3**) that for smaller values of λ, the line becomes symmetric, *e.g.* for $\lambda = 0.2$. More details and references covering Dysonian line shape can be found in the treatise by Poole [6].

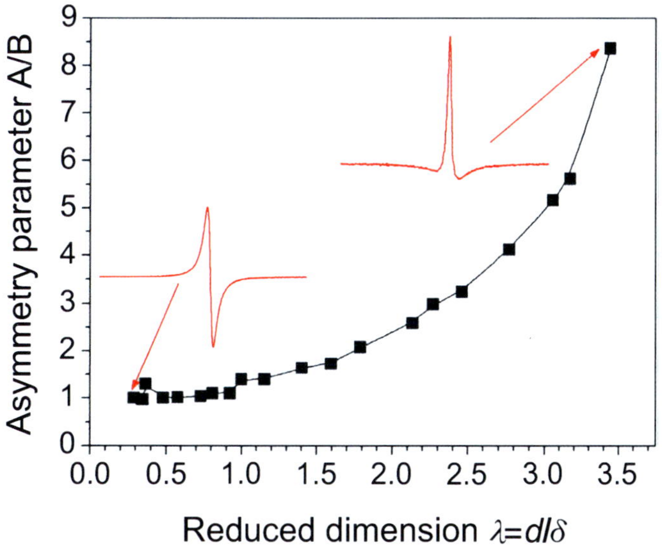

Fig. (3). A plot of the asymmetry parameter(A/B) *versus* the reduced dimension $\lambda = d/\delta$. here d and δ are, respectively, the sample thickness and the skin depth. (Adapted from [5]).

SPIN HAMILTONIAN

For a complete discussion, the reader is referred to Misra [7], the spin Hamiltonian (SH) is expressed as: $\mathcal{H} = \mathcal{H}_1 + \mathcal{H}_2$, where in zero magnetic field

$$\mathcal{H}_1 = \sum_{n=0,2,4,6} \sum_{m=0}^{n} B_n^m \hat{O}_n^m \tag{6}$$

Here n is an even integer, which can take values between 0 and 6 ($n \leq 2S$), for spin $S \leq 7/2$, and m lies in the range $0 \leq |m| \leq n$ for each n. The $\hat{O}_n^m$ are called equivalent spin operators for the electron. They, along with their matrix elements for various spin values ($\leq 7/2$) are listed in [7]. The coefficients are determined by the crystal field symmetry at the site of the transition metal ion. In addition, the

Zeeman terms depending on the external magnetic field, *B*, and the matrix, as well as the HFI terms depending on the HFI tensor are described by:

$$\mathcal{H}_2 = \mu_B(g_x B_x S_x + g_y B_y S_y + g_z B_z S_z) + A_x S_x I_x + A_y S_y I_y + A_z S_z I_z \quad \textbf{(7)}$$

In eq. (7), $g_{\|} = g_z$; $g_{\perp} = g_x = g_y$ and $A_{\|} = A_z$; $A_{\perp} = A_x = A_y$ when there is symmetry about the *z*-axis.

Some special cases relevant to carbon containing materials are described as follows:

(i) For spin S = 1/2, I = 1/2, the EPR spectra are analyzed using the spin Hamiltonian $\mathcal{H}_2$.

(ii) When quadrupole nuclear interactions become important for the case of nuclear spin $I > 1/2$, *e.g.* low-spin Co^{2+} ($S = 1/2, I = 7/2$), the spin Hamiltonian for axial symmetry becomes

$$\begin{aligned}\mathcal{H} = {} & \mu_B(g_x B_x S_x + g_y B_y S_y + g_z B_z S_z) + b_2^0[S_z^2 - S(S+1)/3] + \\ & + b_2^2(S_x^2 - S_y^2) + Q'[I_z^2 - I(I+1)/3] + Q''(I_x^2 - I_y^2).\end{aligned} \quad \textbf{(8)}$$

In the above equation Q', Q'' are the components parallel and perpendicular to the symmetry axis of the quadrupole interaction tensor, μ_n is the nuclear magneton, and g_n is the nuclear *g* factor.

(iii) For $S = 1$; $I > 1/2$, the applicable spin Hamiltonian is:

$$\begin{aligned}\mathcal{H} = {} & \mu_B(g_x B_x S_x + g_y B_y S_y + g_z B_z S_z) + b_2^0[S_z^2 - S(S+1)/3] + \\ & + b_2^2(S_x^2 - S_y^2) + Q'[I_z^2 - I(I+1)/3] + Q''(I_x^2 - I_y^2).\end{aligned} \quad \textbf{(9)}$$

In eq. (9), the parameters b_2^0 and b_2^2 are also denoted as *D* and *E*, respectively.

SIMULATION OF EPR SPECTRA

This section is devoted to simulation of EPR spectra, as adapted from Misra [8], where complete details are given. First, one needs to simulate single-crystal spectrum, which is then exploited to simulate polycrystalline spectrum.

Simulation of Single-crystal Spectrum

Simulation of EPR spectrum requires knowledge of both positions and intensities of EPR lines, as well as the line-shape function, $F(B_{ri}, B)$, where B_{ri} denotes the resonance line positions for the various possible transitions $i' \leftrightarrow i''$, which can be Gaussian, Lorentzian, or a Dysonian appropriate to the sample. It is easier to calculate the line positions for the orientations of *B* parallel to the symmetry axes X, Y, Z by perturbation expressions, but for an arbitrary orientation of *B*, the task becomes more difficult; see [8] for specific details. $P(i,\theta,\phi,v_c)$ is the weight given to the various lines for the ith transition, which takes place between the levels i' and i'', participating in resonance. Here, the mw frequency is v_c and the orientation of *B* over the unit sphere is (θ, ϕ).

Thus, the single-crystal simulated spectrum can be expressed as:

$$S(B,v_c) = \sum_i P(i,\theta,\phi,v_c)F(\omega_i, B) \quad \textbf{(10)}$$

In eq. (10), the summation is over all the lines observed for the particular orientation of *B*.

Transition Probability

The probability $P(i,\theta,\phi,v_c)$ of a transition, *i*, between the energy levels i' and i'' is expressed as follows:

$$P(i,\theta,\phi,v_c) \propto \left| \langle \Phi_{i'} | (B_{1x}S_x + B_{1y}S_y + B_{1z}S_z) | \Phi_{i''} \rangle \right|^2 . \quad \textbf{(11)}$$

In eq. (11), S_α and $B_{1\alpha}$; $(\alpha = x,y,z)$ represent the components of the electron spin operator, *S*, and the modulation r.f. field B_1. $|\Phi_{i'}\rangle$ and $|\Phi_{i''}\rangle$ are the eigenvectors of the spin-Hamiltonian (SH) matrix, $\mathscr{H}$, corresponding to the energy levels $E_{i'}$ and $E_{i''}$ participating in resonance [$\mathscr{H}|\Phi_k\rangle = E_k|\Phi_k\rangle$]. For more details of calculation see [8].

Lineshape Function $F(B_{ri}, B_k)$

The sum in Eq. (10) is performed to calculate the spectrum. $P(i,\theta_j,\phi_j,v_c)$ is considered centered at $B_r(i,\theta_j,\phi_j,v_c)$ and the lineshape function $F(B_{ri}, B_k)$, is taken to be extended over the magnetic-field interval $\pm\Delta B$, as a function of $\Delta B_{1/2}$ (full-width at half-maximum, FWHM), which is characteristic of the lineshape, *e.g.* $\Delta B = \pm 10\Delta B_{1/2}$, in order to obtain a good approximation for the Lorentzian lineshape.) The two most common line shapes are:

(i) Gaussian Lineshape, F_G*:*

$$F_G(B_{ri}, B_k) = K_G \exp\left[-(B_k - B_{ri})^2 / \sigma^2\right], \tag{12}$$

where B_{ri}, σ, and $K_G = (1/B_\Delta)(\ln 2/\pi)^{1/2}$ are the resonant field value for the ith transition, the linewidth and the normalization constant for the lineshape, respectively. In the expression for K_G, $B_\Delta = (1/2)\, B_{1/2}$, where $B_{1/2}$ is HWHM (half-width at half-maximum).

(ii) Lorentzian lineshape, F_L*:*

$$F_L(B_{ri}, B_k) = K_L \Gamma\left[\Gamma^2 + (B_k - B_{ri})^2\right]^{-1} \tag{13}$$

where Γ is the Lorentzian linewidth (HWHM = $(3)^{1/2}\Delta B_{pp}/2$, with ΔB_{pp} being the peak-to-peak first-derivative linewidth.

Simulation of a Polycrystalline Spectrum

In a polycrystalline material one can simulate the EPR spectrum by overlapping single-crystal spectra computed for external magnetic (Zeeman) field, *B*, over a large number of orientations (θ, ϕ) on the unit sphere weighted in proportion of $\sin\theta\, d\theta\, d\phi$, which takes into consideration the distribution of various crystallites, whose principal symmetry axes are oriented about (θ, ϕ) in the interval $d\theta$, $d\phi$. The simulated polycrystalline spectrum is then expressed as:

$$S(B, \nu_c) = \int_{\theta=0}^{\pi/2} \int_{\phi=0}^{2\pi} \sum_i P(i, \theta, \phi, \nu_c) F(B_{ri}, B) d(\cos\theta) d\phi \tag{14}$$

From eq. (14) it is seen the resonant field values for the various transitions must be known, along with their transition probabilities, to compute polycrystalline spectrum. The integrals in Eq. (14) are evaluated by summing over single-crystal spectra for magnetic-field orientations (θ_j, φ_j) over a grid of points covering the entire unit sphere with appropriate weight factors. See Ref [8] for more details.

Calculation of First-derivative EPR Spectrum

The experimental EPR spectrum is most often obtained as the first derivative with respect to the external magnetic field intensity of the absorbed mw power. This first-derivative spectrum is simulated by calculating the derivative with respect to B of $S(B,\nu_c)$, as given by eq. (14), as well as that of the lineshape. In particular, for the Gaussian and Lorentzian lineshapes, given by eqs. (12) and (13), respectively, the first-derivatives are:

$$\partial F_G\left(B_{ri},B_k\right)/\,\partial B_k = -2K_G\left(B_k - B_{ri}\right)\exp\left(-\left(B_k - B_{ri}\right)^2/\sigma^2\right)/\,\sigma^2 \quad \textbf{(15)}$$

$$\partial F_L\left(B_{ri},B_k\right)/\,\partial B_k = -2K_L\Gamma\left[\Gamma^2 + \left(B_k - B_{ri}\right)^2\right]^{-2}\left(B_k - B_{ri}\right) \quad \textbf{(16)}$$

The expression for the simulated first-derivative absorption spectrum, as obtained from eq. (14), is

$$\begin{aligned} F_c(B_k, \nu_c) &= \partial S(B_k, \nu_c)\,/\,\partial B_k = \\ &= C\sum_{i,\theta_j,\varphi_j} P(i, \theta_j, \varphi_j)\partial F(w_i, B_r(i, \varphi_j, \nu_c), B_k)\,/\,\partial B_k \sin\theta_j \end{aligned} \quad \textbf{(17)}$$

From eq. (17), one has for the two lineshapes, using eqs. (12) and (13):

Gaussian lineshape:

$$F_c\left(B_{k,}\ \nu_c\right) = $$

$$\sum_i N\left|\left\langle \Phi_{i'}\right|(B_{1x}S_x + B_{1y}S_y + B_{1z}S_z)\left|\Phi_{i''}\right\rangle\right|^2 \exp\left[-(B_k - B_{ri})^2/\sigma^2\right](B_k - B_{ri})/\sigma^2 \quad \textbf{(18)}$$

Lorentzian lineshape:

$$F_c\left(B_{k,}\ \nu_c\right) = $$

$$\sum_i N\left|\left\langle \Phi_{i'}\right|(B_{1x}S_x + B_{1y}S_y + B_{1z}S_z)\left|\Phi_{i''}\right\rangle\right|^2 \left(B_k - B_{ri}\right)\Gamma\left[\Gamma^2 + \left(B_k - B_{ri}\right)^2\right]^{-2} \quad \textbf{(19)}$$

In eqs. (18) and (19), the normalization constant, N, may be arbitrarily chosen. For example, $|F_c(B_k,\nu_c)|_{max} = 1$, obtained by dividing all values by $|F_c(B_k,\nu_c)|_{max}$, which is the largest magnitude of all the calculated values.

The details given above can be used to write a source code for simulation of a powder (polycrystalline) spectrum. On the other hand, the software Easyspin [9], downloadable from the Internet, is useful for simulations, provided that one has MATLAB available.

PULSE EPR

In this technique, mw pulses are applied over selected finite intervals of time to the sample, so that the orientation of the precessing magnetic moment can be turned by chosen angles about the x, y, or z axes around the external magnetic field (Larmor precession). (Conventionally, the z-axis is chosen to be parallel to the external magnetic field). These angles are usually multiples of 30° or 45°. To this end, one can use a single pulse or more pulses with chosen intervals of time in between, recording a signal referred to as echo signal that is proportional to the magnetization of the sample at appropriate intervals after the application of the last pulse. Consequently, several varieties of pulsed EPR techniques have been developed, *e.g.* electron spin echo (ESE), ESE envelope modulation (ESEEM), electron-electron double resonance (ELDOR), double quantum coherence (DQC), double electron-electron resonance (DEER), also referred to as pulse ELDOR (PELDOR) and ELDOR in ESE, two-dimensional Fourier transform EPR (2D-FTEPR), spin-echo correlated spectroscopy (SECSY), correlation spectroscopy (COSY), hyperfine spin-correlation spectroscopy (HYSCORE). Pulse EPR provides information that is not possible to obtain with CW EPR. For pulse EPR the instrumentation required is much more costly; as well, the analysis of the data is also much more complicated in comparison to that for CW EPR.

Nuclear Modulation Effects Leading to ENDOR and ESEEM

The energy levels of a paramagnetic system with an interacting electron spin (S = 1/2) and a magnetic nucleus of spin I = 1/2 are given in Fig. (**4**) [10, 11]. Two of the energy levels in which the z-component of the electron spin angular momentum component with $m_S = |+1/2\rangle$ is the α-manifold, whereas that with $m_S = |-1/2\rangle$ is the β-manifold, and there are two sublevels in this manifold with the nuclear spin angular momentum components $\pm|1/2\rangle$, and likewise there are two sub levels in the β-manifold, leading to four levels labeled as $|\alpha\alpha\rangle$, $|\alpha\beta\rangle$, $|\beta\alpha\rangle$ and $|\beta\beta\rangle$, the two letters denote the electron and nuclear spins, respectively.

The allowed EPR transitions are two in number indicated by vertical arrows (a) with $\Delta m_S = \pm1$, and $\Delta m_I = 0$, 1↔3 and 2↔5. There are two forbidden EPR transitions (f) with $\Delta m_S = \pm1$, and $\Delta m_I = \pm1$, 1↔4 and 2↔3, the so-called zero-quantum and double-quantum transitions involving the simultaneous flip of the electron and nuclear spins. There are also two transitions involving exclusively nuclear spin flips ($\Delta m_S = 0$), and $\Delta m_I = \pm1$) denoted as ω_α(1↔2) and ω_β(3↔4).

If one denotes the isotropic part of the hyperfine coupling by a_{iso}:

$$a_{iso} = (1/3)[A_1 + A_2 + A_3] \quad \textbf{(20)}$$

where A_i are the principal components of the observed anisotropic hyperfine tensor such that the principal components of the dipolar hyperfine tensor is given by

$$\mathsf{T} = (A_1 - a_{iso}, A_2 - a_{iso}, A_3 - a_{iso}); \text{ and } Tr(\mathsf{T}) = 0 \quad \textbf{(21)}$$

The allowed (ω_{24} and ω_{13}) and the forbidden (ω_{14} and ω_{23}) EPR transitions are noted in Fig. (**4**), and the extreme transitions are seldom seen except in weak-coupling cases where the magnitude of hyperfine coupling at any orientation $|A_\theta| \ll 2\omega_I$, where ω_I is the NMR transition energy of the nucleus. By setting up the 4 x 4 matrix representation of the spin Hamiltonians H_{Zee}, H_{HF} and H_{NZee} within the product spin space |αα>, |αβ>, |βα> and |ββ>, and diagonalizing the matrices one can arrive at the energies of the nuclear(ENDOR) and the electronic (EPR) transitions and their transition probabilities. They are given below:

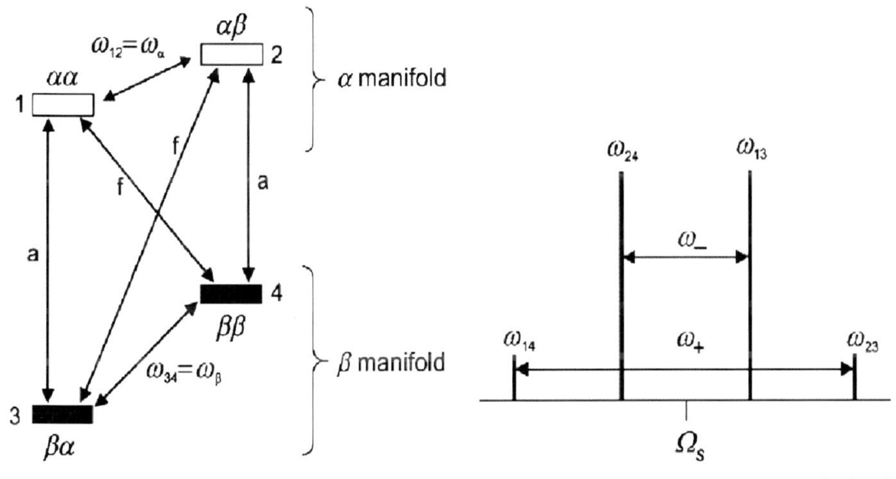

Fig. (4). The schematic EPR spectrum (right) and the corresponding energy level diagram (left) for $S = 1/2$, $I = 1/2$ model system with $|A_s| < |2\omega_I|$ (weak coupling case). Here a: Allowed EPR transitions (1↔3) & (2↔ 4). f: forbidden EPR transitions (1↔4) & (2↔ 3). Nuclear transitions are (1↔2) & (3↔ 4). As for notation, A_s is the hyperfine coupling at an arbitrary orientation. Reprinted from [10] with Permission from John Wiley and Sons.

$$\omega_\alpha = |\omega_{12}| = \left[\left(\omega_I + \frac{A}{2}\right)^2 + \left(\frac{B}{2}\right)^2\right]^{1/2}; \ \omega_\beta = |\omega_{34}| = \left[\left(\omega_I - \frac{A}{2}\right)^2 + \left(\frac{B}{2}\right)^2\right]^{1/2} \quad \textbf{(22)}$$

And, for an axially symmetric hyperfine tensor ($A_1 = A_2 \neq A_3$), A and B above are given by

$$A = a_{iso} + T(3\cos^2\theta - 1); \ B = 3T\sin\theta\cos\theta, \quad \textbf{(23)}$$

where θ is the angle between the electron-nuclear vector and the Zeeman field axis. The probabilities of the EPR transitions for the allowed (a) and forbidden (f) transitions are given by

$$I_a = \cos^2\eta = \frac{\omega_I^2 - \frac{1}{4}\omega_-^2}{\omega_\alpha\omega_\beta}; \ I_\beta = \sin^2\eta = \frac{\omega_I^2 - \frac{1}{4}\omega_+^2}{\omega_\alpha\omega_\beta} \quad \textbf{(24)}$$

Here, 2η is the angle between the nuclear quantization axes in the two electron manifolds (α & β), ω_1 is the nuclear Larmor frequency and $\omega_+ = \omega_\alpha + \omega_\beta$ and $\omega_- = \omega_\alpha - \omega_\beta$. It is clear from the schematics in Fig. (**4**) that the nuclear transitions gain intensity when the forbidden transitions (f) have finite probability. In ESEEM where the mw pulses excite both the allowed and the forbidden EPR transition one can see (indirectly) the nuclear transition frequencies manifest themselves as modulation of the electron spin echo. In ENDOR, on the other hand, the nuclear transition frequencies are excited directly with rf irradiation at the appropriate frequencies; since nuclear transitions are excited between levels whose polarizations are governed by electron spin polarization, one gets an enhancement in transition on the order of the ratio of the respective gyromagnetic factors $\gamma_{\text{electron}}/\gamma_{\text{nucleus}}$.

CW ENDOR: Theory

The following description is basically adapted from Kulik and Lubitz [12], who have discussed this very competently. ENDOR (Electron-nuclear double resonance) was introduced in solid state physics first by Feher [13]; Hyde and Maki [14] later extended it to radicals in solution. One combines the techniques of EPR and NMR spectroscopies in ENDOR, but the two techniques play different roles here. The EPR signal is observed here at a fixed magnetic field, whereas its intensity is varied by the applied scanning rf NMR radiation. ENDOR monitors only the paramagnetic species, *e.g.* those in SiCN ceramics. The addition of NMR enhances the resolution of ENDOR as compared to that obtained by EPR alone. In ENDOR, one studies HFI, (hyperfine interactions), which represent the magnetic interactions of the unpaired electron spin with the spins of magnetic nuclei, which can belong either to (i) the molecule on which the unpaired electron is localized, or (ii) to the surrounding molecules. One detects the electron spin echo (ESE) signal in pulse ENDOR, which limits its application to systems with a sufficiently large transverse electron spin relaxation time (T_2 > 100 ns). This makes it unsuitable to study liquid samples, requiring measurements at low temperatures. CW ENDOR, on the other hand, is free from this limitation. It allows the experiments to be done at ambient temperatures. On the other hand, it requires, for an optimum signal intensity, "fine tuning" of the longitudinal relaxation times of the electron and nuclear spins. Pulse ENDOR is better than CW ENDOR at low temperatures due to the strong temperature dependence of these relaxation rates. The simplest system that can be investigated by ENDOR is a radical with the electron spin $S = 1/2$ associated with a nucleus of spin $I = 1/2$. When the hyperfine coupling between them is isotropic, then if the g-matrix is also isotropic, the spin-Hamiltonian H of this system is (in frequency units):

$$H/h = g\mu_B\, B_0 S_z/h - g_n\mu_n I_z/h + aS\cdot I \qquad \textbf{(25)}$$

The three terms on the right in this equation, respectively, describe the electron Zeeman interaction, the nuclear Zeeman interaction, and the HFI. Here h is Planck's constant, B_0 is the intensity of the external magnetic field, μ_B is the Bohr magneton, g is the electronic g-factor, μ_N is the nuclear magneton, g_n is the nuclear g-value, a is the HFI constant, S and I electron and nuclear spin operators. The z-axis is parallel to the orientation of the constant magnetic field of the EPR spectrometer B_0 (laboratory frame). The spin-Hamiltonian in eq. (25) applies to a radical in a liquid solution, wherein fast rotation averages out all anisotropic interactions. The first term is dominant in eq. (25) in the "strong-field" approximation, The energy levels of the system are characterized by the electronic and nuclear magnetic quantum numbers, $m_S = \pm 1/2$ and $m_I = \pm 1/2$, respectively. The first-order eigenvalues are:

$$E(m_S;m_I)/h = \nu_e m_S - \nu_n m_I + a m_S m_I. \qquad \textbf{(26)}$$

In the above equation, $\nu_e = g\mu_B B_0/h$ is the electron frequency and $\nu_n = g_n\mu_n B_0/h$ is the nuclear Larmor frequency. The energy level diagram, describing eq. (26) is shown in Fig. (**5**). In EPR, the selection rules are $\Delta m_S = \pm 1$ and $\Delta m_I = 0$, which implies that there are possible two allowed EPR transitions in this system The rf field drives also the NMR transitions in an ENDOR experiment with the selection rules: $\Delta m_S = 0$ and $\Delta m_I = \pm 1$. The frequencies of these transitions are:

$$\nu^{\pm}_{ENDOR} = \nu_n \pm a/2. \tag{27}$$

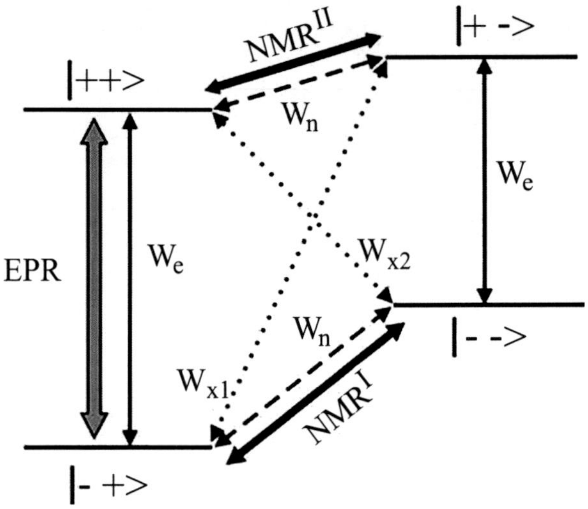

Fig. (5). A plot of the energy levels for coupled electron spin ($S = 1/2$) and nuclear spin ($I = 1/2$) system. The spin functions are shown on the four energy levels of this system. The EPR and NMR transitions, together with the electron spin (W_e), nuclear spin (W_n) and cross-relaxation rates (W_{x1}, W_{x2}) are indicated. The NMR resonances, shown by black arrows, are detected by the change of an irradiated saturated EPR line (gray arrow) simultaneously in a CW ENDOR experiment; for further details, see text (adapted from [12] with Permission from Springer).

Continuous wave ENDOR can be considered as NMR-induced partial desaturation of a saturated EPR line for a stable radical in thermodynamic equilibrium. The spin relaxation processes affecting the $S = 1/2, I = 1/2$ system are shown as dashed lines in Fig. (**5**).

The rate of population relaxation that is the longitudinal spin relaxation, of the electron spin is denoted as W_e, that of the nuclear spin as W_n, and those of the electron-nuclear cross-relaxation are denoted as $W_{x1}$and W_{x2}. The ENDOR transition takes place when the two, mw and rf fields, are, respectively, in resonance with the EPR and NMR transitions and there is a common energy level shared by these transitions. One EPR transition is saturated by mw irradiation in CW ENDOR, as indicated by the thick vertical arrow in Fig. (**5**).

One NMR transition (NMR^{II} or NMR^{I}) is saturated by the rf field simultaneously, opening an alternative relaxation path for thc pumped electron spin. When NMR^{II} is pumped, it can relax *via* a two-step pathway $W_e(|+ ->\leftrightarrow|- ->)$, $W_n(|- ->\leftrightarrow|- +>))$ or directly by $W_{x1}(|+ ->\leftrightarrow|- +>)$. The intensity of the ENDOR signal is determined by the extent to which the EPR line is desaturated by the bypass of the additional relaxation. As a consequence, the number of contributing nuclei is usually not reflected in the intensity of the ENDOR line, in contrast to NMR or EPR. The intensity of the ENDOR signal, *E*, in the limit of the simplifying assumption $W_{x1} = W_{x2} = 0$ and in the presence of strong NMR saturation, is:

$$E = 1/[2(2 + b + b^{-1})]; \qquad \textbf{(28)}$$

where *E* is the 'ENDOR enhancement' and $b = W_n/W_e$. It represents the relative change of the EPR signal. *E* depends strongly on the relaxation properties of the system (Plato *et al.* 1981). To achieve the "matching condition": $W_n = W_e$, corresponding to the maximum ENDOR enhancement, which is $E_{max} = 1/8$, careful optimization of the respective rates, *e.g.*, by variation of temperature, is required. This value might be increased by cross-relaxation. An asymmetry of the ENDOR spectrum is produced by the asymmetric relaxation network Usually $W_{x1}= W_{x2}$.

The situation is qualitatively similar for more complicated systems with $k > 1$ nuclei and with $I = 1/2$, for which Eq. (25) can be easily generalized to:

$$H/h = g\mu_B B_0 S_z/h - \Sigma_i g_n \mu_n I_{zi}/h + \Sigma a_i(S \cdot I_i), \qquad \textbf{(29)}$$

where the index *i* covers all nuclei. When these nuclei are non-equivalent, there are possible 2*k* EPR transitions, associated with only 2*k* ENDOR transitions with the frequencies:

$$\nu_{ENDOR} = |\nu_n(i) \pm a_i/2| \qquad \textbf{(30)}$$

ENDOR spectroscopy thus provides simplification of the spectra as compared to that in EPR. Its sensitivity is many orders of magnitude higher than those of the NMR experiments on paramagnetic systems, although ENDOR is less sensitive than EPR. This is due to the enormous increase in the linewidth as compared to that of NMR on diamagnetic molecules.

ENDOR Instrumentation

The instrumentation for ENDOR experiments is an extension of that for CW or pulse EPR, with the difference that for ENDOR, one requires an rf source and amplifier, whose rf output is introduced into the rf coils, which are placed at the EPR cavity. These coils are placed in such way that the magnetic component of the rf field B_2 is perpendicular to both the static field (B_0) and the excitation field (B_1). See Kurreck *et al.* [15], Kevan and Kispert [16] and Poole [6] for more details on ENDOR instrumentation.

Mims and Davies Pulsed ENDOR Sequences

ENDOR can be performed both in CW and pulsed modes. (For a complete discussion refer to [10]). The CW experiment consists of saturating an EPR transition of the system with coupled nuclear spins. As the EPR transitions lose the intensity due to the reduction in population difference between the levels, a rf sweep is performed to excite the nuclear transitions, which releases the bottleneck in the population distribution and allows the EPR transition to regain the intensity and this is referred to as ENDOR induced EPR. The nuclear transitions are enhanced and occur at frequencies above and below the nuclear Larmor frequency depending on the hyperfine coupling and nuclear quadrupole coupling. Since nuclear resonances are quite narrow, hyperfine splitting which is on the order of EPR line width, and hence unresolved in the EPR spectrum can be evaluated with high precision in ENDOR. This is particularly true for low gamma nuclei such as ^{13}C, ^{15}N, ^{14}N and ^{2}H nuclei, when measured at high spectrometer frequencies. CW ENDOR consists of carefully locking the Zeeman field on a designated EPR transition and then sweeping a range of NMR frequencies keeping the resonator tuned to the changing rf all the time by servo-regulated automatic tuning and matching.

Instrumentation-wise CW ENDOR is quite challenging and requires a critical balance of rf power and relaxation times to achieve optimum sensitivity. Since rf and mw are present simultaneously, a lot of effort is required to reduce interference from low frequency and the associated artifacts. Pulsed ENDOR, on the other hand, is less susceptible to artifacts since the rf and mw power are applied only for short duration and seldom applied simultaneously. Pulsed ENDOR allows the study of the manifestations of electron T_1, T_2 and nuclear T_1 and T_2 as well as electron nuclear cross relaxation with good selectivity. Basically, the nuclear resonance frequencies which contain high-resolution hyperfine information are monitored by creating an enormous polarization between the nuclear sublevels by saturating the connected EPR transition. Upon exciting the nuclear transitions by selective rf pulses, the two-pulse or three-pulse echo gets enhanced when the nuclear resonance frequencies match with the rf frequency.

ENDOR techniques predominantly exploit the electron spin echo (ESE) effect, wherein the echo signal is created by proper mw pulse sequences. The rf pulsc is applied during the "mixing period" of the pulse sequence. It induces transitions among nuclear spin levels, producing a change in the ESE intensity. The pulse ENDOR signal is represented by the amplitude of this change when the rf frequency is scanned.

The two sequences that are used in pulsed ENDOR are those of Davies [17] and Mims [18]. These are shown in Fig. (**6**). The principle of pulse ENDOR is best illustrated for the $S = 1/2$, $I = 1/2$ system.

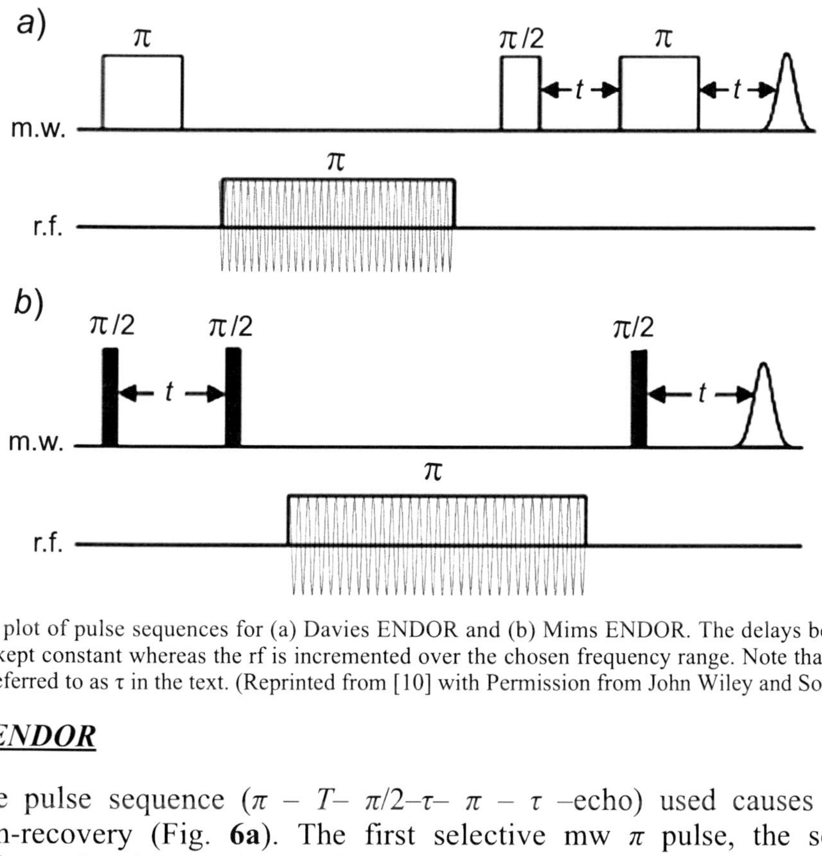

Fig. (6). A plot of pulse sequences for (a) Davies ENDOR and (b) Mims ENDOR. The delays between the pulses are kept constant whereas the rf is incremented over the chosen frequency range. Note that t in these figures is referred to as τ in the text. (Reprinted from [10] with Permission from John Wiley and Sons).

Davies ENDOR

Here the pulse sequence (π – T– $\pi/2$–τ– π – τ –echo) used causes an mw inversion-recovery (Fig. **6a**). The first selective mw π pulse, the so-called preparation pulse, inverts the populations in two connected levels of a particular EPR transition (levels 1↔3, Fig. **7A,B**). Following this, a selective rf π pulse is applied. When the frequency and the band width of this rf pulse corresponds to transitions between nuclear sub-levels connected to the inverted EPR transition, the nuclear polarization also gets inverted (see Figs. **7B,C**), which, in turn, will alter the population of the 'observer' electron spin transition 1↔3. If the rf frequency does not correspond to any of the nuclear hyperfine sublevels, then one does not see any change in the population of the levels 1 and 3 (Fig. **7D**). The amplitude of the mw field B_1 should here be properly adjusted ($B_1 \leq a$) in order to avoid the inversion of the second EPR transition.

As a consequence, Davies ENDOR is found to be useful for systems with large HFIs. When one considers a stable radical in thermal equilibrium, the polarization of the EPR transition is positive initially. This polarization is inverted by the mw π-pulse. During the T interval, during which the rf pulse is applied, the rf pulse changes the population of the nuclear sublevels, which thereby restores the polarization of the EPR transition partially. The change in the electron polarization brought about by a transition of the connected nuclear sublevels can be monitored by appending the conventional Hahn echo sequence $\pi/2 - \tau - \pi - \tau$ (Fig. **6a**) following the selective rf pulse. The echo intensity is monitored as a function of the rf frequency which is incremented stepwise.

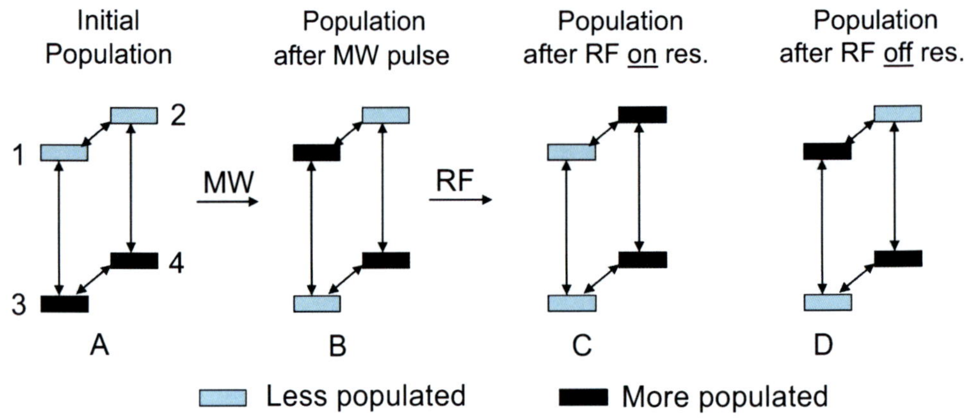

Fig. (7). Energy levels and population in a two spin $S = 1/2$ and $I = 1/2$ system during the Davies ENDOR experiment. (a) equilibrium populations (b) selective mw π pulse inverts the polarization of EPR transition (1,3) and (c) populations after the rf π pulse on resonance with nuclear transitions and (d) off-resonance with the nuclear transitions (no effect). Reprinted from [10] with Permission from John Wiley and Sons.

Mims ENDOR

It essentially uses the same concept as that of Davies ENDOR, except that the nuclear transitions are here monitored by the change in the intensity of the *stimulated echo* as against the Hahn echo. Therefore, the Mims ENDOR sequence (Fig. **6b**) consists of three non-selective $\pi/2$ mw pulses. Further details of Mims ENDOR are as follows. The applied stimulated echo mw produced by the pulse sequence ($\pi/2 - \tau - \pi/2 - T - \pi/2 - \tau$ –echo) excites both EPR transitions. It limits the application of this method to relatively small HFI constants ($B_1 \geq a$). A spin-level population diagram is not suitable for describing Mims ENDOR because the transverse components of the electron spin magnetization (coherencies) become involved in this case.

Mims ENDOR can be qualitatively understood as a partial defocusing of the ESE. Here the rf π-pulse causes a change in m_I, producing a change in the frequency of the Larmor precession of the electron spin. Therefore, the frequency of this precession differs by the value of a during the first and the second s period. Just when the echo is formed, the precessing magnetization acquires the additional phase $\Delta\phi = a\tau$, making the echo intensity proportional to $S_y = \cos(a\tau)$, from which it is seen that when $a\tau = 2\pi n$ no ENDOR effect is observed; here n is an integer number. As a consequence, there appear "blindspots," or regions with severely decreased ENDOR sensitivity for the given s value in Mims ENDOR spectrum around $a = 2\pi n/\tau$. This is a major disadvantage of Mims ENDOR. When the magnitude of the HFI is comparable to or larger than that of the nuclear Larmor frequency, there manifests hyperfine enhancement, both for the cases of CW and pulse ENDOR. It happens due to the effect of the rf field on the electron spin. Because of this, the effective rf field at the sites of the nuclear spins becomes dependent upon (i) m_S and (ii) the strength of HFI. This causes a change in the ENDOR line intensity (more details in [19]).

The preparation pulse pair $\pi/2 - \tau - \pi/2$ generates electron polarization depending on the delay τ. The selective rf pulse that follows this, if it corresponds to a connected nuclear transition, then it will change the electron polarization, and hence the amplitude of the stimulated echo peak generated by the third $\pi/2$ mw pulse followed by the delay time τ. The use of the stimulated echo whose intensity depends on the transverse relaxation time T_2, in Mims ENDOR makes it more efficient than that of Davies ENDOR. (This is because stimulated echo whose intensity is governed by the relatively longer T_2 is more intense than the Hahn echo with intensity being governed by the shorter phase memory time T_M). The ENDOR efficiency also depends on the magnitude of the hyperfine coupling, $A(\theta)$, and the echo time τ. It was shown above that in Mims ENDOR, the maximum efficiency happens when $\tau = (2n + 1)\pi/A(\theta)$ and zero for $\tau = 2n\pi/A(\theta)$ ($n = 0, 1, 2,..$). There will, therefore, appear 'blind spots' just as in ESEEM and HYSCORE which are described below.

Electron Spin Echo Envelope Modulation (ESEEM) and Hyperfine Sublevel Correlation Spectroscopy (HYSCORE)

For a complete discussion refer to [10]. The sequences that are used in ESEEM closely resemble those of pulsed ENDOR, but there are no rf pulses to directly excite nuclear transitions. The basic ESEEM, one-dimensional (1D) sequence and a 2D four-pulse sequence that generates a correlation spectrum known as HYSCORE are shown in Fig. (**8**).

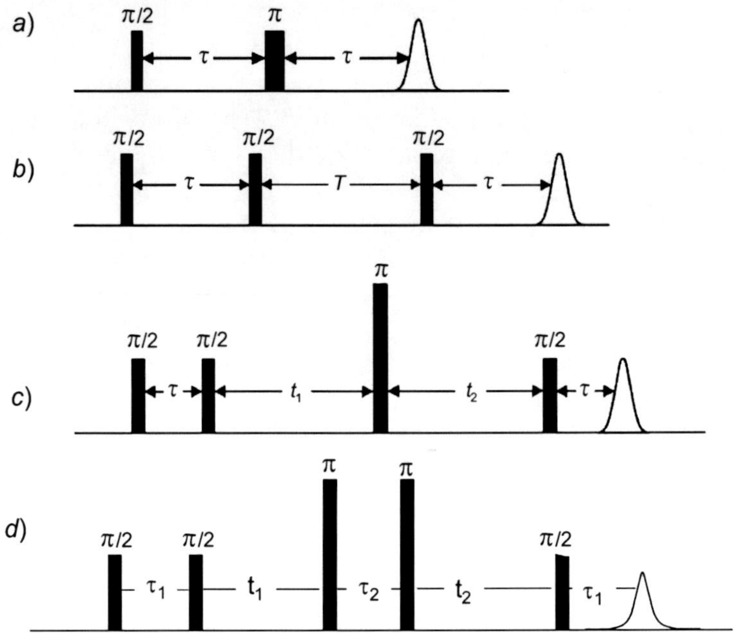

Fig. (8). Pulse sequences for ESEEM: (a) Two-pulse Hahn Echo sequence along with the primary echo; (b) the three-pulse sequence for stimulated echo; (c) four-pulse sequence used in HYSCORE set up; and (d) Five pulse sequence for DONUT-HYSCORE. Please see text for details. Reprinted from [10] with Permission from John Wiley and Sons.

In the two-pulse ESEEM experiment (Fig. **8a**), the intensity of the two-pulse (primary) echo is monitored as a function of the interval τ of the $\pi/2-\tau-\pi-\tau$ sequence. It can be shown that the amplitude of the echo will be modulated, for an $S = 1/2$ and $I = 1/2$ system; it is given by

$$V^{2-pulse}(\tau) = 1 - \frac{k(\theta)}{4}[2 - 2\cos(\omega_\alpha \tau) - 2\cos(\omega_\beta \tau) + \cos(\omega_- t) + \cos(\omega_+ t) \quad \textbf{(31)}$$

with

$$k(\theta) = \left(\frac{B\omega_I}{\omega_\alpha \omega_\beta} \right)^2 \quad \textbf{(32)}$$

where $k(\theta)$ is the orientation dependent modulation depth given in eq. (32). For isotropic hyperfine coupling and along the principal axis of the hyperfine tensor the echo modulation is zero, and it is maximum when $|A(\theta)| = |2\omega_1|$. The two-pulse Hahn echo amplitude is governed by the gross transverse relaxation time T_M which is much shorter than the true transverse relaxation T_2, and, therefore, as in the case of Mims ENDOR, one can employ a three-pulse sequence (Fig. **8b**). In this case, the echo peak amplitude is modulated in terms of the nuclear transition frequencies in the alpha and the beta manifolds (ω_α and ω_β, see eqs. (22) and (23)) and are given as a function of τ and T by,

$$V^{3-pulse}\left(\tau,T\right)=1/2\left[V^{\beta}\left(\tau,T\right)+V^{\alpha}\left(\tau,T\right)\right] \quad \textbf{(33)}$$

$$V^{\alpha(\beta)}(\tau,T)=1-\frac{k}{2}\left[1-\cos(\omega_{\beta(\alpha)}\tau][1-\cos(\omega_{\alpha(\beta)}(\tau+T))\right] \quad \textbf{(34)}$$

When T is varied the echo amplitude is modulated as a function of the two frequencies ω_α and ω_β as well as the sum and difference frequencies ω_+ and ω_- (cf. eq. (24) do not appear, as is the case of the two-pulse ESEEM (cf. eq. 32). The three-pulse ESEEM (as in the Mims ENDOR) exhibits blind spots at $\omega_{\alpha(\beta)}$ when $\tau = 2\pi n/\omega_{\alpha(\beta)}$ (n = 1, 2, 3,...). By performing several experiments at different τ values and coherently summing the results the blind spots can be removed. In order to generate the nuclear transition frequencies, the value of τ in eq. (33) and that of T in eq. (34) are to be systematically incremented and the echo-envelope modulation is traced point by point. This is, therefore, a time-consuming experiment.

Hubrich *et al.* [20] suggested an alternative approach to the two- and three-pulse ESEEM, which allows the entire envelope modulation to be obtained in a single experiment. In this approach, the last refocusing $\pi/2$ pulse is replaced by a weak, long irradiation for a period of time causing an extended time excitation (ETE). Each infinitesimal time segment of this irradiation corresponds to a weak pulse that causes a stimulated echo at the corresponding time after the refocusing pulse and the series of these echoes provide a signals exactly similar to the two-pulse ESEEM sequence, but in a single experiment. The details for the four-pulse sequence for a HYSCORE experiment and five-pulse sequence for DONUT-HYSCORE experiment are not discussed anymore here. They can be found in [10].

CW EPR SPECTROMETERS

High-frequency spectrometers have rendered the technique of EPR a very sensitive and unique technique today. The most commonly used frequency band for EPR is X-band (~9.5 GHz; 1 GHz = 10^9 hertz]), which requires a waveguide resonator of the size ~3.0 cm.

Continuous Wave EPR

The mw field is applied to the sample during the entire time of recording a spectrum in continuous-wave (CW) EPR spectroscopy. The EPR signal is recorded by sweeping the external magnetic field over a chosen range, which is normally over 0.0 – 0.9 Tesla (T; 1 T = 10,000 Gauss) at X-band. There are certain advantages of using X-band spectrometers. However, mw bands different from X-band may be preferable over X-band, depending on the type of investigation. A multi-frequency approach to EPR appears preferable to acquire a complete and unambiguous picture of the interaction of a paramagnetic ion with its environment.

Typical X-band, Low- and High-frequency CW Spectrometers

X-band EPR Spectrometer Design ([10])

X-band spectrometer is most commonly used in studying CB materials. In a continuous-wave (CW) EPR spectrometer, the mw frequency is held constant while the magnetic field is changed linearly with time. This brings about changes in the spacing of the energy levels. Resonance occurs when the spacing of the energy levels participating in resonance becomes equal to the quantum of energy $h\nu$, where ν is the frequency of the mw radiation and h is Planck's constant. There are four components of such a spectrometer:

(i) the source of mw radiation, with provisions to control and monitor the frequency and intensity of the mw radiation;

(ii) the magnetic-field system, to provide a stable, linearly varying, and homogeneous magnetic field of required magnitude;

(iii) the cavity, or resonator, system characterized by a fixed resonant frequency; it holds the sample, directing and controlling the mw waves to and back from the sample;

(iv) the detection system for the signal. Here one modulates the magnetic field with a sinusoidal voltage, which is amplified as the useful EPR absorption signal that is recorded. In the modern design, one uses the cavity in the "reflection mode" in conjunction with a mw circulator, so that the reflected microwaves from the cavity do not go back to the source of radiation. The majority of EPR spectrometers are of the reflection type, wherein one measures the changes in the amount of radiation reflected back from the mw cavity that contains the sample. There is used the same design over the range of frequencies 8-70 GHz.

The schematic of a typical X-band spectrometer is shown in Fig. (**9**). The details of the various components are as follows.

(i) Source of mw radiation. The mw source is usually a klystron, which emits a monochromatic radiation. In a klystron, which is a vacuum tube, mw oscillations are produced which are centered over a small range of frequency. The output voltage, displayed as a function of frequency, is referred to as the *klystron mode.*

One selects the mode with the highest output of the various modes available. To display a mode on an oscilloscope one changes the klystron frequency over the mode's range. It is done by changing the voltage applied to the klystron, which alters the frequency of the output mw radiation.

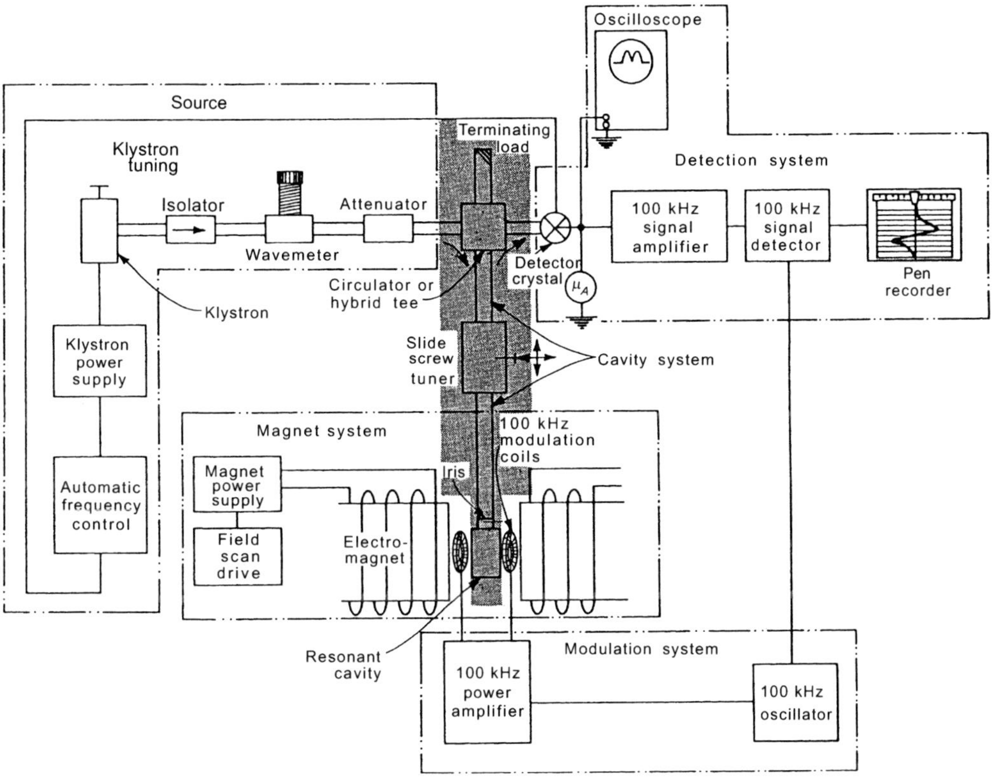

Fig. (9). Schematic diagram of a common X-band EPR spectrometer, using a 100-kHz phase-sensitive detector (Reprinted from [21] with Permission from John Wiley and Sons).

Thus the frequency can be tuned over the range of the mode with the help of the tuning knob on the klystron. When there occurs the cavity resonance a sharp dip appears in the mode. One sets the frequency such that the cavity-resonance dip is at the center of the mode, where the power and stability are maximum. This is done by tuning the klystron. Since the energy density in the resonant cavity depends very strongly on the frequency of the incident radiation, it is necessary to maintain the frequency of the klystron very stable. A wavemeter is used to measure the frequency of the radiation. It is a cylindrical resonant cavity, whose length can be adjusted to an integral number of half wavelengths. This is accomplished by the use of a micrometer. The power detected by the silicon crystal is decreased when there is a match between the resonant frequency of the wavemeter and the frequency of the incident microwaves. This is caused by the resonant absorption of the microwaves by the wavemeter. For greater accuracy in frequency measurement, one uses a frequency meter by coupling it to the mw system. The mw power incident upon the sample is adjusted by an *attenuator*, which consists of an absorptive element.

(ii) Transmission of microwaves. One uses single-mode waveguides for microwave transmission. The size of a waveguide is directly proportional to the wavelength, or inversely proportional to the frequency, of the microwave radiation. For example, the dimensions (cross section) of an X-band (9.8 GHz) waveguide are: 0.9 in x 0.4 in, whereas they are 6.50 x 3.25 in at L-band (1 GHz). The frequency of the klystron is stabilized by an automatic frequency control system. Normally, one locks the klystron frequency to the resonant frequency of the cavity which contains the sample. *Isolator* is a non-reciprocal device that passes the mw radiation in the forward direction readily but strongly attenuates any reflections. By the use of an *isolator*, one avoids feeding the backward reflections of the mw energy from the system to the klystron.

(iii) The cavity (or resonator) system. A resonator enhances the sensitivity of a spectrometer. It stores the mw energy, and its configuration provides for concentration of the mw power at the sample. The most common type of resonators is in the form of cavities. The commonly used cavities are rectangular-parallelepiped and cylindrical cavities. They are constructed by cutting a short section of a rectangular or circular wave guide in which the standing wave is produced. The length of this section is chosen to be an integral number of half wavelengths. It is possible to excite more than one type of standing wave pattern in the cavity, called *modes*. They are denoted as TE (transverse electric) modes, with the subscripts denoting the number of half wave lengths along the various dimensions, *e.g.* TE_{102} for rectangular-parallelepiped and TE_{011} for cylindrical cavities. In the case of a cylindrical cavity, one may tune it by the use of a piston fitted at the end. This technique is also used in a wavemeter, mentioned above,

which is a cylindrical cavity. It is used to measure the wavelength of the mw radiation. For microwaves, one has to take into account both, the electric (E_1) and magnetic (B_1), fields in the cavity.

To observe EPR resonance, one requires that:

(i) the cavity should be in a mode that contains high-energy density;

(ii) one should place the sample at the maximum B_1 density;

(iii) one should set the B_1 field perpendicular to the static field B in order to observe the $\Delta M = \pm 1$ transitions which are the allowed ones; here M is the electron spin magnetic quantum number.

The response of the cavity is determined by its Q-factor, which is the sharpness, or figure of merit of the cavity. It is defined as the ratio of 2π times the maximum mw energy stored in the cavity and the energy dissipated in the cavity per cycle. Q is maximized by heavy silver plating, whose deterioration is prevented by thin gold plating upon it. Further, Q is diminished when a sample with high dielectric constant, *e.g.* an aqueous sample, happens to be introduced in the regions of appreciable electric fields. One uses an *iris* to reduce the fractional reflection of mw energy from the cavity and to increase simultaneously the fractional change in the reflected power. It is a small hole cut into the entrance port to the cavity. This is done by an adjustable screw, whose setting depends upon the size as well as the nature of the sample in the cavity. Discontinuities in the waveguide, or imperfect matching of mw elements from areas outside the cavity, form standing waves, which decrease the sensitivity of the spectrometer. To minimize these, one uses a *slide-screw tuner*, which is a small metallic probe, to produce standing waves, which are chosen to be of such amplitude and phase that minimize the existing standing waves. This is accomplished by varying its depth of insertion and its position along the waveguide. One uses a *circulator* to attenuate microwaves traveling in the reverse direction. It is a non-reciprocal device, which allows the microwaves to pass in the forward direction with little loss, but attenuates strongly the microwaves traveling in the opposite direction, directing the microwaves towards the cavity as well as directing the reflected signal from the cavity to the detector.

(iv) Magnetic-field system. The magnet used at X-band is an electromagnet. It consists of two Helmholtz coils, which have sufficient space between their pole pieces to accommodate the resonator. One requires the magnetic field to be stable and uniform over the sample, whose stability is better than ± 10 mG for organic free radicals in the liquid solution and ~1 G for most inorganic samples. The magnetic field is stabilized by the use of a power supply that drives the magnet,

which is highly regulated. A Hall-effect probe is used to detect directly any fluctuations in the magnetic field, which are then corrected by a feedback system in the power supply so that the magnetic field is maintained at a constant value. The current in the magnet is regulated by a highly reproducible, linear scanning system, which is connected to the power supply, to ensure the variation of the magnetic field in a linear fashion.

(v) Modulation and detection system. One uses the phase-sensitive detection technique for detection of signal. In this technique, a small amplitude modulation of the d.c. magnetic field is introduced to restrict the noise to frequencies in a very narrow range about the modulating frequency. 100 kHz is the modulation frequency used commonly. The modulation is done by the use of Helmholtz coils, which are placed on the two sides of the cavity directed along the axis of the d.c. magnetic field. One uses very thin cavity walls so that the 100-kHz signal reaches the sample without much attenuation. In this manner, the rectified signal reaching the detector becomes amplitude-modulated at the frequency of 100 kHz. A Schottky barrier diode made of a silicon crystal is used most commonly as the detector, which acts as an mw rectifier. The "1/f detector noise", which is inversely proportional to the frequency, becomes appreciably less at 100-kHz modulation as compared to those due to other sources. One can further increase the sensitivity of detection at low mw power by *superheterodyne detection*, wherein the signal is mixed with the output of a local oscillator to generate an intermediate frequency (i.f.), which is then amplified at i.f. and detected. For a sample that requires being irradiated with a very small mw power, the superheterodyne detection proves to be the most sensitive method. Further reduction in noise can be accomplished by phase-sensitive detection, which discards all noise components except those in a very narrow band of ~1 Hz about 100-kHz. In phase sensitive detection, the amplified signal and the output of the modulating 100-kHz oscillator are mixed with each other. When the two have exactly the same phase the output becomes the maximum, if they are exactly opposite in phase the output becomes the minimum. When the amplitude of the modulation used is small compared to the linewidth the detected 100-kHz signal output is almost proportional to the slope of the absorption curve at the mid-point of the modulating field. The output polarity of the phase-sensitive detector is determined by the sign of the slope. This results in the output signal looking like the first derivative of the absorption signal when the modulation amplitude used is small. The noise level of the signal is a function of the time constant of the circuit filtering the output longer time constants reduce the noise.

Design of Low-frequency Spectrometers

The frequency range of these spectrometers, *e.g.* L (1 GHz)- and S (4 GHz)-

bands, is anywhere from ~250 MHz to ~4 GHz. They are not discussed here, as they are not relevant to studying CB materials. They are mostly used for biological systems.

Design of High-frequency (hf) Spectrometers

The various hf frequency ranges used are often referred to as sub-millimeter waves. They are greater than or equal to 95 GHz, with the corresponding wavelengths less than or equal to 2.5 mm. (See Lynch *et al.* [22] for a block diagram of a 1-mm EPR spectrometer). The various components of a hf EPR spectrometer differ from those of an X-band spectrometer as follows.

(i) Sources of radiation.

(ii) Transmission of submillimeter waves. For transmission of hf microwaves, one can use: (a) waveguides, which can be corrugated, or oversized and (b) free space. The attenuation is zero in a vacuum and it is very small in free space. The beam must be contained in this case, as the divergence of radiation is appreciable. To this end, one uses *quasi-optics*, which consists of conventional optics with lenses and elliptical mirrors. Horns are used for conversion from single-mode or corrugated waveguides to propagation in free-space. This is done by using optical elements, such as mirrors, lenses, and polarizers, which manipulate the beam. (See Smith *et al.* [23] for the schematics of a quasi-optical induction-mode spectrometer).

(iii) Resonators and Sensitivity. The resonator used is the Fébry-Perot resonator preferentially in VHF (~250 – 360 GHz) EPR range. To provide for optical excitation it has an open structure.

(iv) Magnetic field. At high frequencies, where one needs magnetic fields extending to 2 – 14 T, superconducting solenoids are commonly used, which requires liquid helium. There have also been developed cryogen-free magnets, which are available commercially.

(v) Detectors. There are mainly two types of detection systems, which are being used currently for sub-millimeter waves – bolometric or heterodyne with Schottky diodes.

EPR IN CARBON-BASED MATERIALS

Small variations in the electronic structure can be rather sensitively detected by EPR, see, e.g [24, 25]. Thus, EPR has been exploited to unravel various types of anthracite, as well as to distinguish anthracite from coal as reported by Tadyszak *et al.* [5]. One requires temperature variation of EPR signal and those of the electrical conductivity and magnetic susceptibility to understand the nature of the EPR signal in anthracites, which consist of a large variety of carbon materials, which are different from each other with regard to their contents of hydrogen and ash, as well as in their degree of metamorphosis. There are found significantly different EPR signals with regard to their shape and linewidth in the materials extracted from different deposits. The first work on an EPR study of anthracite, published in 1956 [26] has been used quite often to characterize various anthracites [24, 27, 28]. EPR is therefore a very powerful tool for investigation of defects in carbon-based (CB hereafter) materials, such as different polymorph forms of carbon: amorphous carbon, diamond, graphite, graphene, fullerenes, fullerites, turbostratic carbon, lonsdaleite, SWCN, MWCN (single and multi-walled carbon nanotubes), carbon nanobuds and carbon nanoribbons, as well different compounds, like SiC, C_3N_4, C_3Al_4, B_4C, CaC_2, WC, Fe_3C, SiCN, which include carbon ions chemically bonded with other ions, or "free" carbon phase. The other systems relevant to carbon-based materials are: radical-like centers, mainly carbon-related dangling bonds; color centers, related to the vacancy; paramagnetic transition metal ions and rare-earth ions; and nanoparticles of magnetically ordered compounds.

X-band investigations of new nanosized carbon-based materials are usually not sufficient for adequate and unequivocal interpretation of EPR spectra and obtaining precise values of spin-Hamiltonian (SH) parameters. Therefore, it is desirable to use several frequencies to get an unambiguous interpretation of EPR spectra. Usually, V-band (50-75 GHz), W-band (75-110 GHz), G-band (110-300 GHz), 250 GHz were used in such investigations of disordered materials. High-frequency (HF) investigations provided a unique possibility to resolve overlapped EPR lines. HF EPR is thus very important for investigation of carbon-based materials to distinguish various EPR signals with slightly different g-values observed near $g = 2.0023$. In addition, Electron Nuclear Double Resonance (ENDOR) is a very sensitive technique to study HFI and local environment of a paramagnetic center in CB materials.

EPR provides very important information on the origin, the location of radical-type carbon-related paramagnetic centers, magnetic properties and structural transformations in carbon-based heat-treated nanomaterials. Recently, nanosized particles of these carbon-related materials have attracted a great interest. These carbon-related paramagnetic centers are usually sp^2- or sp^3-carbon related dangling bonds, which are located on the surface of the nanoparticle, so their density can be very high for nanosized particles. Superparamagnetic nanoparticles of magnetic carbon-based materials have the potential to be used to develop magnetic-sensor devices. Intrinsic magnetism of carbon-based materials, based on a high density of dangling bonds, has recently attracted special attention.

Magnetic composites, as well as electrically conductive ceramics, with conductivity varying from dielectric to semiconductor, can be created using SiC, SiCN, or other carbon materials, as a base. Since SiC and SiCN ceramics also possess excellent piezoelectric properties, they may be exploited to create complex isolator-conductive-material structures and to develop new techniques with SiCN –MEMS-NEMS (Micro- (Nano-) Electro- Mechanical Systems). Consequently, investigation of different polymorph forms of carbon, as well SiC, SiCN ceramics and its conductive and magnetic derivatives is currently of great interest.

Several materials in contemporary electronics are polymorph forms of carbon or some compounds containing carbon ions. Nanosized materials of such carbon-based materials attract great attention due to their potential advantage compared to bulk materials. Therefore, it is very important to investigate different kinds of defects associated with the surface of these nanoparticles. Such defects determine many chemical and physical properties of nanosized compounds. EPR is a powerful tool for investigating such defects. It provides important information on the origin, the location of radical-type carbon-related paramagnetic centers, magnetic properties and structural transformations in heat-treated carbon-based nanomaterials. The various carbon-based materials studied by EPR are as follows.

(i) Single and double-walled carbon nanotubes, turbostratic carbon. Single and double-walled carbon nanotubes are very popular materials for EPR investigations. The g-values in these compounds are similar to those in glassy graphite samples, for which the *g*-value is 2.0027. It is because of the fact that the density of localized states is high at the Fermi level [29]. However, these g-values are far from g = 2.018, the value for the planar graphite phase [29], or from g = 2.020, that for turbostratic carbon [30], or from g = 2.012, that for multiwall nanotubes phase [31].

(ii) Fullerenes. A complete survey from the beginning of fullerene research (1985 up to 1995) was given by Misra and Petkov [32]. They reviewed the various EPR (CW, pulsed and time-resolved) and muon spin resonance (μSR) studies in $C_{60}/C_{70}/C_{82}$ fullerenes. They discussed the use of different EPR frequencies (X-, Q- and L-band) for fullerene investigations. Different techniques for preparing EPR active C_{60}/C_{70} species were also discussed.

(iii) Nanodiamonds; graphite and its transformations. EPR study of diamonds obtained by explosion synthesis started in the middle of 1970s [33]. It was found that ultra-dispersed diamonds (UDD) show strong intrinsic paramagnetism originating from uncoupled spins localized at broken carbon-carbon bonds. All investigated samples demonstrated the same typical EPR features: absence of

characteristic triplet signals from spins located on doped nitrogen atoms; presence of a singlet with a very intense signal at $g = 2.003$, with the linewidth $\Delta B_{pp} \sim 0.7$ - 0.9 mT, which was due to carbon-centered defects; these paramagnetic centers are present in concentrations of the order of 10^{18} - 10^{19} spin/g. By EPR one can control separation of graphite and nanodiamond phases and those of transition-metal ion impurities and dopants. In addition, it is exploited to determine (a) the cause of paramagnetism in pure nanodiamonds and its mesoscopic-related features, *e.g.g*-factor, NMR linewidth, relaxation times; (b) the number of EPR centers related to dangling bonds per particle as well as their precise location; and (c) to track the transformation of nanodiamond to nanographite by heat treatment. In [34] it was demonstrated that EPR is a powerful tool to control para- and ferro-magnetic impurities during UDD purification. EPR was employed in a successful manner to trace the conversion of nanodiamond to nanographite by heat treatment (HT) in Ar atmosphere [35, 36].

(iv) SiC. a-$Si_{1-x}C_x$: H – dangling bonds. It is known that paramagnetic defects in *a*-$Si_{1-x}C_x$:H are mainly associated with carbon incorporation, which leads to the development of "chemical disorder," *i.e.*, homonuclear interatomic bonds (Si-Si, C-C). In a mixed sp^2- and sp^3-coordinated disordered carbon network, there is a variety of possible types of paramagnetic defects: dangling-bonds and unpaired electrons associated with a single sp^2-coordinated carbon atom or with a carbon cluster having an odd number of sp^2-sites.

(v) Color centers in SiC. Color centers in SiC are promising for quantum sensing at ambient conditions because their spin state can be optically initialized, manipulated and readout by ODMR (optically detected magnetic resonance). The color centres in SiC with $S = 3/2$ exhibit unique characteristics, such as optical spin alignment up to the temperatures of 250°C. A giant thermal shift is exhibited in the ZFS (zero-field splitting) of the centres in the excited state, which finds its use in applications to thermometry. The spin and optical properties of the centres remains the same in the SiC nanocrystals allows achieving quantum sensing with high spatial resolution [37, 38].

(vi) SiCN. Pure SiCN ceramic samples have been successfully studied by EPR [39 - 45]. EPR investigations on pure SiCN ceramics derived from non-commercial polymer precursors were reported by Trassl *et al.* [41 - 43]; Erdem *et al.* [44], Li *et al.* [46], whereas investigations on SiBCN ceramics were reported by Berger *et al.* [45]. Dangling bonds in these ceramics were studied by EPR in detail at several frequencies from 9.5 GHz to as high as 170 GHz 4 K to 300 K [39]. An intense EPR line due to carbon-related sp^2-dangling bonds [39], located on the surface of free-carbon phases with $g = 2.0027$ was observed. In addition, a less intense EPR signal was observed at $g = 2.0033$. During ceramization are formed

dangling bonds, followed by breakage of C-H bonds, leading to the formation of free-carbon phase [47]. Tomasella *et al.* [48, 49] obtained confirmation of such origin of this EPR signal recently. EPR signal in UV-irradiated SiCN films due to nitrogen-related dangling bonds was observed by Kobayashi *et al.* [50]. Savchenko *et al.* [40] also investigated SiCN films and found an EPR signal with $g = 2.0033$ related to sp^2-carbon-related dangling bonds. They attributed the EPR signal at $g = 2.009$ to the interface defect due to threefold-coordinated Si dangling bond. The third EPR signal at $g = 2.05$ was deduced to be due to the trapped holes at Si atom.

(vii) Anthracite (Tadyszak *et al.* [5]). Anthracites consist of a large variety of carbon materials, which vary not only in the hydrogen and ash content but also in the degree of their metamorphosis. The EPR signals due to materials from various deposits differ from each other significantly in shape and linewidth. This is because of EPR's strong sensitivity to small variations in electronic structure. The EPR signal due to a thin bulk sample exhibits a single Lorentzian line; on the other hand, the EPR signal of a powder sample consists of two Lorentzian lines with identical *g*-factors and different linewidths. The temperature dependence of the linewidths revealed that the narrow component of the powder spectra corresponds to that of the thin sample. The room-temperature apparent linewidth, ΔB_{pp}, of the two-component EPR signal of anthracite observed is 0.052 mT. The deconvolution of this spectrum is shown in Fig. (**10**).

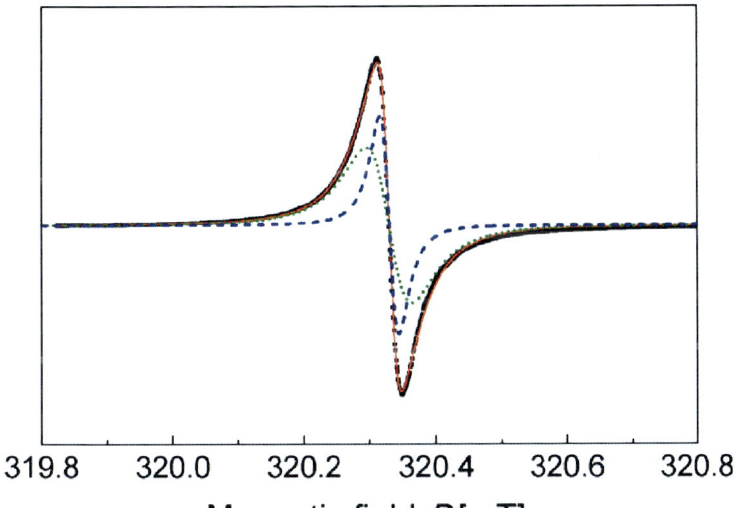

Fig. (10). EPR spectrum of powder anthracite at 10 K. It is deconvoluted into two Lorentzian lines. Low mw power of 1.35mW (attenuation of 20 dB) was used to avoid signal saturation. (Adapted from [5]).

CONCLUSION

This introductory chapter discusses the advantages of using multifrequency EPR in studying carbon-based materials. It includes the details of designing (i) continuous-wave EPR spectrometers operating at different frequencies; and (ii) pulse Electron Nuclear Double Resonance (ENDOR) spectrometers, which are very important for studying hyperfine interactions and local environment of paramagnetic ions in carbon-based materials precisely. Explanations are provided for the analysis of EPR spectra, spin Hamiltonians, evaluation of spin-Hamiltonian parameters, EPR lineshapes, and simulation of single-crystal and powder spectra. In addition, a short review of carbon-based materials studied by EPR is included.

CONSENT FOR PUBLICATION

Not applicable.

CONFLICT OF INTEREST

The author (editor) declares no conflict of interest, financial or otherwise.

ACKNOWLEDGEMENT

Thanks are due Dr. S. Andronenko for his help in organizing this manuscript and providing some useful information on carbon-based materials.

REFERENCES

[1] Ahmad, R.; Kuppusamy, P. EPR Imaging: Theory and Instrumentation. In *Multifrequency Electron Paramagnetic Resonance: Theory and Applications*; Misra, S. K., Ed.; Wiley-VCH Verlag GmbH & Co. KGaA: Weinheim, **2011**; pp. 755-793.

[2] Abragam, A.; Bleaney, B. *Electron paramagnetic resonance of transition ions,* 1st ed; Oxford University Press: Oxford, **1970**.

[3] Ivanshin, V.; Deisenhofer, J.; Krug von Nidda, H.; Loidl, A.; Mukhin, A.; Balbashov, A.; Eremin, M. ESR study in lightly doped$La_{1-x}Sr_xMnO_3$. *Phys. Rev. B,* **2000**, *61*, 6213-6219. [http://dx.doi.org/10.1103/PhysRevB.61.6213]

[4] Popovych, V.; Bester, M.; Stefaniuk, I.; Kuzma, M. Dyson line and modified Dyson line in the EPR measurements. *Nukleonika,* **2015**, *60*, 385-388. [http://dx.doi.org/10.1515/nuka-2015-0068]

[5] Tadyszak, K.; Augustyniak-Jabłokow, M.; Więckowski, A.; Najder-Kozdrowska, L.; Strzelczyk, R.; Andrzejewski, B. Origin of electron paramagnetic resonance signal in anthracite. *Carbon,* **2015**, *94*, 53-59. [http://dx.doi.org/10.1016/j.carbon.2015.06.057]

[6] Poole, C.P., Jr *Electron Spin Resonance— Comprehensive Treatise on Experimental Techniques,* 2nd ed; Wiley Interscience: New York, **1983**.

[7] Misra, S.K. Spin Hamiltonians and Site Symmetries for Transition Ions., **2011**. [http://dx.doi.org/10.1002/9783527633531.ch7]

[8] Misra, S.K. Simulation of EPR Spectra. *Multifrequency Electron Paramagnetic Resonance: Theory and Applications*; Misra, S.K., Ed.; Wiley-VCH Verlag GmbH & Co. KGaA: Weinheim, **2011**, pp. 417-454.
[http://dx.doi.org/10.1002/9783527633531.ch9]

[9] Stoll, S.; Schweiger, A. EasySpin, a comprehensive software package for spectral simulation and analysis in EPR. *J. Magn. Reson.,* **2006**, *178*(1), 42-55.
[http://dx.doi.org/10.1016/j.jmr.2005.08.013] [PMID: 16188474]

[10] Misra, S.K.; Buckmaster, H.A.; Reijerse, E.; Subramanian, S.; Krishna, M.C. In *Multifrequency Electron Paramagnetic Resonance: Theory and Applications*; Misra, S. K., Ed.; Wiley-VCH Verlag GmbH & Co. KGaA: Weinheim, **2011**; pp. 115-228.

[11] Calle, C.; Sreekanth, A.; Fedin, M.; Forrer, J.; Garcia-Rubio, I.; Gromov, I.; Hinderberger, D.; Kasumaj, B.; Léger, P.; Mancosu, B.; Mitrikas, G.; Santangelo, M.G.; Stoll, S.; Schweiger, A.; Tschaggelar, R.; Harmer, J. Pulse EPR Methods for Studying Chemical and Biological Samples Containing Transition Metals. *Helv. Chim. Acta,* **2006**, *89*, 2495-2521.
[http://dx.doi.org/10.1002/hlca.200690229]

[12] Kulik, L.; Lubitz, W. Electron-nuclear double resonance. *Photosynth. Res.,* **2009**, *102*(2-3), 391-401.
[http://dx.doi.org/10.1007/s11120-009-9401-y] [PMID: 19184518]

[13] Feher, G. Observation of nuclear magnetic resonances *via* the electron spin resonance line. *Phys. Rev.,* **1956**, *103*, 834-835.
[http://dx.doi.org/10.1103/PhysRev.103.834]

[14] Hyde, J.S.; Maki, A.H. ENDOR of a free radical in solution. *J. Chem. Phys.,* **1964**, *40*, 3117-3118.
[http://dx.doi.org/10.1063/1.1724957]

[15] Kurreck, H.; Kirste, B.; Lubitz, W. *Electron nuclear double resonance spectroscopy of radicals in solution,* 1st ed; VCH: Weinheim, **1988**.

[16] Kevan, L.; Kispert, L. *Electron spin double resonance spectroscopy,* 1st ed; Wiley: New York, **1976**.

[17] Davies, E.R. A new pulse ENDOR technique. *Phys. Lett. A,* **1974**, *47*, 1-2.
[http://dx.doi.org/10.1016/0375-9601(74)90078-4]

[18] Mims, W.B. Pulsed ENDOR experiments. *Proc. R. Soc. Lond. A Math. Phys. Sci.,* **1965**, *283*, 452-457.
[http://dx.doi.org/10.1098/rspa.1965.0034]

[19] Schweiger, A.; Jeschke, G. *Principles of pulse electron paramagnetic resonance spectroscopy,* 1st ed; Oxford University Press: Oxford, **2001**.

[20] Hubrich, M.; Hoffmann, E.C.; Schweiger, A. Pulse-Train-Detected Electron-Spin-Echo envelope modulation. *J. Magn. Reson. A,* **1995**, *114*, 271-273.
[http://dx.doi.org/10.1006/jmra.1995.1139]

[21] Misra, S.K. In *Multifrequency Electron Paramagnetic Resonance: Theory and Applications*; Misra, S. K., Ed.; Wiley-VCH Verlag GmbH & Co. KGaA: Weinheim, **2011**; pp. 1-22.

[22] Lynch, W.; Earle, K.; Freed, J. 1-mm wave ESR spectrometer. *Rev. Sci. Instrum.,* **1988**, *59*, 1345-1351.
[http://dx.doi.org/10.1063/1.1139720]

[23] Smith, G.M.; Le Surf, J.C.G.; Mitchell, R.H.; Riedl, P.C. Quasi-optical cw mm-wave electron spin resonance spectrometer. *Rev. Sci. Instrum.,* **1998**, *69*, 3924-3927.
[http://dx.doi.org/10.1063/1.1149200]

[24] Konchits, A. A.; Shanina, B. D.; Valakh, M. Y.; Yanchuk, I. B.; Yukhymchuk, V. O.; Alexeev, A. D.; Vasilenko, T. A.; Molchanov, A. N.; Kirillov, A. K. Local structure, paramagnetic properties, and porosity of natural coals: spectroscopic studies. *J. Appl. Phys.,* **2012**, *112*, 043505-1-9.
[http://dx.doi.org/10.1063/1.4745015]

[25] Bratek, K.; Bratek, W.; Gerus-Piasecka, I.; Jasien'ko, S.; Wilk, P. Properties and structure of different rank anthracites. *Fuel,* **2002**, *81*, 97-108. [http://dx.doi.org/10.1016/S0016-2361(01)00120-X]

[26] Garif'yanov, N.S.; Kozyrev, B.M. Paramagnetic resonance in anthracite and other carbonaceous substances. *Sov. Phys. JETP,* **1956**, *3*, 255-258.

[27] Brach, I.; Zanchetta, J.V.; Giuntini, J.C. Study of a peranthracite by conductivity and electron spin resonance. *Fuel,* **1995**, *74*, 779-781. [http://dx.doi.org/10.1016/0016-2361(95)00002-M]

[28] Fedaruk, R.; Adashkevich, S.V. Manifestation of strong coupling between microwave photons and spins in the EPR spectrum of anthracite. *J. Appl. Spectrosc.,* **2014**, *81*, 355-359. [http://dx.doi.org/10.1007/s10812-014-9936-3]

[29] von Bardeleben, H.J.; Cantin, J.L.; Zeinert, A.; Racine, B.; Zellama, K.; Hai, P.N. Spins and microstructure of hydroganated amorphous carbon: a multiple frequency electron paramagnetic resonance study. *Appl. Phys. Lett.,* **2001**, *78*, 2843-2845. [http://dx.doi.org/10.1063/1.1370980]

[30] Coleman, J.N.; Dalton, A.B.; Curran, S.; Rubio, A.; Davey, A.P.; Drury, A.; McCarthy, B.; Lahr, B.; Ajayan, P.M.; Roth, S.; Barklie, R.C.; Blau, W.J. Phase separation of carbon nanotubes and turbostratic graphite using a functional organic polymer. *Adv. Mater.,* **2000**, *12*, 213-216. [http://dx.doi.org/10.1002/(SICI)1521-4095(200002)12:3<213::AID-ADMA213>3.0.CO;2-D]

[31] Beuneu, F.; l'Huillier, C.; Salvetat, J-P.; Bonard, J-M.; Forro, L. Modification of multiwall carbon nanotubes by electron irradiation: An ESR study. *Phys. Rev. B,* **1999**, *59*, 5945-5949. [http://dx.doi.org/10.1103/PhysRevB.59.5945]

[32] Misra, S.K.; Petkov, V. Electron paramagnetic and muon spin-resonance studies in fullerenes. *Appl. Magn. Reson.,* **1995**, *8*, 277-310. [http://dx.doi.org/10.1007/BF03162792]

[33] Vlasova, M.V.; Kakazei, N.G.; Savvakin, G.I. Some properties of ultradispersion diamonds prepared under high-temperature explosion synthesis conditions, *Izv. Akad. Nauk SSSR, Neorg. Mater,* **1979**, *15*, 1303-1305.

[34] Shames, A.I.; Panich, A.M.; Kempiński, W.; Alexenskii, A.E.; Baidakova, M.V.; Dideikin, A.T.; Osipov, V.Yu.; Siklitski, V.I.; Osawa, E.; Ozawa, M.; Vul', A.Ya. Defects and impurities in nanodiamonds: EPR, NMR and TEM study. *J. Phys. Chem. Solids,* **2002**, *63*, 1993-2001. [http://dx.doi.org/10.1016/S0022-3697(02)00185-3]

[35] Andersson, O.E.; Prasad, B.L.V.; Sato, H.; Enoki, T.; Hishiyama, Y.; Kaburagi, Y.; Yoshikawa, M.; Bandow, S. Structure and electronic properties of graphite nanoparticles. *Phys. Rev. B,* **1998**, *58*, 16387-16395. [http://dx.doi.org/10.1103/PhysRevB.58.16387]

[36] Prasad, B.L.V.; Sato, H.; Enoki, T.; Hishiyama, Y.; Kaburagi, Y.; Yoshikawa, M.; Rao, A.M.; Eklund, P.C.; Endo, M. Heat treatment effect on the nanosized graphite π-electron system during diamond to graphite conversion. *Phys. Rev. B,* **2000**, *62*, 11209-11218. [http://dx.doi.org/10.1103/PhysRevB.62.11209]

[37] Vainer, V.S.; Il'in, V.A. Electron spin-resonance of exchange-coupled vacancy pairs in hexagonal silicon carbide. *Soviet Physics: Solid State,* **1981**, *23*, 2126-2133.

[38] Koehl, W.F.; Buckley, B.B.; Heremans, F.J.; Calusine, G.; Awschalom, D.D. Room temperature coherent control of defect spin qubits in silicon carbide. *Nature,* **2011**, *479*(7371), 84-87. [http://dx.doi.org/10.1038/nature10562] [PMID: 22051676]

[39] Andronenko, S. I.; Stiharu, I.; Misra, S. K. Synthesis and characterization of polyureasilazane derived SiCN ceramics. *J. Appl. Phys.,* **2006**, *99*, 113907-1-5. [http://dx.doi.org/10.1063/1.2202291]

[40] Savchenko, D.; Kulikovsky, V.; Vorliček, V.; Lanček, J.; Kiselov, V.; Kalabukhova, E. Optical and magnetic resonance study of a-SiC_xN_y films obtained by magnetron sputtering. *Phys. Status Solidi, B Basic Res.*, **2014**, *251*, 1178-1185. [http://dx.doi.org/10.1002/pssb.201451041]

[41] Trassl, S.; Motz, G.; Rossler, E.; Ziegler, G. Characterization of the free-carbon phase in precursor-derived SiCN ceramics. *J. Non-Cryst. Solids*, **2001**, *293-295*, 261-267. [http://dx.doi.org/10.1016/S0022-3093(01)00678-0]

[42] Trassl, S.; Motz, G.; Rossler, E.; Ziegler, G. Characterization of the Free-Carbon Phase in Precursor-Derived Si-C-N Ceramics: I, Spectroscopic Methods. *J. Am. Ceram. Soc.*, **2002**, *85*, 239-245. [http://dx.doi.org/10.1111/j.1151-2916.2002.tb00072.x]

[43] Trassl, S.; Kleebe, H-J.; Stormer, H.; Motz, G.; Rossler, E.; Ziegler, G. Characterization of the Free Carbon Phase in Precursor-Derived SiCN Ceramics Part II: Comparison of Different Precursors. *J. Am. Ceram. Soc.*, **2002**, *85*, 1268-1275. [http://dx.doi.org/10.1111/j.1151-2916.2002.tb00256.x]

[44] Erdem, E.; Mass, V.; Gembus, A.; Schulz, A.; Liebau-Kunzmann, V.; Fasel, C.; Riedel, R.; Eichel, R-A. Defect structure in lithium-doped polymer-derived SiCN ceramics characterized by Raman and electron paramagnetic resonance spectroscopy. *Phys. Chem. Chem. Phys.*, **2009**, *11*(27), 5628-5633. [http://dx.doi.org/10.1039/b822457a] [PMID: 19842480]

[45] Berger, F.; Muller, A.; Albinger, F.; Muller, K. Solid state NMR investigation on Si-B-C-N ceramics derived from boron-modified poly(allylmethylsilazane). *Z. Anorg. Allg. Chem.*, **2005**, *631*, 355-363. [http://dx.doi.org/10.1002/zaac.200400259]

[46] Li, Y.; Yu, Y.; San, H.; Han, Q.; An, L. Effect of Pyrolysis Temperature on the Electrical Behavior of Polymer-Derived SiOCN Ceramic. *J. Mater. Sci. Chem. Eng*, **2015**, *3*, 9-16.

[47] Kleebe, H-J.; Störmer, H.; Trassl, S.; Ziegler, G. Thermal stability of SiCN ceramics studied by spectroscopy and electron microscopy. *Appl. Organomet. Chem.*, **2001**, *15*, 858-866. [http://dx.doi.org/10.1002/aoc.243]

[48] Tomasella, E.; Rebib, F.; Dubois, M.; Cellier, J.; Jacquet, M. Structural and optical properties studies of sputtered a-SiCN thin films. *J. Phys. Conf. Ser.*, **2008**, *100*, 082045. [http://dx.doi.org/10.1088/1742-6596/100/8/082045]

[49] Tomasella, E.; Spinelle, L.; Bousquet, A.; Rebib, F.; Dubois, M.; Eypert, C.; Gaston, J.P.; Cellier, J. Structural and Optical Investigations of Silicon Carbon Nitride Thin Films Deposited by Magnetron. *Plasma Process. Polym.*, **2009**, *6*, S11-S16. [http://dx.doi.org/10.1002/ppap.200930103]

[50] Kobayashi, K.; Yokoyama, H.; Endoh, M. Leakage current and paramagnctic defects in SiCN dielectrics for copper diffusion barriers. *Appl. Surf. Sci.*, **2008**, *254*, 6222-6225. [http://dx.doi.org/10.1016/j.apsusc.2008.02.144]

CHAPTER 2

Resolution of EPR Signals in Graphene-based Materials from Few Layers to Nanographites

Francesco Tampieri and **Antonio Barbon***

Department of Chemical Sciences, University of Padova, Padova, Italy

Abstract: The focus of this chapter is set on the application of EPR methods to carbon-based materials, from nanographites to graphene-based materials, for the resolution and characterization of the different signals, related to the presence of specific species, or structures. Because of the intrinsic heterogeneity of the samples, this goal is not simple: most of the signals coming from different types of structures have similar spectroscopic features and are overlapping in the cw-EPR spectra with very different relative intensities. It is then necessary to use all possibilities that EPR offers, from the cw-EPR techniques to pulse EPR methods, to disentangle ideally all contributions. Our analysis of the EPR spectra considers the presence of three types of paramagnetic contributions: conduction electrons, edge states and molecular states. This interpretation framework has been shown to be effective for the considered materials, characterized by the presence of finite-dimension graphene layers, eventually stacked one above the other. In our analysis, we investigated different experimental parameters, like the variation in the temperature of the EPR intensity, the values of the g-tensors and the homogeneous lineshapes of the spectra to obtain further structural information. Pulse EPR methods were used to study and characterize species with long relaxation times (molecular states). Echo-detected EPR enabled to obtain their spectral lineshapes. Hyperfine spectroscopies, ESEEM, ENDOR and HYSCORE, determined the electron hyperfine couplings of unpaired electrons with magnetic nuclei, thus allowing the evaluation of the extent of the π-system and the presence of different types of nuclei.

Keywords: Conduction electrons, Dysonian lineshape, Edge states, Echo-detected EPR (ED-EPR), ENDOR, EPR, ESEEM, Expanded graphite, FT EPR, Graphene defects, Graphite, Hyperfine interaction, HYSCORE, Lorentzian lineshape, Molecular states, Nanographite, Natural graphite, π-systems, Reduced graphene oxide (RGO), Synthetic graphite.

* **Corresponding author Antonio Barbon:** Department of Chemical Sciences, University of Padova, Padova, Italy; Tel: +39 049 8275151; E-mail: antonio.barbon@unipd.it

Dariya Savchenko & Abdel Hadi Kassiba (Eds.)

INTRODUCTION

Electron paramagnetic resonance (EPR) spectroscopy has proven to be a valuable technique to study graphitic materials and recently also graphenic systems. In his paper on graphite, published in 1960 [1], G. Wagoner thoroughly characterized the EPR signals of mobile charge carriers in a single graphite crystal. Nevertheless, this fascinating argument, almost sixty years later, is still demanding attention [1, 2]. Beside single crystals, the interest has been risen by graphite powders [3, 4], nanographitcs [5 - 10], activated carbon fibers [11, 12], and carbon nanotubes [13, 14]. Many EPR techniques have been exploited to study graphitic systems under different experimental conditions: different temperatures [15], different microwave frequencies [16] and pulsed experiments [17].

For non-ideal graphitic materials, the magnetic properties exhibit certain variability, mostly due to the presence of *defects*: theoretical studies show how imperfections, particle terminations and other sources of crystal faults determine the magnetic properties of graphene-based materials [18], inferring them antithetic anti/ferromagnetic properties [19, 20]. The EPR technique is particularly convenient because of its spectral resolution, enabling to distinguish contributions from different paramagnetic species. Moreover, pulse techniques have been used to isolate species with long relaxation times [8, 17, 21], or identify species with unresolved hyperfine interactions [6 - 8, 21, 22, 23].

It is rather surprising that the potentialities of this spectroscopy have not been fully exploited yet to study the properties of graphene and related materials. After the seminal EPR paper on graphene obtained by the scotch-tape technique in 2009 by L. Ćirić et al. [24], more publications on the topic followed, sometimes reporting contradictory results [18, 19, 22, 25, 26], indicating that the studied systems are complex, and require a more general new approach for their characterization.

The focus of this chapter is on the EPR studies conducted on different types of nanographitic and graphene-like materials, produced by completely different methods, with the aim of finding analogies that can be attributed to the presence of similar structures and/or, at the same time, to give insights into the variability in the properties within materials with very different properties. The interpretation of the characteristic response of the different materials will be offered in terms of the presence of different types of *active sites,* responsible for the macroscopic behavior of the material. All the examples are taken either from published and unpublished materials that have been investigated in our laboratory.

Bulk Material, Edges and Defects

In the following, we introduce the studies conducted in our laboratory on molecular structures, produced by different methods ,determining the different EPR signals in carbon-based materials.

When approaching this subject, first of all, we note that all the structures generating an EPR response are based on sp^2-hybridized carbon atoms, organized in hexagons to form 2D networks (graphenic layer) of a variable extent that can eventually stack together in more complex 3D structures.

The 2D structures are stabilized by σ-bonds and π-bonds. They are formed by overlapping of either $sp2$ orbitals (σ-bonds), or p_z orbitals (π-bonds). Only a π-system can host unpaired electrons, detectable by EPR. The C – C bond length in graphene is 1.42 Å; the unit cell contains two carbon atoms and has an area of 5.2 Å^2; the density of graphene is therefore 0.77 mg m^{-2}.

3D structures are further stabilized by π-π interactions between stacked layers.

Within the single graphene layer, the delocalized π-system spreads over the entire layer, and it is energetically separated into two bands; electrons at 0 K completely fill the lower band (valence band), whereas the higher band (conduction band) is completely unoccupied. The energy gap between the two bands is zero (zero-band gap semiconductor, see Scheme **1a**). Conduction electrons are obtained by the promotion of electrons to the conduction band.

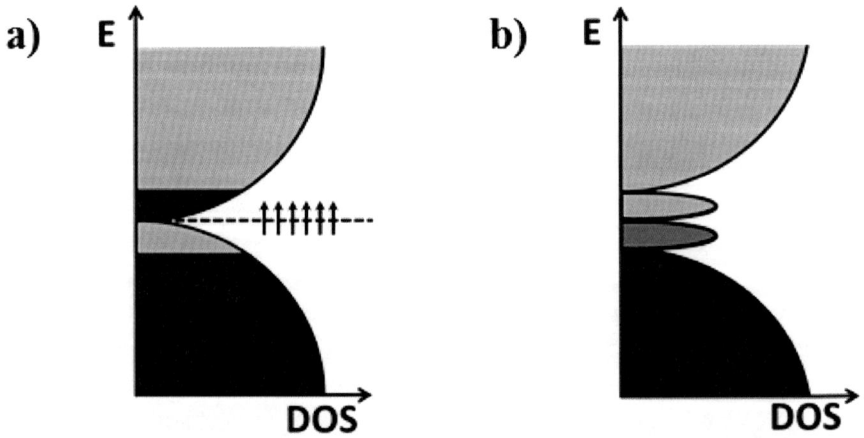

Scheme 1. a) sketch of the density of states for a graphitic or graphenic system in the presence of edge states for $T > 0$ K. Adapted from [27]. b) Sketch of Density of states with edge states split by magnetic exchange interaction at 0 K. Adapted from [28]

Edge states are formed because the π system is limited in extension, and at the boundaries there might be a localization of electrons (classified then as edge states).

Finally, we note that the dimensions of the π-system can be limited, as due to the breaking of the graphenic *motif* (carbon atoms with sp^3 hybridization, vacancies, etc.). The consequent effect, besides the formation of edges, is that the distribution in energy of the levels is altered. In particular, when particles of small dimension are formed, the bands are composed of a discrete number of levels, and the system is better described in terms of molecular states; we will consider the electrons present in these structures as confined within molecular states.

We will interpret the EPR data obtained in nanographenes and graphene-like materials in terms of contributions coming from conduction electrons, edge states and molecular states.

In this section, we will first discuss with more details the general characteristics of these different contributions then we will show how to identify their fingerprint, and finally discuss, when possible, the variability of the signals coming from these contributions in different types of materials.

Conduction Electrons

Within zero-bandgap semiconductors, like graphenic and graphitic systems, electrons are thermally promoted from the valence (bonding) band to the higher energy conduction (antibonding) band. G. Wagoner [1] has unequivocally shown that these are the electrons (together with the generated holes) responsible for the EPR signals in a graphite single crystal, based on the observation that the lineshape is Dysonian, and on the temperature dependence of the EPR intensity. The thermal population of the states was used by McClure [29] for a theoretical estimation of the magnetization, for both graphene and graphite, as a function of the temperature. The experimental trend of the magnetization as a function of the temperature is linear with positive slope, in stark contrast with a common Curie paramagnetic behavior. At very low temperature, a residual magnetization temperature independent Pauli magnetization contribution prevents the vanishing of the magnetization.

Fig. (**1**) reports the variation in the temperature of the magnetization of graphite. For graphene, the trend is similar (see [24]). In defective samples, the magnetization measured by standard methods varies in more complex ways because of the presence of several contributions. In these cases, the EPR spectrum can show the contribution of more overlapping signals. The temperature dependence of their intensities gives information on the different contributions to the overall magnetization of the sample. The electron-electron interaction is responsible for a short spin relaxation time [30], generating intrinsic broad linewidths.

The presence of narrow EPR lines can be attributed to a partial or complete averaging of the principal values of the *g*-tensor. This effect can be, for example, due to a fast jump of electrons between different particles inside conductive samples composed of isotropically distributed particles [31]. A modified sample with different inter-particle interactions should show a different EPR behavior.

Some authors attribute the averaging of the *g*-values of conduction electrons to their fast switching back and forth between the conduction and the valence band [17]. In this case, it is implicitly assumed that there must be a strong variation of the *g* principal values as a function of the energy in order to have an efficient averaging. Readers are referred to Chapter 3 for insights on the argument.

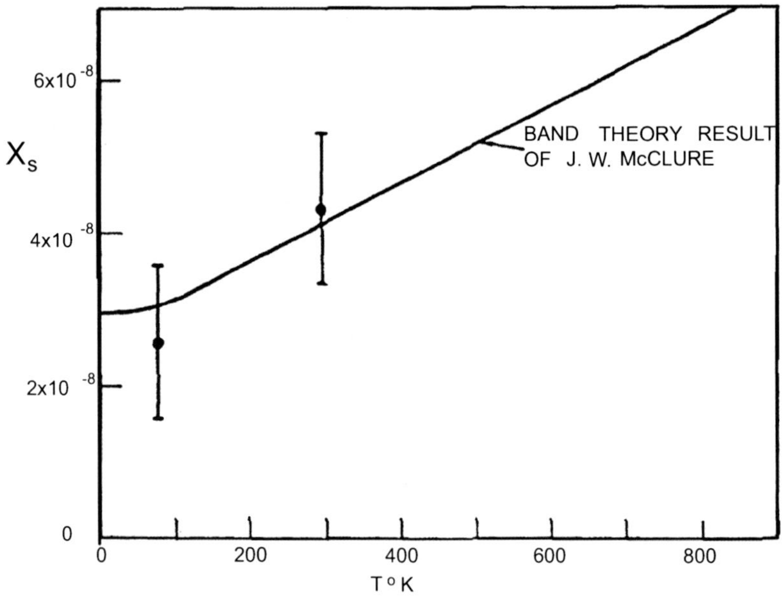

Fig. (1). Magnetization as a function of the temperature for a single crystal of graphite (Reprinted with Permission from [1]. Copyright (1960) by the American Physical Society).

Edges and Edge States

A non-ideal graphene layer has edges at its termination or where the π-system is interrupted because of the presence of defects.

From a topological point of view, these edges are in principle generated by the cut of an infinitely extended graphene layer. Cuts can be realized by bisecting two opposite bonds within the hexagons (thus separated by two bonds), or it can be realized by cutting two bonds separated by one bond (see Scheme **2**).

The structures of the edges obtained by the two methods are topologically different and are named respectively *zigzag* and *armchair* edges (Scheme **2**). Zigzag-edged and armchair-edged finite size graphene have different properties.

The interruption of the π-system in finite size graphenes gives rise to localized electrons in *edge states* which are energetically localized at the Fermi levels (see Scheme **1a**), and are singly occupied [32].

These states have the same nature of those found in non-Kekulé structures [33], where not all electrons can pair to form bonds. A relatively simple way to determine the presence of unpaired electrons can be found in Ref [34]. An instructive example is offered by the case of phenalenyl derivatives [35], used as graphene-like fragments, and showing magnetic properties.

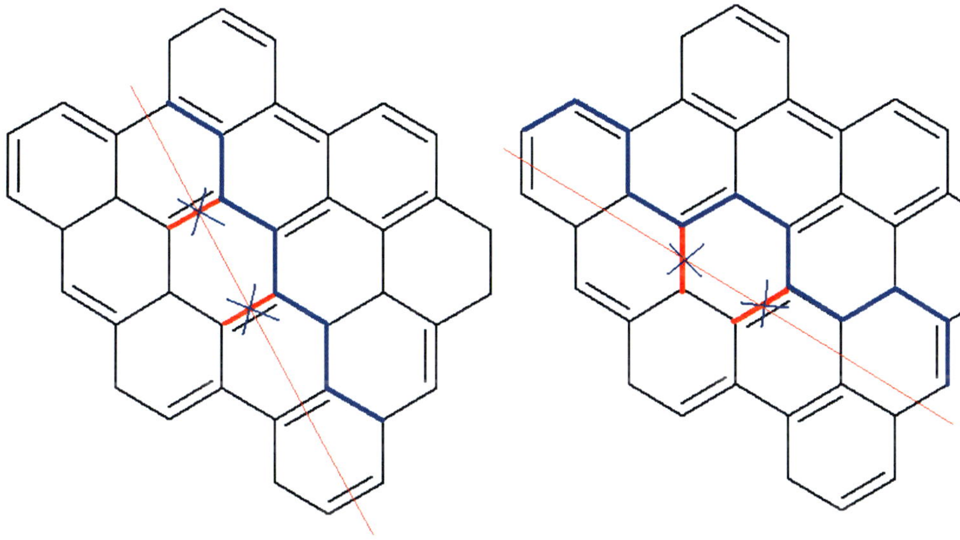

Scheme 2. Formation of the zigzag edges (left, blue color) or of the armchair edges (right, blue color) as resulting from the cutting of two different non adjacent bonds of the graphene hexagons.

Another known effect of the presence of finite size graphenes is related to the stabilization of open shell structures below the close-shell ground state [36].

Edge states are characterized by orbitals with very narrow extension toward the interior of the material, in particular for zigzag edges, where the wavefunction is practically defined in terms of a combination of p_z orbitals of the carbon atoms forming the edge [7]. The wavefunction describing these electrons is similar to those found in polyenes [37] and their energy (close to Fermi level) corresponds to the energy of the p_z carbon orbital [32], like the case of non-Kekulé molecules.

Conduction electrons interact with edge electrons [38] due to their mobility and delocalization over the whole π-system. The mechanism is analogous to that of paramagnetic impurities in metals, where the Kondo effect has been found, i.e. the antiferromagnetic pairing into singlets between impurities and conduction electrons below a given temperature (T_K). The specificity of graphite and graphene is that they are in general zero-band gap semiconductors (for graphene-like materials the band gap can be non-zero), with the consequence that T_K should be likely very low.

Conduction electrons play another important role for edges, that of being able to act as a mediator for the exchange interaction between nearby edges, the so-called Ruderman-Kittel-Kasuya-Yoshida (RKKY) coupling [39, 40]. This coupling determines the presence, in graphite or in graphene-like materials, of contributions to the magnetization, which can be either ferromagnetic or antiferromagnetic. Saremi [41] has shown that RKKY coupling between magnetic impurities is ferromagnetic only on the same sublattice but antiferromagnetic on different sublattices.

To conclude, we mention that dangling bonds might remain at the termination of the structure, and eventually, they can rearrange to couple electrons into new-forming bonds and minimize the energy. It is also possible, in a reactive environment, to have flake terminations passivated, typically, with hydrogen or oxygen atoms.

Defects and Molecular States

Like for edges, imperfections and defects are very common in carbon-based materials. They change deeply the chemical and physical properties of graphite- and graphene-based materials. Sometimes defects come from imperfections in the crystal lattice as due to reactions or to mechanical events, sometimes from the synthesis of the materials and sometimes they are introduced on purpose to modify and tune the reactivity of the systems.

Defects in three-dimensional crystals are called *intrinsic* if the crystalline order is perturbed without the introduction of foreign atoms. Within this category, several different types of defects are responsible for an interruption of the π-system, and the reader is referred to Ref. [42] for details.

Some of these defects are associated to the presence of magnetic moments. It has been shown, for example, that the creation of vacancies by bombardment with ions is associated with a rise of the magnetization of the material [18].

Besides the formation of holes or dislocations, the produced voids are responsible for an interruption of the π-system and this can create small islands of carbon atoms with their p_z orbitals that are not communicating with the larger π-system. The same effect can be obtained by the mechanical breaking of the material, with the separation of pieces of very small dimensions.

π-systems formed by a limited number of carbon atoms are better described by using tools that are appropriate for molecular systems. Indeed, the molecular approach is the most appropriate to discuss spectroscopic results, for example, when hyperfine EPR methods are applied.

Oxygen

Here we comment on the presence of oxygen in carbon-based samples. EPR spectroscopists work, as a rule, under deaerated conditions, or with samples kept under inert gas, to avoid the spin relaxation produced by the paramagnetism of oxygen gas. This is mandatory for solutions, but often EPR spectra of solid (crystalline) samples are obtained in the presence of air.

For carbon-based materials it is a good practice to carefully remove oxygen from samples, first of all, because, often, the signals are due to edge states, and then they are associated with paramagnetic centers located on the surface, in direct contact with oxygen. On top of this, oxygen molecules can be adsorbed on the graphite/graphene layers [43] and have shown to induce swelling of nanopores in activated carbon fibers.

Oxygen not only changes the relaxation properties [11], but also the magnetic properties [44, 45]; a charge transfer to oxygen has even been found with the production of superoxide [46]. It appears that the interaction between oxygen and graphite or graphenic materials is weak, most likely a simple physisorption. Indeed cycles of removing of oxygen and reoxygenation of the samples show a reversible quenching/reappearance of the EPR signals.

The only advantage of working under the aerated condition is the thermal contact that the gas offers with the environment, favoring the dissipation of the resistively heating of the sample, that is particularly important when working at high microwave power [3]. In general, then, it is advisable to carefully remove the oxygen, and this often requires the samples to remain under dynamic vacuum overnight.

CONTINUOUS WAVE EPR SPECTRA

The cw-EPR spectra of carbon-based materials show the overlapping signals of all the paramagnetic centers present in the samples. The disentangling of the spectra is possible when some features, like linewidths or *g*-factors, are relevantly different. When the spectra are a superposition of signals from different types of local structures, the separation of the different contributions can be a difficult task, and we are able to separate two or three different contributions at most. We will discuss in the following the main features of the cw-EPR spectra obtained from some different types of characteristic graphitic systems taken as examples, highlighting analogies and differences, and stressing the information that can be obtained from the measurements.

Subsequently, in the next section, we will discuss as the disentangling of the signals can be done by exploiting the differences in electron spin relaxation.

Lineshape

Anisotropies in the magnetic interactions result in a modification of the homogeneously broadened lineshape, which is of two fundamental types: Lorentzian or Dysonian. In the presence of anisotropic interactions, the resulting spectrum is inhomogeneously broadened, given by a convolution of homogeneously broadened lines as described in Chapter 1, where the total spectrum line profile is given by eq. (5).

The first feature to discuss when considering nanographites and graphene-based materials is the homogeneous lineshape, as samples are often conductive, and this implies that the lineshape can be Dysonian. We then consider the consequences of the conductivity either on the lineshape or on the *g*-values.

In Fig. (**2**) we show the cw-EPR spectra of four different samples having different lineshapes. We have two graphites, natural Madagascar graphite, GMad, and the synthetic graphite, GPyr (100 Mesh, Aldrich). We also have two materials derived from these by mechanical treatment (nanographite G90) or by chemical treatment (expanded graphite EH).

Sample EH is expanded graphite, which is obtained by intercalation with acids of GMad, followed by a strong microwave treatment [8]. The material is characterized by a consistent reduction of the density due to a separation of the graphenic layers. Sample G90 is obtained by ball milling of the synthetic graphite GPyr for 1.5 h [6, 7].

We begin by comparing the lineshape of the spectra of the two graphite samples GMad and GPyr that have been used to prepare nanographites of graphene-like materials. The spectrum of sample GMad is very asymmetric, giving a clear indication of a mixing of absorption and dispersion components, typical of

Dysonian lineshapes, in line with what accurately documented by G. Wagoner [1] for bulk crystalline graphite.

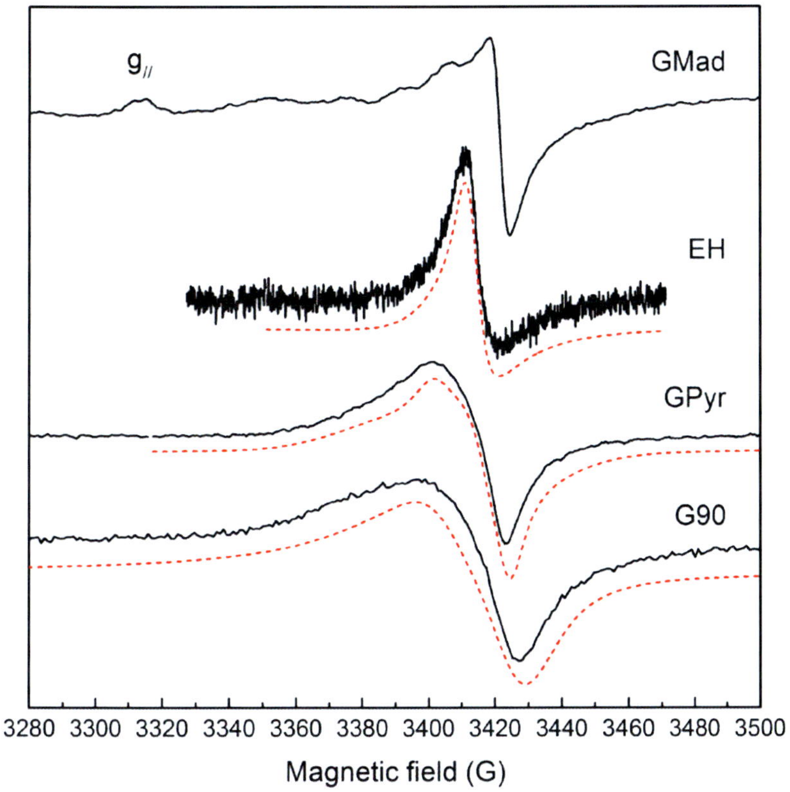

Fig. (2). X-band cw-EPR spectra of high purity Madagascar graphite (Gmad), and expanded graphites [8] obtained by treatment with acids (EH), GPyr is a commercial pyrolitic powder graphite (100 Mesh), and G90 is after ball milling for 1.5 h [6]. The spectra are recorded at 260 K (black lines), and are overlapped to their simulations (red dashed lines). The parameters obtained from the simulation are collected in Table **1**.

The spectrum of GMad could be simulated neither as a single crystal nor as a powder sample, as it is composed of relatively large crystallites resulting in some evident features within the spectrum. Nevertheless, the $g_{\parallel}$ ($g_{\parallel} = 2.06$) component (shown in Fig. **2**) can be distinguished. This value is in line with an estimate of the $g_{\parallel}$ value typical of macroscopic single crystals of graphite taken from the literature [1, 47], moreover, because the line asymmetries of the two spectra are also similar (this can be proved also by the integration of the spectra), we then deduce that the lineshape for GMad is also Dysonian.

For the other graphite, GPyr, the simulation of the powder spectrum (see the best fit parameters in Table **1**) shows that two components are resolved, both with a

Dysonian homogeneous lineshape, as expected on the base of the dimension of the particles: the dimension of the mesh fixes the largest particle dimension to 5-6 μm. In this case, because the skin depth [48] is of the order of few μm in graphite [1], the particles are not completely penetrated by microwaves, and, according to theory [49], the lineshape is Dysonian.

Table 1. Relevant simulation parameters of the cw-EPR spectra at 260 K reported in Figs. (2 and 3). For each contribution, the linewidth and the principal components of the *g*-tensors are reported (isotropic–starred- or parallel, $g_{i\parallel}$ and perpendicular, $g_{i\perp}$) and α is the asymmetry parameter for the Dysonian lineshape [50]. The relative abundance of the species (C_i) is indicated.

Sample	% C_i	$g_{1\parallel}$	$g_{1\perp}$	Γ_1 (G)	$g_{2\parallel}$	$g_{2\perp}$	Γ_2 (G)	α[a]
GPyr	80	2.016	2.007 2.0023	16[a]	2.020*	-	30[a]	-0.4 -1[e]
EH	89.2	-	2.0050	14.5[b]	-	2.0032	6.60[a]	3.7
G90	45	2.020	2.007 2.004	35 [a]	2.020*	-	35[d]	-1
G1200[f]	96	2.013	2.0034 2.0023	25[c]	2.0031	2.0030 2.0029	4[c]	-
EHK	73	2.0099	2.0029	11.2[c]	2.0070	2.0025	5.28[c]	-
RGO	79	2.0035*	-	12.8	2.0035*	-	4.47	-
MAPTES[g]	55 30 15	2.0037*	-	1.5 [c] 0.6 [c] 4.1 [c]	-	-	-	-

[a]General Dysonian lineshape [50]
[b]Bulk graphite lineshape [1]
[c]Lorentzian lineshape
[d]Gaussian lineshape
[e]Relative to the first and second species
[f]From Ref [6], $T = 270$ K
[g]From Ref [51]; three components with the same *g*-value but different linewidth, $T = 290$ K

Two components were used for the simulation of GPyr sample, having a general phenomenological Dysonian expression proposed by Bath [50] that in derivative form reads as:

$$\Gamma(B - B_0(\Omega)) \propto \frac{\alpha}{4[B - B_0(\Omega)]^2 - \Gamma^2} - \frac{8[B - B_0(\Omega)\cdot\{\Gamma + \alpha[B - B_0(\Omega)]\}]}{\{4[B - B_0(\Omega)]^2 - \Gamma^2\}^2} \quad \textbf{(1)}$$

where Γ is the homogeneous linewidth, and $B_0(\Omega)$ the resonance field at a given orientation Ω:

$$B_0(\Omega) = \frac{h\nu_{\exp}}{g_{eff}(\Omega)\beta} \tag{2}$$

Sample EH also shows the presence of two components, one of these having the same Dysonian lineshape of GPyr, eq. (1) with strong asymmetry (α = 3.7, see Table **1**) and the second relative to the typical bulk graphite lineshape [1]

$$Int \propto \Gamma^2 B \frac{\left[\frac{(B-B_0)^2}{\Gamma} - 1\right]^2 - 2}{\left[1 + \frac{(B-B_0)^2}{\Gamma^2}\right]^2} = \Gamma^2 B \frac{\left[\frac{(B-B_0)^2}{\Gamma^2} - 2\left(\frac{(B-B_0)}{\Gamma} - 1\right)\right]}{\left[1 + \frac{(B-B_0)^2}{\Gamma^2}\right]} \tag{3}$$

The presence of this component indicates that the expansion process has left some parts of the sample slightly affected by the treatment, with a structure very much similar to that of the pristine graphite.

We deduce that the sample is composed of regions with relatively high conductivity, and these regions are larger than the skin depth δ.

The attribution of the signal observed in the EH sample as due to conduction electrons is further supported by an evident weak intensity associated with a strong decrease upon lowering of the temperature.

The spectrum of sample G90 is the superposition of two main components, a Dysonian contribution and a Gaussian contribution, this last with a relatively high *g*-value, but a small anisotropy, that we attribute to the field-induced orientation of some crystallites with the *c*-axis along the magnetic field [4, 6]. A Gaussian contribution appears when distributions of small interaction are present. In this case a Gaussian profile can be obtained from a *g*-strain of the parallel component of the *g*-tensor.

We note that when the magnetic field is along the *c*-axis also for big graphite crystals the lineshape can be Lorentzian [47].

Grinding of the material for 90 min (1.5 h) does not affect much the dimension of the crystallites, and, with respect to sample GPyr, only the linewidth of the first component is clearly different.

A clear change of the lineshape can be obtained with heavier treatment of the materials, either by further grinding or by further chemical exfoliation. Fig. (**3**) reports the spectra of samples G1200 and EHK, which are obtained by 20 h grinding (G1200), or by a second chemical expansion (EHK). For these samples, the lineshape is Lorentzian, meaning that the particle size is smaller than the skin width.

If the grinding is much longer than 1.5 h, the energy delivered to the sample is sufficient for either breaking the graphite crystallites into pieces smaller than the skin depth and/or for introducing defects in the lattice. It is plausible to suppose that grinding produces breaking, shearing, exfoliation and re-aggregation of the graphene layers and creating π-system with limited extension, systems with the increased disorder and with a larger concentration of defects and edges. Indeed, in the spectrum of G1200 it is observable a narrow, low-intensity band (starred in Fig. **3**) which is attributed to a signal coming from edge states. The intensity gets stronger when the sample is cooled to cryogenic temperatures, in contrast to intensities due to conduction electrons.

We have just seen that grinding of Gpyr for an increasing time produced particles with smaller and smaller dimensions; the same has been observed if multiple treatments are applied to GMad: EH, for instance, was further chemically exfoliated by intercalation of potassium followed by reaction with ethanol [8] to produce EHK. The lineshape of the cw-EPR spectrum of the material so obtained is Lorentzian (see Fig. **3** and Table **1**). The quantity of sample penetrated by the microwaves is larger, increasing the number of electrons interested in the resonance, giving a stronger signal for the sample.

Chemically expanded graphites, unfortunately, have not been able to provide a total exfoliation of graphite to obtain single layer graphene [8]. In order to investigate this latter type of material we selected graphene produced by Hummers method [52], followed by chemical reduction (reduced graphene oxide, RGO, commercial name 'Single layer graphene' from ACS). The characteristic of this sample is the presence of a single Lorentzian line with an isotropic *g*-value = 2.0035, like that found by L. Ćirić et al [24], in the seminal paper showing the EPR spectrum of an oriented graphene sample obtained by scotch-tape method [53]. As discussed later a practically isotropic *g*-tensor for this material is the effect of a weak spin-orbit coupling along the direction perpendicular to the plane (no stacking of layers, see further).

The last example is taken from graphitic materials included inside SiO_2 matrix and obtained by the thermal annihilation of sol-gel prepared silica (from methacryloxypropyl-triethoxysilane, MAPTES). It has been found that calcination of bulk material produces carbon nanoclusters inside the silica matrix [6]. Recent literature classifies these nanometric particles as *graphene quantum dots* [54, 55].

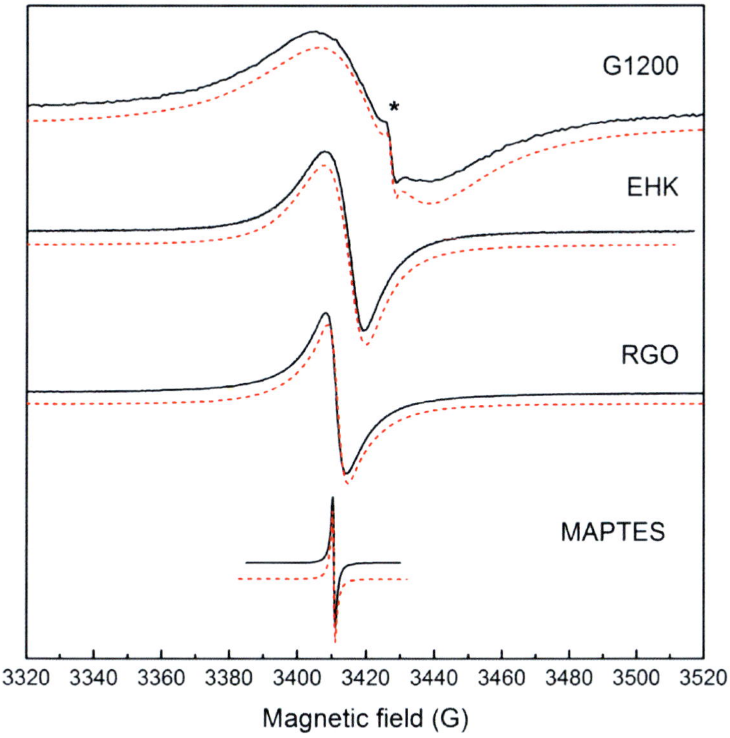

Fig. (3). X-band cw-EPR spectra at 260 K of an expanded graphite [8] (black lines) obtained by treatment with acids and followed by intercalation with potassium and reaction with ethanol (EHK). RGO is a commercial reduced graphene oxide ('Single layer graphene', ACS). MAPTES is a sample obtained by calcination in the oven of a methacryloxypropyl-triethoxy silane at 1300 K. The spectra are overlapped to their simulations (red dashed lines). The parameters obtained from the simulation are collected in Table **1**.

The spectrum is meaningfully different from all the others present in the figure, and it is composed by an overlap of very narrow Lorentzian lines with linewidth ranging from 0.6 to 4.1 G. Because of the limited dimension of the particles (the estimate of the dimension is 5 Å ca [51].), the system is likely closer to a molecular system, where the signal is mostly due to edge states. Indeed a strong electron spin echo (ESE) signal was observed.

In the next section we will see how a temperature change can modify, in some cases dramatically, the shape of the cw-EPR spectra.

Temperature Variation of the Spectra and Separation of the Components

So far, we have shown the variety of signals with different characteristics, ranging from Dysonian lineshape, to Lorentzian, and in some cases Gaussian. The temperature is a critical parameter that can strongly influence the cw-EPR spectrum.

The cw-EPR spectrum of another exfoliated graphite, EKS (obtained by intercalation of potassium, followed by reaction with ethanol and sonicated for 90 minutes), is shown in Fig. (**4**). The spectrum has been simulated as a sum of two components (both with Lorentzian lineshape), and has a behavior very similar to that of sample EHK. The component with an isotropic *g*-value (component 2) shows a monotonous decrease of the *g*-value with the temperature from 2.012 at room temperature to 2.004 at 10 K. The principal values of the *g*-tensor of the other component (component 1, giving the largest contribution at room temperature) are rather stable from room temperature down to 30 K ca, and then the parallel component drops quickly from 2.010 to 2.007 at 4 K.

The intensity of the two components as a function of the temperature is shown in Fig. (**4b**).

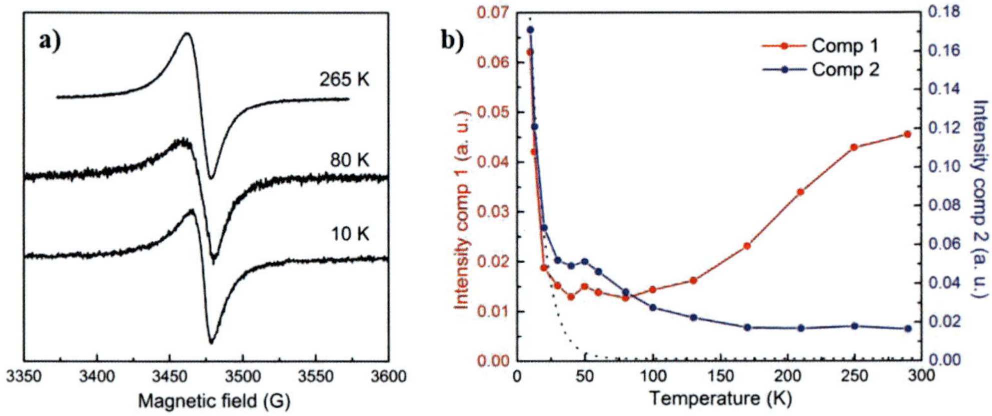

Fig. (4). a) cw-EPR spectrum of the EKS sample as a function of temperature. The sample has characteristics very similar to EHK, and it is simulated in all temperature range as a superposition of two Dysonian components. At room temperature component 1 has $g_{1\parallel} = 2.010$ and $g_{1\perp} = 2.002$, whereas component 2 has isotropic value of 2.012 and its value decreases constantly with the temperature down to 2.004 at 10 K b) EPR intensities of the two components as a function of the temperature. The dotted curve is the calculated typical Curie type trend.

On the base of what was reported in the first section of this chapter, component 1 in the range 300-50 K ca. is attributed to conduction electrons, as the intensity decreases with the lowering of the temperature. Below 50 K the intensity grows like for paramagnetic contributions, and at the same time the *g*-anisotropy quickly reduces. Therefore, in this last temperature range, the signal must be attributed to other structures (edges and/or molecular states), with paramagnetic characteristics. As it will be shown in the section of pulse-EPR methods, these components provide also spin echoes, and ESEEM and ENDOR spectra evidence hyperfine interaction with nuclei. According to the classification given before, these signals derive from molecular states.

Component 2 grows mildly from room temperature down to 150 K, and it is attributed to edge states interacting with conduction electron. Indeed edge states can deeply modify the temperature dependence of the magnetization as a function of the temperature, depending on the particle size [56].

Fig. (**5**) is an even more emblematic example of the variability of the spectrum profile on lowering of the temperature. In this figure three different contributions are resolved with different lineshape: Dysonian, Gaussian and Lorentzian.

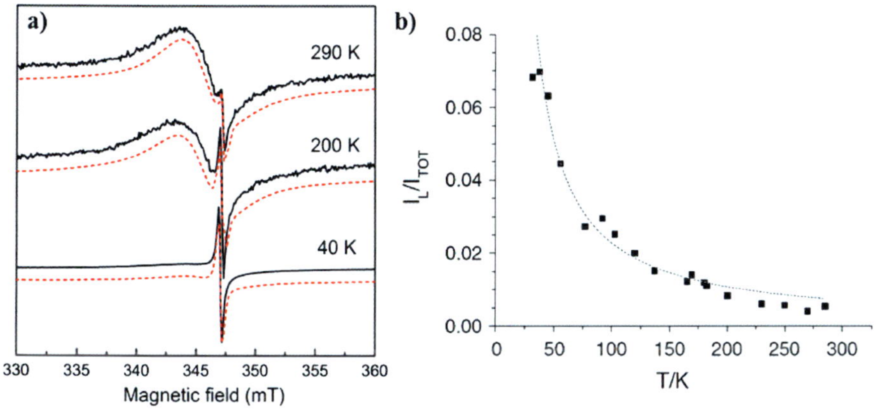

Fig. (5). (left) cw-EPR spectra of a graphite (GPyr) milled for six hour (G360, black: experimental, dashed grey: simulated). The EPR spectra have been simulated as the superposition of a Dysonian ($g = 2.013$), a Gaussian ($g = 2.018$) and a Lorentzian ($g = 2.0033$) [6]. For 200 K the LW were 74, 90 and 2.8 G respectively, with relative intensities 45%, 45% and 10%. For 40 K the parameters are similar to those used for the simulation at 7 K. The simulation parameters are unchanged under 40 K. (right). Relative intensity variation of I_L as a function of the temperature.

The narrow Lorentzian line is attributed to edge states [57], and it is particularly narrow because exchange interactions with mobile electrons are effective, and they are able to wash out all hyperfine interactions (cxchange-narrowing *régime*) [39, 57]. With respect to sample EKS, the number of electrons per particle is expected to be larger, being the dimensions of the particles larger, and this allows edge states to have a larger number of interacting events per unit timc.

The Dysonian component is attributed to conduction electrons in large particles, or in particles with good contacts, where they can easily diffuse, whereas the Gaussian component is attributed to conduction electrons in small particles having weak contacts among them.

The dependence of the intensities of the different components on the temperature is complex and not easily explained. The three components raise their intensity from room temperature down to 50 K ca, and then they start to decrease. Beside this general behavior, a deeper analysis on the weight of the Lorentzian band with respect to the total cw-EPR intensity (see Fig. **5**) shows that it raises its contribution relative to the other two components on lowering the temperature in a way that is analogous to a Curie-Weiss behavior [6].

This complex behavior shows the variety of situations that can be found within these materials that cannot be explained simply by the distribution of energy levels reported in Scheme (**1a**). A possible qualitative explanation can be offered by the distribution of states in the work by K. Tanaka et al [28] and sketched in Scheme (**1b**). Because edge states can exchange couple, two edge bands are formed near the Fermi level, separated by the exchange interaction, thus providing a Curie-Weiss behavior for the edge state. The presence of nearby valence and conduction band states can complicate further the population scheme of these states.

g-Values

Another significant parameter in the EPR spectra of nanographitic and graphenic-like materials is the *g*-tensor. The *g* principal values vary in these materials for a number of different effects.

L. S. Singer et al [3] have shown that the averaged *g*-tensor depends on the dimension of the particles. K. Matsubara et al [15] have shown that the anisotropy of the *g*-tensor depends on the type and number of states at the Fermi level, where edge states are located: the concentration of the edge states can influence the observed *g*-value. Again, the orientational disorder in the stacking of graphene layers can also cause a variation of *g*-values [4].

Most of the theoretical work done so far has been focused on the calculation of *g*-values for extended structures [15, 29, 58]. In summary, the results of these works are that in-plane components of the *g*-tensor have principal values close to g_e, whereas the principal value of the components perpendicular to the plane is affected by a spin-orbit mixing which rises the value.

If in the single layer the deviation from g_e value is rather small (up to $g = 2.004$ ca

[24]), in multilayer systems the admixture brings the *g*-value up to 2.01-2.02 [47].

The extent of anisotropy and the average value of the *g*-tensor is then a further indication of the extent of the stacking. The broad components, mostly Dysonian, in the examples of EPR spectra presented so far, as nanographites (G90, G360 and G1200) have $g_{\|}$ = 2.013-2.020, whereas in expanded graphites, with a limited number of stacked planes, and possibly with variable interplanar distance (EHK and EKS) the values are lower, around 2.010.

The particles with a lower dimension (RGO and MAPTES) have the smallest values of 2.0027 and 2.0037, with no anisotropy.

Another source of the variability of the principal values of the *g*-tensor is due to the conductivity of the sample. In fact, in conducting materials an averaging of the *g*-values can be produced by the jumping of the electrons between particles differently oriented but in good electrical contact [3]. The effect is similar to that of a reorientational motion averaging the *g* tensor.

For sample EH, a highly conducting sample showing Dysonian lineshape, the unusual isotropic value of 2.0050 is the average [4] of the three principal values of the *g*-tensor ($g_{av} = (g_a + g_b + g_{b'})/3$) of the parent sample EHK.

Edge states are often detected as narrow lines, with no *g*-anisotropy because of exchange interactions with itinerant electrons, which causes an averaging of the tensor [39]. As the concentration of edge states depends on the defectivity of the sample [7], and the linewidth on the local concentration of mobile electrons, narrow lines are seen more often in graphitic systems than in graphenic one.

PULSE EPR SPECTRA

Pulse EPR methods allow the separation of the signals of different species, depending on their spin relaxation. Species with short relaxation time, as conduction electrons, are undetectable by pulse EPR, whereas edge and molecular states can be detected [59].

Free Induction decay (FID) is suitable for narrow-linewidth components, and generally with small inhomogeneity. On the other hand, spin echo methods (ESE) are more suitable for the detection of signals broadened by resolved or unresolved anisotropic interactions, such as *g*-tensor or hyperfine interactions. Consequently, we expect molecular states to be detected by ESE-based techniques.

In the following we will present examples of spectra obtained by both FID based (FT-EPR) and ESE based (ED-EPR) techniques.

FT- and ED-EPR

Echo-detected EPR (ED-EPR) spectra are obtained by recording the intensity of a 2-pulse (2p) echo (2p: $\pi/2$ - τ-π-τ - *echo*) as a function of the magnetic field. Because the echo intensity is recorded at 2|-from the first pulse, as a rule of thumb only species having a T_2 value (more precisely, the phase memory time T_M [59]), longer than 2| can provide detectable echo intensity. In principle, for $\tau = 0$ the same spectrum as the cw-EPR should be obtained, but we are experimentally limited by the instrument dead-time, that for commercial X-band spectrometers it is of the order of 10^2 ns.

By increasing the value of τ, fast relaxing species reduce their contribution to the echo with respect to slow relaxing species. Thus, τ acts as a kind of filter for selection of the species observed.

FT-EPR spectra are obtained from the Fourier Transform (FT) of the Free Induction Decay (FID), resulting from the precession of the magnetization in the *xy* plane (perpendicular to the magnetic field) after a $\pi/2$ pulse. Only species whose relaxation time is longer T_2 than the instrument dead time are detectable.

We start presenting the results obtained from sample MAPTES, where both FT-EPR and ED-EPR spectra could be obtained.

In Fig. (**6**) the spectra FT-EPR and ED-EPR in comparison with the integral form of the cw-EPR. This is a particularly instructive sample, as different components are observed by applying cw- FT- or ED-EPR methods.

The cw-EPR has three major pure Lorentzian components overlapping; the narrower one has a linewidth of 0.6 G, which corresponds to a T_2 of about 100 ns. This component is expected to be barely visible by pulse methods, whereas the other components, corresponding to species relaxing faster, cannot be detected.

The FT-EPR spectrum is broader than the cw-EPR spectrum, from which we immediately deduce that, in contrast with components in the cw-EPR spectrum, the dominant species observed by FT-EPR have small anisotropies in the magnetic parameters. In fact, the homogeneous linewidths must be smaller than around 0.6 G (T_2 of about 100 ns), which is a necessary condition for the FID to be observed. We can then estimate these anisotropies by simulating the spectra. A simulation, using a Gaussian profile, gave two major components with an inhomogeneous linewidth of 1.2 and 3.8 G. From these values, we estimate the narrower component ($\Gamma = 1.2$ G) to have unresolved anisotropies broadening the spectrum by 0.5-1 G, and the other in the range 3.1-3.6 G.

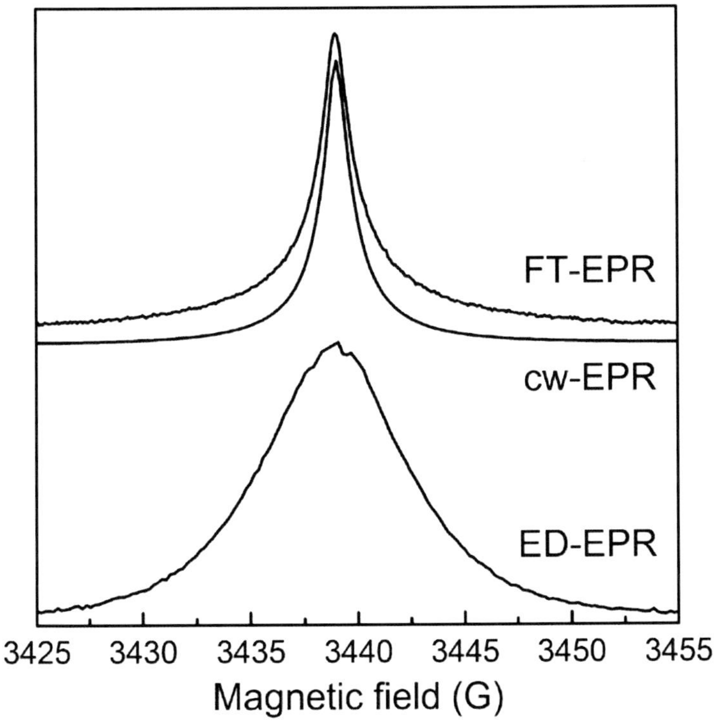

Fig. (6). Top) FT-EPR spectrum *vs* the integrated cw-EPR spectrum of MAPTES at room temperature. Bottom) normalized ED-EPR is taken with a high resolution (narrow bandwidth) echo sequence of 276-20--572 ns.

The narrower spectrum is consistent with two possible structures: a) pure graphenic layers (no *g*-anisotropy) in the absence of paramagnetic nuclei around the unpaired electron, or b) edge states with almost complete averaging of the magnetic interactions. We are in favor of this second hypothesis, as we expect some left protons to be bound to the particles (see also next section) creating larger anisotropy.

As for the broader component having a linewidth of 3.8 G, this component has been observed in the ED-EPR with 2.0037.

The nature of the inhomogeneity of this band has been investigated by hyperfine spectroscopy and it will be discussed in the next section. We anticipate here our attribution of this band to molecular states.

Let us consider now the results for the graphenic material RGO. In this case, the comparison between pulse EPR and cw-EPR spectrum tells a different story. For this material, the cw-EPR spectrum is much broader than the ED-EPR one. The two types of spectra, i.e. the integrated cw-EPR and the ED-EPR at 20K are reported together in Fig. (**7**). As one can notice, the two types of spectra are different not only for their linewidth, but also for their *g*-factor.

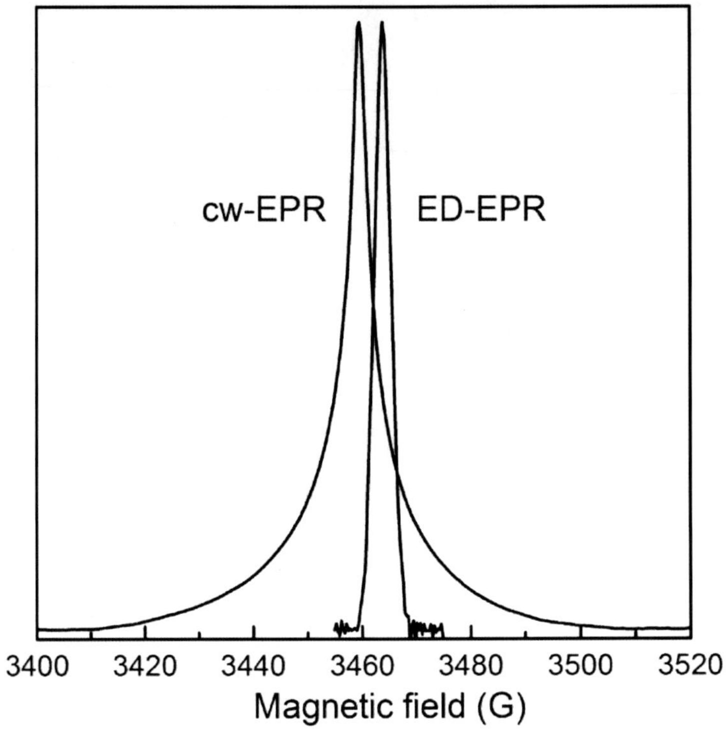

Fig. (7). ED-EPR spectrum *vs* the integrated cw-EPR spectrum of RGO at 20 K. The 2p-pulse sequence is with 16 ns and 140 ns. Adapted from [21] with the Permission from AIMS Press.

The ED-EPR spectrum could be simulated with a single Gaussian ($g = 2.0038$), indicating the absence of strong hyperfine interactions with paramagnetic nuclei. On the other hand, the cw-EPR spectrum can be simulated by the superposition of two components.

The component with the narrowest line has a homogeneous linewidth of 5 G ca, that is $T_2 \approx 10$ ns.

We can conclude that, in this case, the cw-EPR spectrum shows the presence of paramagnetic species with short spin relaxation times, not detectable by Pulse

EPR. On the other hand, the ED-EPR shows the presence of another paramagnetic species, with a smaller *g* factor, which should give a contribution also to the cw-EPR spectrum, but with a much lower intensity with respect to the main cw-EPR spectrum. Therefore, its cw-EPR signal is buried below the more intense components.

In conclusion, the two types of spectra give complementary information: the dominant contributions observed by cw-EPR cannot be observed by pulse methods, and by pulse methods we can detect and identify low-intensity components that could not be resolved in the cw-EPR spectrum.

The nanographite G1200, for which the results are analogous to those found for sample RGO, provides the third example (see Fig. **8**).

A narrow spectrum is obtained by ED-EPR, with $g = 2.003$ and with a Gaussian profile; the narrow linewidth indicates the absence of strong hyperfine interactions. The broad lines present in the cw-EPR spectrum are not observed with pulse-EPR.

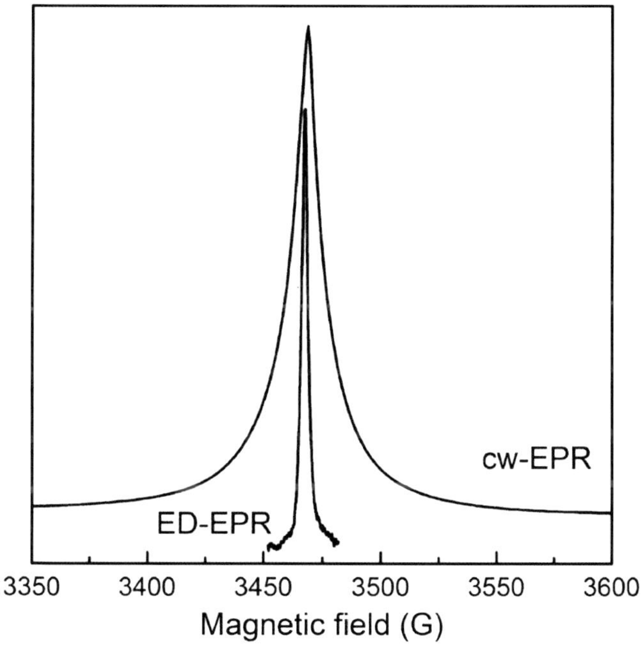

Fig. (8). ED-EPR spectrum *vs* the integrated cw-EPR spectrum of G1200 at 21 K. The 2p-pulse sequence is with $\pi/2$ = 16 ns and τ = 200 ns.

We conclude this section pointing up the similarity of ED-EPR spectra of different types of materials. They all are characterized by a narrow linewidth, with common features, like the width of 2-5 G ca, and *g*-values close to 2.004. Clearly, these spectra should correspond to similar structures. We have further characterized these signals by hyperfine spectroscopies, to have an insight of their structure, as reported in the following.

Hyperfine Spectroscopies

To further characterize the ED-EPR signals described above, we have applied the so-called hyperfine spectroscopies, to obtain information on the presence of hyperfine interactions. For this purpose, we have performed an electron spin echo envelope modulation (ESEEM), electron-nuclear double resonance (ENDOR) [59] and hyperfine sublevel correlation (HYSCORE) spectroscopies.

For ESEEM and ENDOR spectroscopies, the spectra obtained show signals centered at the Larmor frequency of the coupled nuclei, and are eventually split by the value of their hyperfine interaction. We expect in general the presence of the paramagnetic nuclei ^{13}C (with a relative abundance of ca 1% depending on the type of graphite) and ^{1}H as some materials have been prepared in strong chemical reductive systems. For sample MAPTES we might expect to find also ^{29}Si, which belongs to the silica matrix. Unfortunately, oxygen is undetectable as its natural isotopic abundance of paramagnetic isotopes is too low. The free atom Larmor frequencies of the three isotopes at 3500 G (typical magnetic field for X-band EPR spectroscopy) are 3.75, 14.90 and 2.96 MHz for ^{13}C, ^{1}H and ^{29}Si, respectively.

ESEEM spectra are obtained from the Fourier Transform of the modulation present in the 2p-ESE decay, or from the stimulated 3p-ESE decay. The sequences are, respectively, $\pi/2$ - τ-π-τ-*echo* and $\pi/2$ - τ-$\pi/2$ - T-$\pi/2$ - τ-*echo*. The echo intensity is measured as a function of the time delay between pulses: for the 2p-ESE and for the 3p-ESE. Mims ENDOR spectra are obtained by using the 3p-ESE sequence. During delay between the second and the third pulse, a further radiofrequency pulse is sent to induce nuclear transitions. The radiofrequency is scanned to measure echo decrease when, during the scan, a nuclear transition reaches the resonance condition (see Chapter 1).

In Fig. (**9**) the 2p-ESEEM spectra of sample MAPTES at room temperature, and of samples G1200 and EKS at 20 and 10 K, respectively. Spectra of MAPTES and G1200 exhibit a rather similar ratio between the ^{13}C and the proton band, with a stronger intensity of the ^{13}C. For MAPTES the ^{29}Si band is barely observed close to the ^{13}C band, despite the much higher natural abundance. Sample MAPTES exhibit the free ^{13}C Larmor peak, and satellite peaks due to a hyperfine interaction of 5 MHz [51] (see the HYSCORE spectrum discussed further).

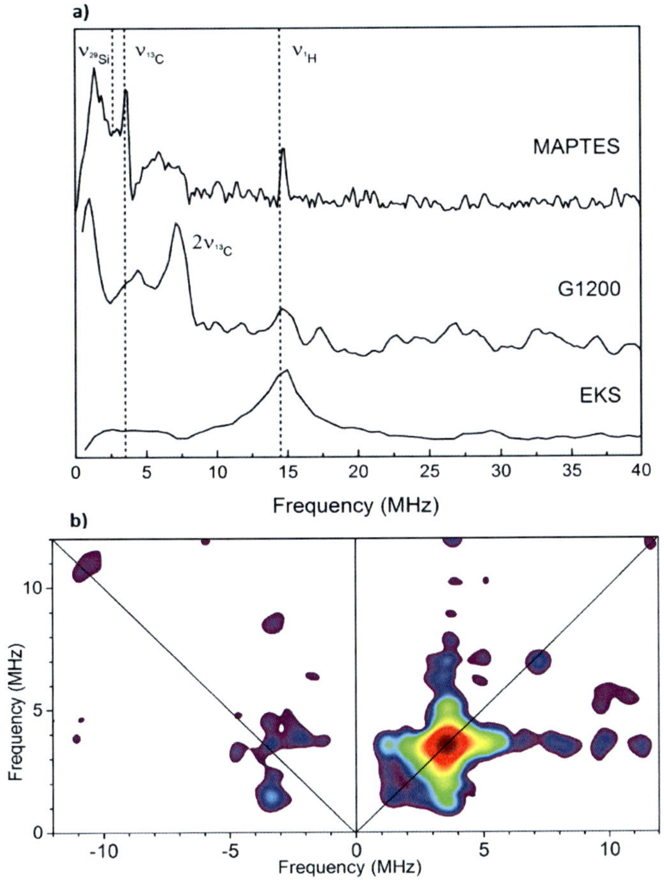

Fig. (9). a) The 3p-ESEEM spectra of sample MAPTES recorded at room temperature (140 ns), and 2p-ESEEM of samples G1200 and EKS recorded at 20 and 10 K respectively. The Larmor frequencies of ^{1}H, ^{13}C and ^{29}Si nuclei are indicated. For sample G1200 the strong peak at 7.9 MHz is attributed to twice the Larmor frequency of ^{13}C. b) The HYSCORE powder spectrum of MAPTES_1000 at room temperature taken with 136 ns (left) and its simulation (right) by using two sets of coupled nuclei: a ^{13}C with hyperfine principal values of -1.5, -3.5 and 5 MHz and an isotropic interaction of 5 MHz. The simulation of the real part is reported in Ref [51].

Sample G1200 displays a rather intense double frequency of the ^{13}C, observable in the 2p-ESEEM experiments [59], that likely overlap an analogous band as found for MAPTES relative to the hyperfine interaction of few MHz.

For very small hyperfine interactions, the width of the bands is related to the hyperfine coupling constants (hcc), which are of particular value as in such π-type systems they are directly connected to the distribution of the spin density across the graphenic layer. For protons, saturating the dangling bonds of the edge carbon atoms, the well-known McConnell theory [60] links the hcc value to the spin density at the carbon atom of the π-system to which they are bound. In general the widening of the dimension of the π-system induces a reduction of the hcc value, and *vice versa*, from the value of the hcc is related to the extent of the layer.

The spectra of MAPTES and G1200 exhibit proton lines which are 1-1.5 MHz wide (0.35-0.5 G ca). Following the discussion presented in Ref [21], we make a comparison of the experimental widths with the hcc value of protons in reference molecules, namely coronene and circumcoronene to have a rough estimate of the extension of the π-system. An analogous discussion can be conducted for the calculation of the extent by using the ^{13}C hcc. In this case we take as reference material fullerene derivatives [51], where the hcc are in the range 2-7 MHz, which are values close to that found for MAPTES. To conclude, the analysis of the ^{13}C and ^{1}H signal agree with the attribution of these signals as due to fragments bearing more than 50 C atoms, whose edges are terminated by protons.

To further support the ESEEM data for this sample, HYSCORE was obtained with the aim of separating and isolating of the components thanks to the better spectral resolution of this 2D spectroscopy.

The spectrum display ridges centered at the ^{13}C frequency, and no visible peaks relative to ^{29}Si are present. The ridges obtained from a 5 MHz isotropic hcc constant are consistent with what found by ESEEM spectroscopy. Off-diagonal peaks are present at about the same positions in the imaginary-part quadrant. The simulation of the spectrum is reported in Ref [51].

Finally we analyze the spectrum of EKS that exhibit a much larger proton band (4 MHz ca, 1.4 G) and a larger ^{13}C band (7 MHz ca, 2.5 G). The estimated proton hcc is much closer to that of coronene, and ^{13}C interaction is stronger than in the other two samples. We then conclude that the aromatic fragments in sample EKS are smaller, and much closer to a molecular system.

For RGO we discuss the hyperfine interactions as determined by the Mims ENDOR spectrum, which is shown in Fig. (**10**) as it was not possible to obtain the ESEEM spectra due to a too little modulation depth in the 2p-ESE decay time traces. In RGO the proton band is higher than that of the ^{13}C, but in ENDOR there do not gain in intensity at low frequencies as for ESEEM. Consequently, the proton band is better resolved than the ^{13}C band, which is characterized by a narrow component (indicative of the presence of distant matrix carbon atoms), and likely a badly resolved broader band centered at the same frequency.

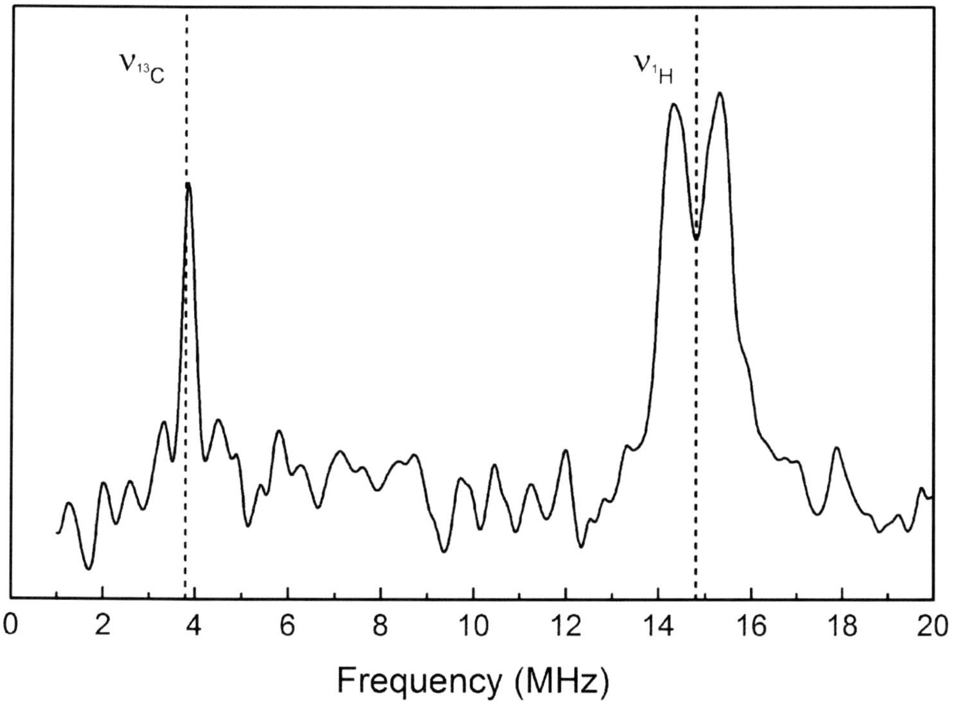

Fig. (10). Mims ENDOR spectrum of sample RGO, adapted from Ref [21] with the Permission from AIMS Press.

For this sample we then concentrate on the information coming from the proton signals, which nicely exhibit a band with a maximum width is 4 MHz ca. On the base of this value, we deduce that also for RGO the fragments are relatively small (smaller than C_{54}), and they very much saturated by protons at the edges.

To conclude this section, we note that the common characteristic of the materials is the presence of small fragments with a relatively small extension of the π-system. These fragments are produced by the disruption of the structure into pieces with various methods, like the grinding for hours of graphite, or its chemical exfoliation. In the case of MAPTES, they represent impurities in the silica host, where clusterization of the carbon atoms is observed.

CONCLUSION

By studying graphene-like materials by EPR techniques, it is possible to detect similar features in their spectra, only slightly variable, useful for a characterization of the materials.

We have identified three types of structures: bulk graphitic/graphenic structures providing conduction electrons, edge states, providing often narrow signals, and molecular defects which are graphenic fragments of different dimensions, decorated with protons.

By pulse methods it is possible to isolate molecular states that are overlooked in the cw-EPR spectra, as they are of small intensity.

CONSENT FOR PUBLICATION

Not applicable.

CONFLICT OF INTEREST

The author (editor) declares no conflict of interest, financial or otherwise.

ACKNOWLEDGEMENT

The authors acknowledge Prof. M. Brustolon who first stimulated the beginning of this experimental work on the study of carbon-based materials.

REFERENCES

[1] Wagoner, G. Spin Resonance of Charge Carriers in Graphite. *Phys. Rev.,* **1960**, *118*, 647-653. [http://dx.doi.org/10.1103/PhysRev.118.647]

[2] Huber, D.; Urbano, R.; Sercheli, M.; Rettori, C. Fluctuating field model for conduction electron spin resonance in graphite. *Phys. Rev. B.,* **2004**, *70*(125417), 1-5. [http://dx.doi.org/10.1103/PhysRevB.70.125417]

[3] Singer, L.; Wagoner, G. Electron Spin Resonance in Polycrystalline Graphite. *J. Chem. Phys.,* **1962**, *37*, 1812-1817. [http://dx.doi.org/10.1063/1.1733373]

[4] Chehab, S.; Guérin, K.; Amiell, J.; Flandrois, S. Magnetic properties of mixed graphite containing both hexagonal and rhombohedral forms. *Eur. Phys. J. B,* **2000**, *13*, 235-243. [http://dx.doi.org/10.1007/s100510050028]

[5] Osipov, V.; Shames, A.; Enoki, T.; Takai, K.; Endo, M.; Hayashi, T.; Kaburagi, Y.; Vul', A. Magnetic and EPR studies of edge-localized spin paramagnetism in multi-shell nanographites derived from nanodiamonds. *Diamond Related Materials,* **2009**, *18*, 220-223. [http://dx.doi.org/10.1016/j.diamond.2008.09.015]

[6] Barbon, A.; Brustolon, M. An EPR Study on Nanographites. *Appl. Magn. Reson.,* **2011**, *42*, 197-210. [http://dx.doi.org/10.1007/s00723-011-0285-6]

[7] Tommasini, M.; Castiglioni, C.; Zerbi, G.; Barbon, A.; Brustolon, M. A joint Raman and EPR spectroscopic study on ball-milled nanographites. *Chem. Phys. Lett.,* **2011**, *516*, 220-224. [http://dx.doi.org/10.1016/j.cplett.2011.09.094]

[8] Tampieri, F.; Silvestrini, S.; Riccò, R.; Maggini, M.; Barbon, A. A comparative electron paramagnetic resonance study of expanded graphites and graphene. *J. Mater. Chem. C Mater. Opt. Electron. Devices,* **2014**, *2*, 8105-8112. [http://dx.doi.org/10.1039/C4TC01383B]

[9] Osipov, V.; Shames, A.; Enoki, T.; Takai, K.; Endo, M.; Kaburagi, Y.; Vul', A. Interaction between

edge-localized spins and molecular oxygen in multishell nanographites derived from nanodiamonds. *Diamond Related Materials,* **2010**, *19*, 492-495. [http://dx.doi.org/10.1016/j.diamond.2010.01.027]

[10] Kausteklis, J.; Cevc, P.; Arčon, D.; Nasi, L.; Pontiroli, D.; Mazzani, M.; Riccò, M. Electron paramagnetic resonance study of nanostructured graphite. *Phys. Rev. B.,* **2011**, *84*(125406), 1-5. [http://dx.doi.org/10.1103/PhysRevB.84.125406]

[11] Joly, V.J.; Takai, K.; Enoki, T. Oxygen guest molecules in the nanopores of nanographite-network. *J. Phys. Chem. Solids,* **2010**, *71*, 575-578. [http://dx.doi.org/10.1016/j.jpcs.2009.12.040]

[12] Shibayama, Y.; Sato, H.; Enoki, T.; Endo, M. Disordered magnetism at the metal-insulator threshold in nano-graphite-based carbon materials. *Phys. Rev. Lett.,* **2000**, *84*(8), 1744-1747. [http://dx.doi.org/10.1103/PhysRevLett.84.1744] [PMID: 11017615]

[13] Dinse, K.; van Tol, J.; Ozarowski, A.; Corzilius, B. Multi-Frequency EPR and DC Conductivity of Itinerant Spins in Single-Wall Carbon Nanotubes. *Appl. Magn. Reson.,* **2009**, *37*, 595-603. [http://dx.doi.org/10.1007/s00723-009-0084-5]

[14] Likodimos, V.; Glenis, S.; Guskos, N.; Lin, C. Magnetic and electronic properties of multiwall carbon nanotubes. *Phys. Rev. B.,* **2003**, *68*(045417), 1-6. [http://dx.doi.org/10.1103/PhysRevB.68.045417]

[15] Matsubara, K.; Tsuzuku, T.; Sugihara, K. Electron spin resonance in graphite. *Phys. Rev. B Condens. Matter,* **1991**, *44*(21), 11845-11851. [http://dx.doi.org/10.1103/PhysRevB.44.11845] [PMID: 9999320]

[16] Sercheli, M.; Kopelevich, Y.; Ricardo da Silva, R.; Torres, J.; Rettori, C. Evidence for internal field in graphite: a conduction electron spin-resonance study. *Solid State Commun.,* **2002**, *121*, 579-583. [http://dx.doi.org/10.1016/S0038-1098(01)00465-3]

[17] Augustyniak-Jabłokow, M.; Tadyszak, K.; Maćkowiak, M.; Lijewski, S. ESR study of spin relaxation in graphene. *Chem. Phys. Lett.,* **2013**, *557*, 118-122. [http://dx.doi.org/10.1016/j.cplett.2012.12.018]

[18] Nair, R.; Sepioni, M.; Tsai, I.; Lehtinen, O.; Keinonen, J.; Krasheninnikov, A.; Thomson, T.; Geim, A.; Grigorieva, I. Spin-half paramagnetism in graphene induced by point defects. *Nat. Phys.,* **2012**, *8*, 199-202. [http://dx.doi.org/10.1038/nphys2183]

[19] Matte, H.; Subrahmanyam, K.; Rao, C. Novel Magnetic Properties of Graphene: Presence of Both Ferromagnetic and Antiferromagnetic Features and Other Aspects. *J. Phys. Chem. C,* **2009**, *113*, 9982-9985. [http://dx.doi.org/10.1021/jp903397u]

[20] Augustyniak-Jabłokow, M.; Tadyszak, K.; Maćkowiak, M.; Yablokov, Y. EPR evidence of antiferromagnetic ordering in single-layer graphene. *Phys. Status Solidi Rapid Res. Lett.,* **2011**, *5*, 271-273. [http://dx.doi.org/10.1002/pssr.201105234]

[21] Barbon, A.; Tampieri, F. Identification of slow relaxing spin components by pulse EPR techniques in graphene-related materials. *AIMS Mater. Sci.,* **2017**, *4*, 147-157. [http://dx.doi.org/10.3934/matersci.2017.1.147]

[22] Rao, S.; Stesmans, A.; Kosynkin, D.; Higginbotham, A.; Tour, J. Paramagnetic centers in graphene nanoribbons prepared from longitudinal unzipping of carbon nanotubes. *New J. Phys.,* **2011**, *13*(113004), 1-9. [http://dx.doi.org/10.1088/1367-2630/13/11/113004]

[23] Marciano, O.; Gonen, S.; Levy, N.; Teblum, E.; Yemini, R.; Nessim, G.D.; Ruthstein, S.; Elbaz, L. Modulation of Oxygen Content in Graphene Surfaces Using Temperature-Programmed Reductive

Annealing: Electron Paramagnetic Resonance and Electrochemical Study. *Langmuir,* **2016**, *32*(44), 11672-11680.
[http://dx.doi.org/10.1021/acs.langmuir.6b02987] [PMID: 27728764]

[24] Ćirić, L.; Sienkiewicz, A.; Náfrádi, B.; Mionić, M.; Magrez, A.; Forró, L. Towards electron spin resonance of mechanically exfoliated graphene. *Phys. Status Solidi, B Basic Res.,* **2009**, *246*, 2558-2561.
[http://dx.doi.org/10.1002/pssb.200982325]

[25] Panich, A.; Shames, A.; Aleksenskii, A.; Dideikin, A. Magnetic resonance evidence of manganese–graphene complexes in reduced graphene oxide. *Solid State Commun.,* **2012**, *152*, 466-468.
[http://dx.doi.org/10.1016/j.ssc.2012.01.005]

[26] Ney, A.; Papakonstantinou, P.; Kumar, A.; Shang, N.; Peng, N. Irradiation enhanced paramagnetism on graphene nanoflakes. *Appl. Phys. Lett.,* **2011**, *99*(102504), 1-3.
[http://dx.doi.org/10.1063/1.3628245]

[27] Wakabayashi, K. Electronic and Magnetic Properties of Nanographites.*Carbon-based Magnetism – An overview of the magnetism of metal free carbon-based compounds and materials*; Makarova, T.; Palacio, F., Eds.; Elsevier: Amsterdam, **2016**, pp. 279-304.

[28] Tanaka, K.; Koike, T.; Yamabe, T.; Yamauchi, J.; Deguchi, Y.; Yata, S. Electron-spin-resonance studies of pristine and heavily doped polyacenic materials. *Phys. Rev. B Condens. Matter,* **1987**, *35*(16), 8368-8373.
[http://dx.doi.org/10.1103/PhysRevB.35.8368] [PMID: 9941183]

[29] McClure, J.; Yafet, Y. Theory of the g-factor of the current carriers in graphite single crystals. *Proceedings of the Fifth Conference on Carbon,* **1962**, , pp. 22-28.
[http://dx.doi.org/10.1016/B978-0-08-009707-7.50008-X]

[30] Zhou, Y.; Wu, M. Electron spin relaxation in graphene from a microscopic approach: Role of electron-electron interaction. *Phys. Rev. B.,* **2010**, *82*(085304), 1-8.

[31] Fung, A.W.; Wang, Z.H.; Dresselhaus, M.S.; Dresselhaus, G.; Pekala, R.W.; Endo, M. Coulomb-gap magnetotransport in granular and porous carbon structures. *Phys. Rev. B Condens. Matter,* **1994**, *49*(24), 17325-17335.
[http://dx.doi.org/10.1103/PhysRevB.49.17325] [PMID: 10010914]

[32] Fujita, M.; Wakabayashi, K.; Nakada, K.; Kusakabe, K. Peculiar Localized State at Zigzag Graphite Edge. *J. Phys. Soc. Jpn.,* **1996**, *65*, 1920-1923.
[http://dx.doi.org/10.1143/JPSJ.65.1920]

[33] Kahn, O. *Magnetism: A Supramolecular Function*; , **1996**.
[http://dx.doi.org/10.1007/978-94-015-8707-5]

[34] Enoki, T.; Takai, K. The edge state of nanographene and the magnetism of the edge-state spins. *Solid State Commun.,* **2009**, *149*, 1144-1150.
[http://dx.doi.org/10.1016/j.ssc.2009.02.054]

[35] Morita, Y.; Suzuki, S.; Sato, K.; Takui, T. Synthetic organic spin chemistry for structurally well-defined open-shell graphene fragments. *Nat. Chem.,* **2011**, *3*(3), 197-204.
[http://dx.doi.org/10.1038/nchem.985] [PMID: 21336324]

[36] Kertesz, M.; Choi, C.H.; Yang, S. Conjugated polymers and aromaticity. *Chem. Rev.,* **2005**, *105*(10), 3448-3481.
[http://dx.doi.org/10.1021/cr990357p] [PMID: 16218558]

[37] Tommasini, M. Department of Chemistry, Materials and Engineering, Politecnico di Milano, Milan, IT. Personal communication. *January,* **2017**.

[38] Fritz, L.; Vojta, M. The physics of Kondo impurities in graphene. *Rep. Prog. Phys.,* **2013**, *76*(032501), 1-15.

[http://dx.doi.org/10.1088/0034-4885/76/3/032501]

[39] Joly, V.; Takahara, K.; Takai, K.; Sugihara, K.; Enoki, T.; Koshino, M.; Tanaka, H. Effect of electron localization on the edge-state spins in a disordered network of nanographene sheets. *Phys. Rev. B,* **2010**, *81*(115408), 1-6. [http://dx.doi.org/10.1103/PhysRevB.81.115408]

[40] Black-Schaffer, A. RKKY coupling in graphene. *Phys. Rev. B,* **2010**, *81*(205416), 1-8. [http://dx.doi.org/10.1103/PhysRevB.81.205416]

[41] Saremi, S. RKKY in half-filled bipartite lattices: Graphene as an example. *Phys. Rev. B,* **2007**, *76*(184430), 1-6.

[42] Banhart, F.; Kotakoski, J.; Krasheninnikov, A.V. Structural defects in graphene. *ACS Nano,* **2011**, *5*(1), 26-41. [http://dx.doi.org/10.1021/nn102598m] [PMID: 21090760]

[43] Giannozzi, P.; Car, R.; Scoles, G. Oxygen adsorption on graphite and nanotubes. *J. Chem. Phys.,* **2003**, *118*, 1003-1006. [http://dx.doi.org/10.1063/1.1536636]

[44] Lijewski, S.; Wencka, M.; Hoffmann, S.; Kempinski, M.; Kempinski, W.; Sliwinska-Bartkowiak, M. Electron spin relaxation and quantum localization in carbon nanoparticle: Electron spin echo studies. *Phys. Rev. B,* **2008**, *77*(014304), 1-8.

[45] Boukhvalov, D.; Moehlecke, S.; da Silva, R.; Kopelevich, Y. Effect of oxygen adsorption on magnetic properties of graphite. *Phys. Rev. B,* **2011**, *83*(233408), 1-4. [http://dx.doi.org/10.1103/PhysRevB.83.233408]

[46] Shames, A.; Osipov, V.; Vul', A.; Kaburagi, Y.; Hayashi, T.; Takai, K.; Enoki, T. Spin–spin interactions between π-electronic edge-localized spins and molecular oxygen in defective carbon nano-onions. *Carbon,* **2013**, *61*, 173-189. [http://dx.doi.org/10.1016/j.carbon.2013.04.082]

[47] Stankowski, J.; Waplak, S.; Bednarski, W. The anisotropy and temperature dependence of g-factor in graphite. *Solid State Commun.,* **2000**, *115*, 489-491. [http://dx.doi.org/10.1016/S0038-1098(00)00208-8]

[48] Poole, C. *Electron spin resonance: a comprehensive treatise on experimental techniques,* 2nd ed; Dover Publications: New York, **1996**.

[49] Dyson, F. Electron Spin Resonance Absorption in Metals. II. Theory of Electron Diffusion and the Skin Effect. *Phys. Rev.,* **1955**, *98*, 349-359. [http://dx.doi.org/10.1103/PhysRev.98.349]

[50] Joshi, J.P.; Bhat, S.V. On the analysis of broad Dysonian electron paramagnetic resonance spectra. *J. Magn. Reson.,* **2004**, *168*(2), 284-287. [http://dx.doi.org/10.1016/j.jmr.2004.03.018] [PMID: 15140439]

[51] Barbon, A.; Gross, S.; Tondello, E.; Brustolon, M. Carbon paramagnetic defects in silica sol-gel prepared materials. *Mol. Phys.,* **2007**, *105*, 2177-2183. [http://dx.doi.org/10.1080/00268970701747207]

[52] Hummers, W.; Offeman, R. Preparation of Graphitic Oxide. *J. Am. Chem. Soc.,* **1958**, *80*, 1339-1339. [http://dx.doi.org/10.1021/ja01539a017]

[53] Novoselov, K.S.; Geim, A.K.; Morozov, S.V.; Jiang, D.; Zhang, Y.; Dubonos, S.V.; Grigorieva, I.V.; Firsov, A.A. Electric field effect in atomically thin carbon films. *Science,* **2004**, *306*(5696), 666-669. [http://dx.doi.org/10.1126/science.1102896] [PMID: 15499015]

[54] Tang, L.; Ji, R.; Li, X.; Teng, K.; Lau, S. Size-Dependent Structural and Optical Characteristics of Glucose-Derived Graphene Quantum Dots. *Part. Part. Syst. Charact.,* **2013**, *30*, 523-531. [http://dx.doi.org/10.1002/ppsc.201200131]

[55] Wang, Y.; Hu, A. Carbon quantum dots: synthesis, properties and applications. *J. Mater. Chem. C,* **2014**, *2*(6921), 1-19.
[http://dx.doi.org/10.1039/C4TC00988F]

[56] Ćirić, L.; Sienkiewicz, A.; Djokić, D.; Smajda, R.; Magrez, A.; Kaspar, T.; Nesper, R.; Forró, L. Size dependence of the magnetic response of graphite oxide and graphene flakes - an electron spin resonance study. *Phys. Status Solidi, B Basic Res.,* **2010**, *247*, 2958-2961.
[http://dx.doi.org/10.1002/pssb.201000448]

[57] Andersson, O.; Prasad, B.; Sato, H.; Enoki, T.; Hishiyama, Y.; Kaburagi, Y.; Yoshikawa, M.; Bandow, S. Structure and electronic properties of graphite nanoparticles. *Phys. Rev. B,* **1998**, *58*, 16387-16395.
[http://dx.doi.org/10.1103/PhysRevB.58.16387]

[58] Dresselhaus, G.; Dresselhaus, M. Spin-Orbit Interaction in Graphite. *Phys. Rev.,* **1965**, *140*, A401-A412.
[http://dx.doi.org/10.1103/PhysRev.140.A401]

[59] Brustolon, M.; Barbon, A. Pulsed EPR of Paramagnetic Centers in Solid Phases., **2003**.
[http://dx.doi.org/10.1007/978-1-4757-5166-6_2]

[60] Weil, J.; Bolton, J. *Electron Paramagnetic Resonance: Elementary Theory and Practical Applications,* 2nd ed; Wiley-Interscience: Hoboken, N.J., **2007**.

CHAPTER 3

Study of Electron Spin Lifetime of Conducting Carbon Nanomaterials

Bálint Náfrádi[1,*], **Mohammad Choucair**[2] and **László Forró**[1]

[1] *Laboratory of Physics of Complex Matter (LPMC), Ecole Polytechnique Fédérale de Lausanne, CH-1015 Lausanne, Switzerland*

[2] *School of Chemistry, University of Sydney, Sydney, New South Wales 2006, Australia*

Abstract: The usefulness of electron spins in quantum information technologies such as spintronics or quantum computation is determined by the spin-lattice (T_1) and spin-spin (T_2) relaxation times. These relaxation times should be long relative to the characteristic times required for spin control in order to allow for controlled information manipulation. Despite the central importance of T_1 and T_2 in modern information technologies, direct experimental access to these quantities is scarce. Electron spin resonance (ESR) spectroscopy is one of the few-experimental methods, offering direct access to both T_1 and T_2 of electrons. In this chapter, we present recent advancements in pulsed and continuous wave ESR spectroscopy of conducting carbon nanomaterials that have emerged with the potential for practical applications.

Keywords: Conducting carbon materials, Conduction ESR (CESR), Disordered onion-like carbon nanospheres (DOLCNS), Electron Spin Resonance (ESR), ESR linewidth, ESR signal amplitude, *g*-shift, Graphene, High-field ESR, Pulsed ESR, Rabi oscillations, Saturation methods, Skin depth, Spin dynamics, Spin-lattice relaxation time, Spin lifetime, Spin relaxation process, Spin-spin relaxation time, Spitronics, Synthetic graphene.

INTRODUCTION

This chapter is focused on paramagnetism in conducting carbon materials and the analysis of this weak magnetism by electron spin resonance (ESR) experiments. Paramagnetism is a property of any assembly of permanent magnetic dipoles if they are free to change their orientations and are in thermal equilibrium [1].

* **Corresponding author Bálint Náfrádi:** Laboratory of Physics of Complex Matter, Ecole Polytechnique Fédérale de Lausanne, Lausanne, CH-1015, Switzerland; Tel: +41 21 69 34515; Fax: +41 21 693 4470; E-mail: nafradi@yahoo.com

Dariya Savchenko & Abdel Hadi Kassiba (Eds.)

Nearly all strongly paramagnetic substances contain transition metals or elements with unpaired inner group electrons, and they give rise to temperature dependent paramagnetism [2]. Weakly paramagnetic metals can contain conduction electrons which are relatively free to move throughout a sample [3]. When the spin moments of conduction electrons experience an applied magnetic field, they show temperature-independent paramagnetic behavior [1, 4 - 7]. It is also worth noting for later discussions on the interaction with molecular oxygen (O_2) and carbon surfaces, that O_2 is paramagnetic [7].

Spin Dynamics and Electron Spin Resonance for Spintronics

Studies on the spin dynamics of solid-state materials in conventional spintronic devices, for example, magnetic random access memory (MRAM) were not central to the phenomena probed. However, it is now necessary to understand the spin dynamics in new materials that exploit spin coherence phenomena in the next generation of spintronic devices. One of the most important spin coherence properties is the precession of the magnetization in nanomaterials [6, 8 - 11], as these new materials are able to facilitate the transport of spin information.

The utility of conducting materials for spintronic and spin quantum computing applications relies on two spin parameters: T_S and l_S, the spin relaxation time and the spin diffusion length respectively, which are connected by the equation:

$$l_S = \sqrt{DT_S} \tag{1}$$

where D is the diffusion coefficient. Increasing T_S permits the manipulation of coherent spin, while long l_S enables the transport of information without significant loss. ESR spectroscopy is an ideally suited contactless method to determine spin transport parameters. A detailed description of ESR, in general, is given in chapter 1, but briefly, the ESR experiment measures the transitions of unpaired electrons between its energy states in the presence of a static magnetic field B and a microwave excitation field.

ESR measurements are commonly performed on unpaired electron spin systems which are localized on separate atoms. However, ESR can provide information on the delocalized – or conduction – electrons in metals. The resonance ESR condition for an unpaired delocalized electron in a metal is the same as that for an unpaired localized electron on an atom:

$$\hbar\omega_Z = g\mu_B B \tag{2}$$

where $\hbar$ is the reduced Planck constant, ω_Z the Zeeman frequency, μ_B is the Bohr magneton. The value of g, or the g-factor, is usually close to 2, although it depends on the details of the electron spin wavefunctions [12].

At resonance, the microwave B-field excites electrons to higher energy states. This excess energy is lost to the lattice by a process termed as (*electron*) *spin relaxation*.

The spin-orbit coupling interaction (SOC) leads to electron spin relaxation *via* several mechanisms like the D'yakonov-Perel' [13] and thc Elliot-Yafet mechanism [14, 15]. A detailed account of the mechanisms of electron spin relaxation has been addressed in detail elsewhere [16, 17]. Briefly, the energy relaxation processes of electron spins are described by a time constant T_1 which unavoidably also leads to the loss of quantum coherence described by a time constant T_2. The energy relaxation is due to a spin-phonon coupling which is mediated by the SOC interaction, and as a result, dictates the value of T_1. If the effect of the nuclear field on the electron spin coherence is completely suppressed, the spin-orbit interaction would limit T_2 to a value of $2T_1$. Moreover, the effect of COS interaction can be observed through a deviation of the g-factor of electrons from the spin-only value of 2.

The T_1 and T_2 relaxation times of electrons can be acquired to a very high approximation from continuous wave ESR experiments employing saturation methods. The saturation methods are based on the fact that the ESR linewidths and ESR signal amplitudes are differently influenced by T_1 and T_2 at different microwave power levels.

Both T_1 and T_2 can be calculated following the procedure outlined by Poole and Farach [18]. Briefly, T_2 is obtained from the ESR linewidth by:

$$T_2 = \frac{\Delta H^0 2\text{h}}{g\mu_B} \tag{3}$$

where ΔH^0 is the limiting linewidth and is determined from the low power limit.

The microwave power dependence (P) of ΔH is given by:

$$\Delta H = \frac{\Delta H^0}{\sqrt{\frac{1+Pg\mu_B}{\text{h}T_1T_2}}} \tag{4}$$

While the microwave power dependence of the ESR signal amplitude, α, has a form:

$$\alpha = \alpha^0 \sqrt{\left(\frac{1+Pg\mu_B}{hT_1T_2}\right)^3} \quad \textbf{(5)}$$

where α^0 is the limiting amplitude determined far from saturation.

In metals that the condition $T_1 = T_2$ is often met. This condition represents the longest period of time that in-phase precessing electron spins and magnetization can propagate as a uniform mode [19]. For simplicity, we will refer to T_1 and T_2 as T_S when the condition $T_1 = T_2$ is met.

This width of the resonance in conduction ESR (CESR) is also determined by the spin lifetime similarly to insulators [17]. The major spin relaxation path in metals is due to the scattering of conduction electrons from lattice vibrations and/or lattice imperfections. This involves the SOC thus, there is a strong correlation between the spin relaxation rate and the electrical resistivity of the metal [15]. Due to the delocalized nature of the conduction electrons, the shape of the CESR differs from the shape of the ESR line of the same atom. In ESR, the fine- and hyperfine-interactions usually dominate the lineshape. For CESR, these effects are usually absent because the itinerant electrons average out these effects due to their movement in the crystal that results in a single motionally narrowed line. Instead, the shape of the CESR is influenced by the electronic conductivity of the metal. An asymmetric Dysonian lineshape is frequently observed rather than a Lorentzian-type lineshape. This effect becomes dominant when the skin-depth becomes comparable to sample size [20]. Since we consider nanosized carbon in this chapter, where the size of the individual particles is much smaller than the skin-depth, we neglect this effect in the following discussions.

The nanometer size of the metallic particles, however, has an important effect on the CESR observables of *g*-shift, T_S and spin susceptibility. Once the size of the conducting particle becomes comparable to the electron wave functions, quantum size effects must be considered. With quantum size effects, the discreetness of electron energy level spectrum becomes apparent [3]. Based on Kawabata's theory [4], when quantum size effects are important, the observed linewidth of CESR is inversely proportional to the size of metallic particles given by:

$$\Delta H \sim \frac{T_S \mathrm{h}\omega_Z}{(\gamma_e \delta) d} = \frac{V_F (\Delta g_\infty)^2 \mathrm{h}\omega_Z}{\delta \gamma_e d} \quad \textbf{(6)}$$

where V_F is the conduction electrons' Fermi velocity, Δg_∞ is the bulk metal g shift from the electron g-factor $g_e = 2.0023$, δ is the electronic energy level spacing, γ_e is the magnetogyric ratio, ω_Z is the Zeeman frequency, d is the particle diameter and $T_S = T_1 = T_2$ is the electron spin relaxation time.

In the small particle size limit described above, we would expect, according to Kawabata [4], that the g-shift has the form [21]

$$\Delta g = (1 - f_S)\Delta g_\infty + f_S \Delta g_S \quad \textbf{(7)}$$

where f_S is the fraction of atoms at the surface and Δg_S is the contribution of the surface atoms to the g-shift.

The spin susceptibility, which corresponds to the integrated intensity of the CESR, assumes a Curie-Weiss like temperature dependence that emerges from the temperature independent Pauli susceptibility.

Coherent Control of the Quantum State of Electron Spin

The most well-known approach to coherently drive Zeeman-transitions between the split electron energy levels is ESR. If we consider a spin-½ electron system between two Zeeman-split energy levels (*i.e.* +1/2 and –1/2) in an oscillatory external field, the probability of the electron being in the excited energy level at a given time t is found from the Bloch equations [22] and is given by [23]:

$$P_{1\to 2}(t) = \left(\frac{\omega_1}{\Omega}\right)^2 \sin^2\left(\frac{\Omega t}{2}\right) \quad \textbf{(8)}$$

where $\Omega = \sqrt{(\omega - \omega_0)^2 + \omega_1^2}$; and $\omega_0 = \gamma B_0$, $\omega_1 = \gamma B_1$, and γ is the gyromagnetic ratio.

This is called Rabi oscillation. This shows that the probability of finding the system in an excited state when continuously driven by the microwave excitation is a finite, oscillatory. The maximum amplitude of oscillation is achieved at frequency $\omega = \omega_0$ which is the condition for resonance. At resonance, the transition probability is then given by:

$$P_{1\to 2}(t) = \sin^2\left(\frac{\omega_1 t}{2}\right) \tag{9}$$

where ω_1 is the Rabi frequency. Thus the detected Rabi oscillations of several cycles can be taken as evidence that the system allows the deliberate preparation of any superposition of a two-level spin-½ system. For example, to go from one state (ground state) to the other (excited state) we can adjust the time t during which the rotating field acts such that:

$$\frac{\omega_1 t}{2} = \frac{\pi}{2}, \tag{10}$$

which gives

$$t = \frac{\pi}{\omega_1} \tag{11}$$

this is called a π pulse. If a time intermediate between 0 and $\frac{\pi}{\omega_1}$ is chosen, *e.g.* in the case for, we obtain a pulse, and this results in a superposition of both states.

$t = \frac{\pi}{2\omega_1}$ we obtain a $\frac{\pi}{2}$ pulse, and this results in a superposition of both states.

ESR AND SPIN DYNAMICS OF SYNTHETIC GRAPHENE

Intrinsically, a pure graphene lattice has a zero magnetic moment and at 0 K behaves as an insulator. Above 0 K graphene is conducting and is magnetic due to the presence of thermally excited spin-½ itinerant electrons. This weak form of magnetism is referred to as Pauli-paramagnetism. Graphene also shows a large diamagnetism similar to that of graphite [24, 25]. In reality, graphene can only exist if the sheets are of finite size [26], and hence must contain chemical or structural defects. Synthetic graphenes, for example, those prepared by exfoliation [27], dispersion [28], chemical reduction [29], epitaxial growth [30], and chemical synthesis [31] exhibit a number of vacancies, lattice mismatching, adatoms, unpaired electrons, and mechanical strains. These changes essentially modify not only the chemical properties of the material but also the electron spin-dependent physical properties.

Of particular focus in this chapter is graphene material synthesized *via* the solvothermal method [31]. The solvothermal method is used to produce graphene in the absence of a hexagonal template material or step edges, which are often present in chemical vapor deposition [1]. Rather, bulk gram-scale quantities of graphene material are directly produced by rapid pyrolysis of clathrate-like alkoxide salts. In this scenario, there is strong theoretical [32] and experimental [33, 34] evidence to support that the formation of (metallic) paracrystalline domains are preferentially nucleated over crystalline graphene as metastable intermediates which then anneal to form graphene, and that the nucleation process is dominated by the swift transition of prolonged carbon chains towards an extended sp^2 carbon-bonded network. A representative TEM image is shown in Fig. (**1**).

Graphene has been suggested as a favorable material for spintronics applications due to theoretical predictions of long T_S in the microsecond range [35]. This long T_S arises from the weak hyperfine interactions in the absence of nuclear spins for the main ^{12}C isotope and from the small spin-orbit coupling of the atomically lightweight carbon atoms.

However, the current measured T_S values for graphene obtained from spin transport experiments are shorter than predicated microsecond values, ranging between 0.2 to 12 ns [36, 37]. The intrinsic T_S values in spin transport measurements for graphene remain subject to strong debate, as spin transport devices have notable drawbacks in measurement with the use of traditional metallic ferromagnetic contacts [38]. It has been proposed that sample preparation has an important effect on the measured T_S [37]. Alternative explanations of the small experimental T_S of graphene have included the effects of external effects likc substrate [39], impurities [40], finite-sized flakes [41], and ripples [42, 43].

Continuous wave ESR was used to determine that the conduction electron density in synthetic graphene obtained *via* solvothermal methods [31] closely resembled that of 'pristine' graphene cleaved from natural graphite [5, 7, 44]. This contactless measurement also provided a lower bound and also record-value of ~50 ns lifetime below 50 K for a clearly resolved itinerant electron spin in the synthetic graphenes [5, 44]. For the observation of the CESR, high-frequency ESR techniques [45, 46] were required due to the minuscule *g*-factor anisotropy of powdered carbon. The ~50 ns spin lifetime value was comparable to a conservative estimate of spin relaxation time resulting from the Elliot-Yafet mechanism of spin relaxation, where mid-range estimates put the spin-relaxation time exceeding microseconds [47]. By comparison, the spin lifetime values obtained for graphene using spin-valve device measurements are of the order 0.2-12 ns at 300 K [37, 47 - 52]. Hence, the discrepancy between experimentally and theoretically obtained values of spin lifetime is yet to be fully reconciled [35]. Current observations of the itinerant electron spin lifetime in graphene indicate that although spin relaxation could be described by a number of intrinsic mechanisms, extrinsic interaction mechanisms must play an essential role.

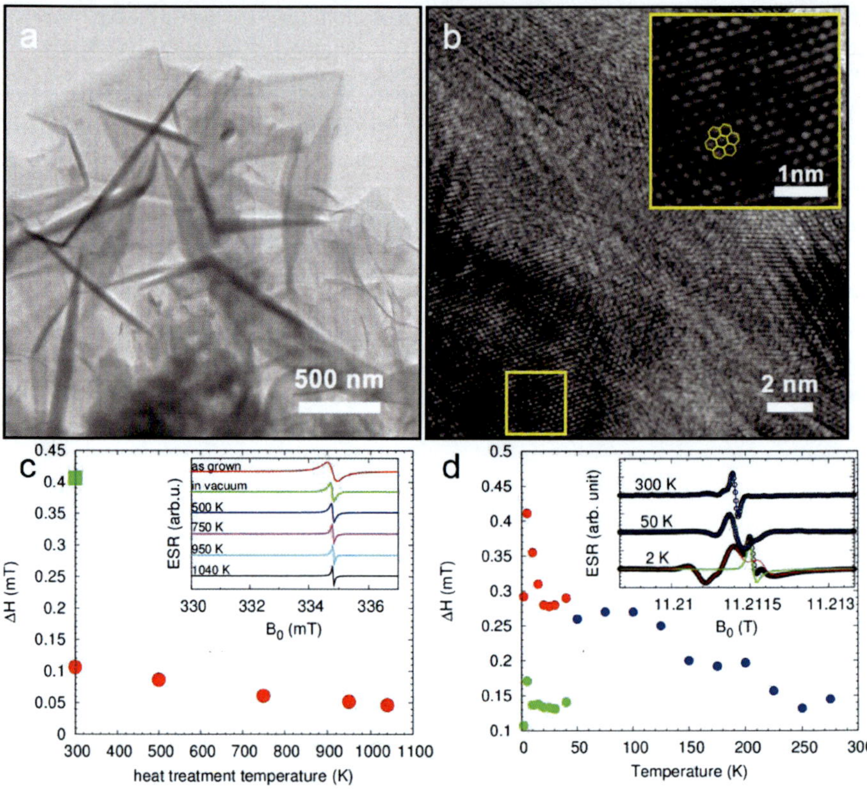

Fig. (1). (**a**) TEM images of typical regions in the graphene sample showing the extended assembly of graphene sheets, and (**b**) a high-resolution image of layered graphene sheets with no interplanar correlation. Sheet corrugation was also visible. (Reused from [7] with the Permission from John Wiley and Sons). Inset of (**b**) shows a magnified region of the hexagonally arranged graphene lattice with hexagons drawn as a guide to the eye. (**c**) Evolution of the ESR linewidth, ΔH, at T = 300 K, and ν = 9.4 GHz of samples undergone various heat treatments under dynamic vacuum, inset in (**c**) shows the ESR spectrum at T = 300 K after subsequent heat treatments with the spectra normalized for comparison. (**d**) The temperature dependence of the ESR line-width measured at 315 GHz Larmor frequency. At high-temperatures above 50 K a single motional narrowed line is observed (blue symbols). Below 50 K (red symbols) represent the paramagnetic line-width, a CESR line appears at the same temperature with temperature independent line-width (green symbols). The inset shows a typical spectrum recorded at 300 K and at 50 K above the transition with the blue lines tracing fits to a Lorentzian profile. The lower trace in the inset shows spectrum recorded at 2 K with fits to the narrow CESR line shown in green and paramagnetic powder spectra shown in red. Reproduced with permission, [5]. Copyright 2014, Elsevier.

An attempt to clarify the origin of the long T_S in solvothermally produced graphene, measurements were performed using continuous wave ESR spectroscopy in the exceptionally broad 9.4–420 GHz frequency-range and in the 2–300 K range [7]. The ESR signal of the as-prepared graphene material at 300 K temperature and 9.4 GHz frequency had a highly symmetric Lorentzian shape. The position at g = 2.0137 is typical of graphitic material. The width of the resonance was ΔH = 0.41 mT. The ESR signal was significantly different after the evacuation of the sample overnight in a dynamic vacuum of 10^{-5} mbar. The ESR line width decreased by more than a factor of 4 and reached ΔH = 0.106 mT. The g-factor was, however, unchanged, within experimental accuracy. Subsequent heat treatments up to 1040 K under dynamic vacuum further changed the ESR width at T = 300 K significantly. The ESR line width decreased monotonically as a function of increasing heat treatment temperature. It decreased to ΔH = 0.046mT after heat treatment at 1040 K; this is an order of magnitude smaller than that observed in the as-prepared graphene exposed to air. The intensity of the ESR line of a heat-treated sample under vacuum at 1040 K was $\chi = 3.7\times10^{19}$ spin g^{-1}. This value is approximately 25 times higher than that of the as-prepared sample in air. The line-width of ΔH = 0.046mT at 300 K nominally gives a spin lifetime of 160 ns corresponding to paramagnetic localized spins.

The record-long electron spin lifetime detected by multifrequency ESR in the solvothermal graphene allowed for the spectroscopic monitoring of O_2 to within 1.4 nm of the graphene surface and demonstrated a clear correlation between the measured spin lifetime and O_2 proximity [7]. A model was put forward where dipole-dipole interactions between the graphene surface and localized paramagnetic O_2 spins destroys spin coherence. The loss of coherence manifests as an increase in the observed ESR line width. Given the sensitivity of spin lifetime to oxygen molecules only a few nanometers away, it was noted that future attempts to assemble graphene-based spintronic devices will require the special consideration of surrounding chemical environments.

ESR AND SPIN DYNAMICS OF CONDUCTING CARBON NANOSPHERES

It is not straightforward to classify spherically shaped carbon particles definitively. An initial distinction of the overall morphology of spheres can be whether they are solid or hollow, however. More generally, carbon nanospheres (CNS) have been classified according to three proposed categories: concentric, radial, or randomly oriented carbon layers [53]. These three categories have been further divided into groups according to their size: the C_n family (e.g. C_{60} and C_{70}) and graphitic onion-like carbon with diameters in the range 2-20 nm, less graphitic onion-like carbon with diameters in the range of 20-50 nm, less graphitic carbon spheres 50-1 μm, and carbon beads with diameters from 1-10 μm [54].

There are several pathways have been proposed regarding the formation of spherical carbon bodies [53, 55 - 59] that highlight the interfacial energy as the major governing factor in determining the critical radius of a nucleated carbon sphere and the atomic-texture of spherical carbon bodies at the stage of their formation. The degree of graphitization of CNS is commonly evaluated by spectroscopic means and by transmission electron microscopy (TEM) [6, 54, 58, 60].

As outlined in the previous section, conduction electrons in extended graphene sheets can have T_S = 50 ns at 50 K, however, the electron spin lifetime is inherently fragile due to the interaction of large open surfaces with the environment [5, 7]. Closed surfaces could, therefore, stabilize the long spin lifetime. A spherical morphology would minimize the free surface for a given volume and reduce the influence of O_2 absorption on the spin dynamics [7]. Furthermore, inspired by Kawabata's theory [4], confining conduction electrons to small volumes containing finite sized graphenic fragments which comprise the carbon nanospheres could be advantageous to the spin lifetime. The combination of these two phenomena has led to the development of disordered onion-like carbon nanospheres (DOLCNS) with prolonged room temperature T_S [6].

ESR measurements uncovered that the DOLCNS were conducting and metallic-like with an itinerant spin lifetime (T_S) of ~175 ns at room temperature [6] and were a feasible option for use as quantum bits. These DOLCNS were synthesized in gram-scale quantities by the combustion of naphthalene, which has proved highly reproducible and repeatable [6, 58, 61]. The DOLCNS material is produced free of metal impurities and other carbon allotropes, (Fig. **2**). The structural and chemical properties of the DOLCNS were a fundamental key to the observation of long T_S. Furthermore, the coherent control of the quantum state of the electron spin confined on the DOLCNS at 300 K by ESR was also achieved, (Fig. **3**) [6].

TEM images of DOLCNS showed the extensive formation of spherical carbon across micron scales, (Fig. **2**). The as-prepared DOLCNS form conglomerates of carbon spheres. After sonication in solvent-based suspensions, the spheres could appear as individual particles. In comparison with other nanoparticle quantum dots used as qubits, for example, Mn-doped PbS (Ref [62]), the DOLCNS are relatively uniform with a size distribution of 37±7 nm estimated from TEM images. TEM tomography revealed that a number of DOLCNS were also observed to have an asymmetric shape, which results from the formation of joint graphitic layers of contacting particles.

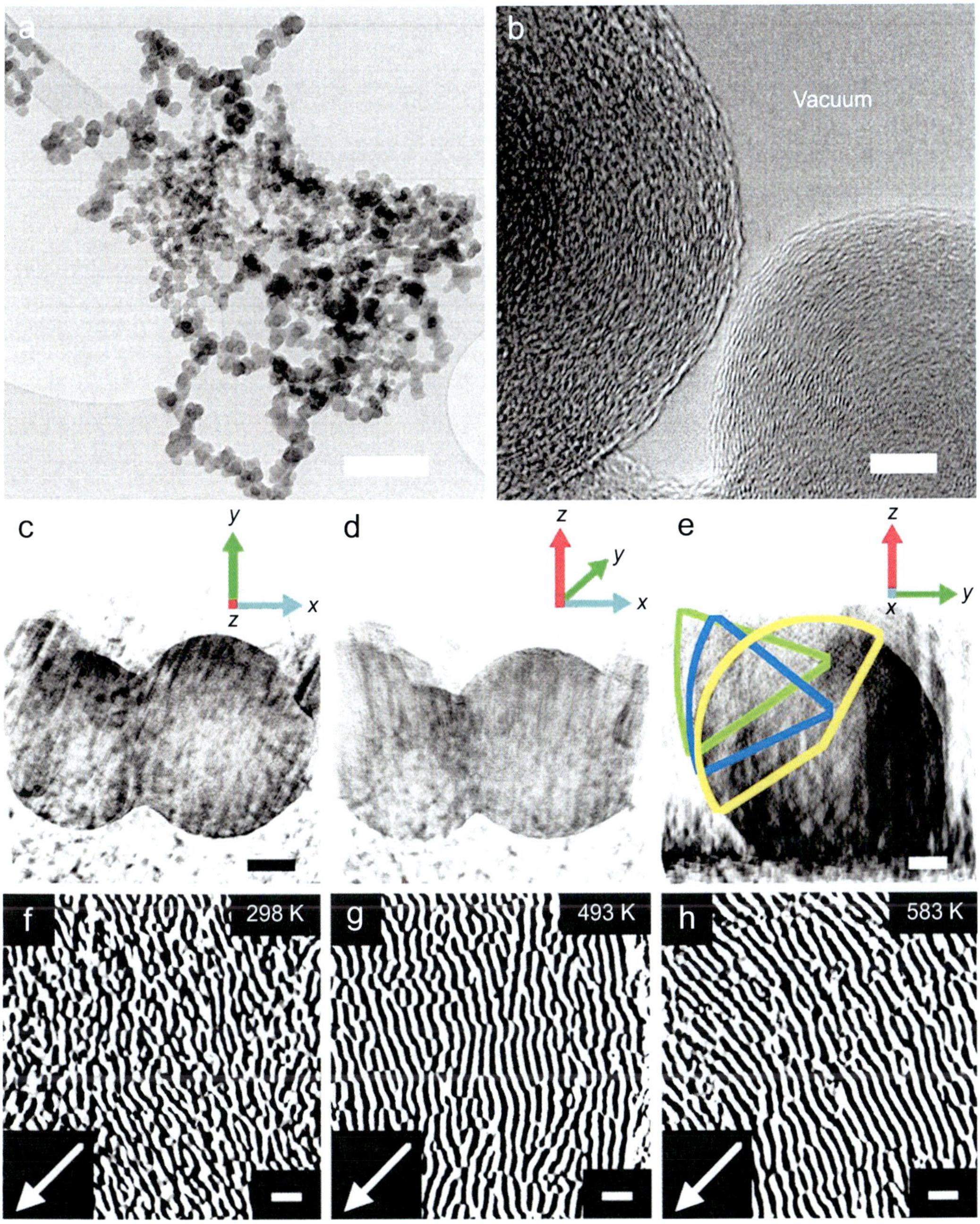

Fig. (2). (**a**) TEM image of carbon nanospheres on a carbon support. (**b**) High-resolution TEM of discrete carbon nanospheres. (Reused from [6] with the Permission from Springer Nature).

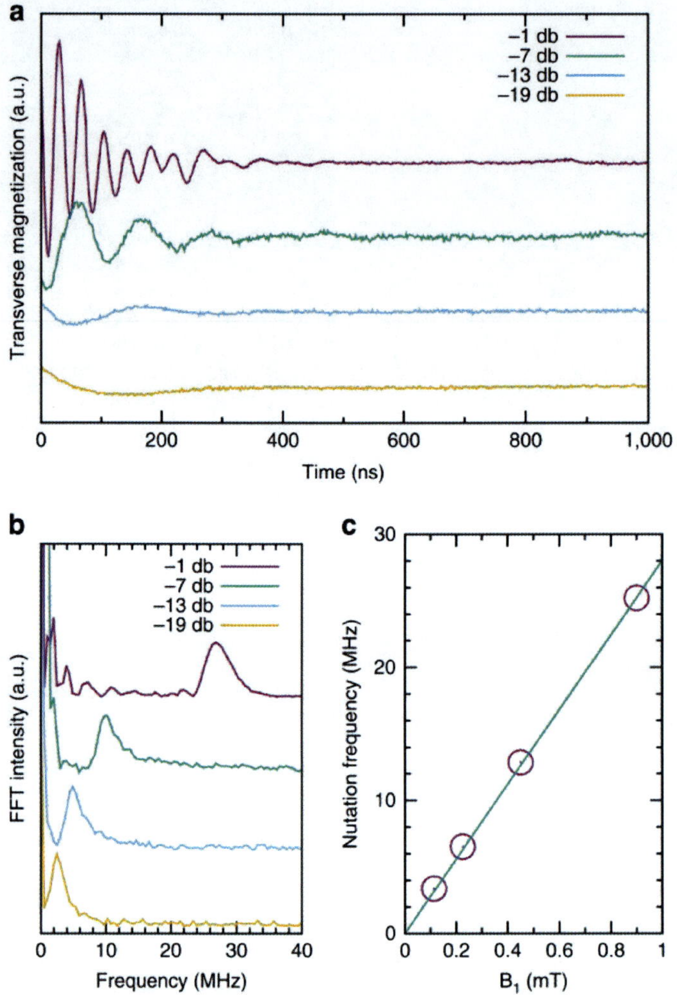

Fig. (3). (**a**) Rabi oscillations of the DOLCNS electron qubits at 300 K and B_0 = 337 mT for different microwave powers. (**b**) Fourier transform of the Rabi oscillations, with the signal shifted vertically for clarity. (**c**) Rabi frequency is proportional to the square root of the power. The value 25 MHz observed for the maximum power is consistent with the previously determined maximum B_1 field in the dielectric cavity of 0.9 mT, indicating the presence of an effective spin S = 1/2. (Reused from [6] with the Permission from Springer Nature).

High-resolution TEM images showed short graphitic fragments that comprise the DOLCNS followed the curvature of a sphere, creating many open edges. It has to be noticed, however, that the individual graphitic fragments in the DOLCNS are neither curved nor they resemble the curvature in nanotubes or fullerenes. Rather, the flakes exhibit an intricate array of interplanar bonding all the way to the center of the nanosphere even when heated to temperatures of 583 K [6]. The DOLCNS show a continuation of the closed cage structure towards the center they are not hollow [6]. Moreover, X-ray photoelectron spectroscopy revealed the presence of non-bonding π and σ orbitals as a result of fragmented sheets that contained carbon arranged in a distorted hexagonal network [6]. As a result of fragmented sheets and the non-bonding π and σ orbitals, the electron density of states is significantly enhanced and conducting state results [63].

The T_S in the carbon nanospheres is remarkably a two order of magnitude enhancement over that found in conducting crystalline graphene [6, 64]. Part of this increase in T_1 and T_2 was attributed to quantum confinement effects readily observed in conducting nanostructures [6, 65]. This assessment is further supported by the fact that T_S was enhanced by a factor of 3 relative to synthetic graphene sheets [5]. The observation of long T_1 and T_2 even at 300 K despite the highly defective structure of carbon nanospheres highlights the intrinsically weak spin-orbit interaction of carbon [6].

Temperature-dependent ESR measurements were performed to further support that conduction electrons are confined within the DOLCNS [6]. The *g*-factor was found temperature independent. This observation is in good agreement with the general behavior of metals with weak spin-orbit coupling and of graphitic nanoparticles. The observed increase in T_1 and T_2 as the temperature was decreased is characteristic of metals, where the SOC is responsible for the temperature-induced shortening of T_1 and T_2 due to electron-phonon scattering due. Furthermore, the spin susceptibility showed temperature-dependent behavior, following Curie-Weiss dependence, as one may expect for nanoparticles of metals [3, 6].

A key observation of Náfrádi *et al.* [6] was the demonstration of coherent control of the spins of DOLCNS, which has broad applicability for room temperature qubit operation, (Fig. **2**). Pulsed ESR at 300 K revealed oscillation of the transverse magnetization, the so-called Rabi oscillations [6]. More generally, the work demonstrated that a long itinerant electron spin lifetime in a magnetically homogenous conducting material can be achieved at room temperature [6]. Furthermore, it was demonstrated the facile preparation of a carbon material using common laboratory reagents like naphthalene, combined with the use of well-established ESR manipulation measurements at room temperature, has effectively reduced many of the technological barriers to realizing practical quantum computing and spintronics using solid-state materials [6].

PERSPECTIVES & OUTLOOK

A potential solution to practical spintronics and quantum computing may involve the use of electrically conducting carbon materials with long room temperature electron spin lifetimes. Until recently, this solution represented a materials dichotomy. Our breakthrough findings detailed in this chapter demonstrated for the first time that it is feasible to operate the electron spins with long room temperature T_S in conducting carbon materials as quantum bits. Methods to synthesize electrically conducting solid-state materials with prolonged T_S at room temperature have however only just emerged.

For solid-state materials, the strong spin-orbit interaction imposed by heavy nuclei in materials like semiconductors [66] and metals [67] requires the pursuit of structural perfection and spin manipulation at low temperatures (40 mK to 100 K). This has resulted in well-defined (highly crystalline) materials that currently set the benchmark for long T_1 relaxation times that exceed seconds [68]. At room temperature, however, a number of intrinsic spin-lattice relaxation processes and phonon modes reduce the spin coherence and diminish T_1 and T_2 even for structurally perfect crystals [17]. Therefore, for metal nanoparticles [65] T_1 and T_2 shorten to 10–40 picoseconds and for bulk semiconductors [69] to 1–4 nanoseconds. Hence, the use of solid-state materials containing light atomic weight nuclei was found to fundamentally reduce the detrimental effects of spin-orbit coupling on electron spin lifetimes [6].

With structural perfection in mind, attempts to prolong the electron spin lifetime have been made using crystalline carbon materials; e.g., quantum dots based on carbon nanotube have been fabricated with spin lifetimes of ~3 ns at millikelvin temperatures [70]. The shortening of spin lifetimes in carbon nanotube based quantum dots is attributed to an increase in the spin-orbit coupling induced by the high curvature of the carbon sheets [42] which is not as prevalent in DOLCNS [6]. Graphene, on the other hand, is flat (or at best not curved) [26] and experiments on single- and few-layer graphene have demonstrated T_1 and T_2 of 0.5–12 ns at room temperature [48]. However, in recent studies, it was found that the lower-boundary room temperature T_1 and T_2 spin lifetime in synthetic graphene approached a record-long 50 ns at room temperature [5, 7]. Even so, T_1 and T_2 values are still well below microsecond expectations [35]. Hence, any improvement in the room temperature electron spin lifetime of graphene would be of great importance.

Efforts continue to be made along disparate, competing pathways to long spin lifetime materials. Molecular materials with confined spins can be produced on an industrial scale and are readily processed [71]. These spin-bearing organic molecules often have weak spin-orbit coupling, possess many non-degenerate spin transitions, and can be chemically modified. However, the dipole-dipole interactions between spins of neighboring molecules require dilution of the spin-containing molecules. The hyperfine interaction of the molecules necessitates isotope engineering of the host system to enhance spin coherence. It also needs low temperatures (< 100 K) in order to dampen molecular vibrations which cause spin decoherence [72]. Non-carbon based materials have been developed, including heterometallic molecular nanomagnets comprised of isotopically engineered and diamagnetically diluted Cr_7M (M = Ni, Mn, Zn) that are able to prolong the electron spin lifetime, but only at low temperature (5 K) [9]. These possible solutions demonstrate that there are clear trade-offs to be considered regarding the feasibility of applying an electron-spin material system. Thus, despite progress to date, there are currently limited solutions to the problem of unifying the disparate fields of inorganic and molecular materials for electron spin manipulation; therefore new materials are needed.

Attempts to combine the advantages of molecular and inorganic quantum materials resulted in electronically insulating carbon-based systems like nitrogen-vacancy centers in nanodiamonds [8] and N@C_{60} fullerenes [73]. In these materials, long microsecond T_1 and T_2 of localized electrons can be achieved. Moreover, manipulation at room-temperature was demonstrated. Magnetic inhomogeneity in these materials is originating from the nuclear spins of ^{13}C and N nuclei. Moreover, the localized electron spins themselves generate fluctuating magnetic fields, posing limitations on material spin density. This limitation is particularly to localized electron spins, but it is absent for conduction electrons where the motional narrowing process contravenes this effect [5]. Thus, conduction electron spins are favored in spintronics applications as they provide greater robustness against decoherence compared to localized spin states.

In our recent work outlined in this chapter, we show that conducting carbon materials can be prepared with long T_S without the need for isotopic engineering, spin dilution, cryogenic temperatures, well-defined crystal structures, or the use of metals, which effectively reduces many of the technological barriers to realizing practical spintronics using solid-state materials. However, our knowledge of the mutual dependence of defects, adatoms, and electron interactions and the degree to which they affect carbon materials' magnetic properties remains limited. Concerted efforts to understand the structural and chemical origins of this magnetic behavior in carbon nanomaterials are of fundamental importance, and continue to have broad applicability.

CONSENT FOR PUBLICATION

Not applicable.

CONFLICT OF INTEREST

The author (editor) declares no conflict of interest, financial or otherwise.

ACKNOWLEDGEMENT

Declare none.

REFERENCES

[1] Náfrádi, B.; Choucair, M.; Forró, L. Electron Spin Dynamics of Two-Dimensional Layered Materials. *Adv. Funct. Mater.*, **2017**, *27*, 1604040. In Press [http://dx.doi.org/10.1002/adfm.201604040]

[2] Bailyn, M. Magnetism in Transition Metals. *Phys. Rev.*, **1965**, *139*, A1905-A1919. [http://dx.doi.org/10.1103/PhysRev.139.A1905]

[3] Kubo, R. Electronic Properties of Metallic Fine Particles. I. *J. Phys. Soc. Jpn.*, **1962**, *17*, 975-986. [http://dx.doi.org/10.1143/JPSJ.17.975]

[4] Kawabata, A. Electronic Properties of Fine Metallic Particles. III. E.S.R Absorption Line Shape. *J. Phys. Soc. Jpn.*, **1970**, *29*, 902-911. [http://dx.doi.org/10.1143/JPSJ.29.902]

[5] Náfrádi, B.; Choucair, M.; Forró, L. Spin lifetime of itinerant electrons in chemically synthesized graphene multi-layers. *Carbon*, **2014**, *74*, 346-351. [http://dx.doi.org/10.1016/j.carbon.2014.03.046]

[6] Náfrádi, B.; Choucair, M.; Dinse, K.; Forró, L. *Room temperature manipulation of long lifetime spins in metallic-like carbon nanospheres.,* Nature Communications volume 7, Article number: 12232, **2016**. [http://dx.doi.org/10.1038/ncomms12232]

[7] Náfrádi, B.; Choucair, M.; Southon, P.D.; Kepert, C.J.; Forró, L. Strong interplay between the electron spin lifetime in chemically synthesized graphene multilayers and surface-bound oxygen. *Chemistry*, **2015**, *21*(2), 770-777. [http://dx.doi.org/10.1002/chem.201404309] [PMID: 25394656]

[8] Childress, L.; Gurudev Dutt, M.V.; Taylor, J.M.; Zibrov, A.S.; Jelezko, F.; Wrachtrup, J.; Hemmer, P.R.; Lukin, M.D. Coherent dynamics of coupled electron and nuclear spin qubits in diamond. *Science*, **2006**, *314*(5797), 281-285. [http://dx.doi.org/10.1126/science.1131871] [PMID: 16973839]

[9] Moro, F.; Kaminski, D.; Tuna, F.; Whitehead, G.F.; Timco, G.A.; Collison, D.; Winpenny, R.E.; Ardavan, A.; McInnes, E.J. Coherent electron spin manipulation in a dilute oriented ensemble of molecular nanomagnets: pulsed EPR on doped single crystals. *Chem. Commun. (Camb.)*, **2014**, *50*(1), 91-93. [http://dx.doi.org/10.1039/C3CC46326E] [PMID: 24187686]

[10] Awschalom, D.D.; Bassett, L.C.; Dzurak, A.S.; Hu, E.L.; Petta, J.R. Quantum spintronics: engineering and manipulating atom-like spins in semiconductors. *Science*, **2013**, *339*(6124), 1174-1179. [http://dx.doi.org/10.1126/science.1231364] [PMID: 23471400]

[11] Koppens, F.H.L.; Buizert, C.; Tielrooij, K.J.; Vink, I.T.; Nowack, K.C.; Meunier, T.; Kouwenhoven, L.P.; Vandersypen, L.M.K. Driven coherent oscillations of a single electron spin in a quantum dot. *Nature*, **2006**, *442*(7104), 766-771. [http://dx.doi.org/10.1038/nature05065] [PMID: 16915280]

[12] Eisberg, R.; Resnick, R. *Quantum physics of atoms, molecules, solids, nuclei and particles,* 2nd ed; John Wiley: New York, **1985**.

[13] Dyakonov, M.; Perel', V. Spin Orientation of Electrons Associated with the Interband Absorption of Light in Semiconductors. *Sov. Phys. JETP*, **1971**, *33*, 1053-1059.

[14] Yafet, Y. g-factors and spin-lattice relaxation of conduction electrons. In Solid State Physics Vol. 14;

Seitz, F.Turnbull, D., Ed.; Academic Press: New York, 1963; pp. 1-98.

[15] Elliott, R. Theory of the Effect of Spin-Orbit Coupling on Magnetic Resonance in Some Semiconductors. *Phys. Rev.,* **1954**, *96*, 266-279. [http://dx.doi.org/10.1103/PhysRev.96.266]

[16] Hanson, R.; Kouwenhoven, L.; Petta, J.; Tarucha, S.; Vandersypen, L. Publisher's Note: Spins in few-electron quantum dots. *Rev. Mod. Phys.,* **2007**, *79*, 1455-1455. [http://dx.doi.org/10.1103/RevModPhys.79.1455]

[17] Boross, P.; Dóra, B.; Kiss, A.; Simon, F. *A unified theory of spin-relaxation due to spin-orbit coupling in metals and semiconductors.,* 3, Article number: 3233, **2013**. [http://dx.doi.org/10.1038/srep03233]

[18] Poole, C.P., Jr; Farach, H.A. *Relaxation in Magnetic Resonance*; Academic Press: New York, **1971**, pp. 17-29. [http://dx.doi.org/10.1016/B978-0-12-561450-4.50008-9]

[19] Jánossy, A. Resonant and nonresonant conduction-electron-spin transmission in normal metals. *Phys. Rev. B,* **1980**, *21*, 3793-3810. [http://dx.doi.org/10.1103/PhysRevB.21.3793]

[20] Dyson, F.J. Electron Spin Resonance Absorption in Metals. II. Theory of Electron Diffusion and the Skin Effect. *Phys. Rev.,* **1955**, *98*, 349-359. [http://dx.doi.org/10.1103/PhysRev.98.349]

[21] Buttet, J.; Car, R.; Myles, C. Size dependence of the conduction-electron-spin-resonancegshift in a small sodium particle: Orthogonalized standing-wave calculations. *Phys. Rev. B,* **1982**, *26*, 2414-2431. [http://dx.doi.org/10.1103/PhysRevB.26.2414]

[22] Bloch, F. Nuclear Induction. *Phys. Rev.,* **1946**, *70*, 460-474. [http://dx.doi.org/10.1103/PhysRev.70.460]

[23] Le Bellac, M.; Forcrand-Millard, P. *A short introduction to quantum information and quantum computation,* 1st ed; Cambridge University Press: Cambridge, **2006**. [http://dx.doi.org/10.1017/CBO9780511755361]

[24] Li, Z.; Chen, L.; Meng, S.; Guo, L.; Huang, J.; Liu, Y.; Wang, W.; Chen, X. Field and temperature dependence of intrinsic diamagnetism in graphene: Theory and experiment. Phys. Rev. B 2015, 91, 094429(1-5). [http://dx.doi.org/10.1103/PhysRevB.91.094429]

[25] Sepioni, M.; Nair, R.; Rablen, S.; Narayanan, J.; Tuna, F.; Winpenny, R.; Geim, A.; Grigorieva, I. Limits on Intrinsic Magnetism in Graphene. *Phys. Rev. Lett.,* **2010**, *105*, 207205(1-4). [http://dx.doi.org/10.1103/PhysRevLett.105.207205]

[26] Meyer, J.C.; Geim, A.K.; Katsnelson, M.I.; Novoselov, K.S.; Booth, T.J.; Roth, S. The structure of suspended graphene sheets. *Nature,* **2007**, *446*(7131), 60-63. [http://dx.doi.org/10.1038/nature05545] [PMID: 17330039]

[27] Notley, S.M. Highly concentrated aqueous suspensions of graphene through ultrasonic exfoliation with continuous surfactant addition. *Langmuir,* **2012**, *28*(40), 14110-14113. [http://dx.doi.org/10.1021/la302750e] [PMID: 22985170]

[28] Li, D.; Müller, M.B.; Gilje, S.; Kaner, R.B.; Wallace, G.G. Processable aqueous dispersions of graphene nanosheets. *Nat. Nanotechnol.,* **2008**, *3*(2), 101-105. [http://dx.doi.org/10.1038/nnano.2007.451] [PMID: 18654470]

[29] Tran, D.; Kabiri, S.; Losic, D. A green approach for the reduction of graphene oxide nanosheets using non-aromatic amino acids. *Carbon,* **2014**, *76*, 193-202. [http://dx.doi.org/10.1016/j.carbon.2014.04.067]

[30] Hao, Y.; Bharathi, M.S.; Wang, L.; Liu, Y.; Chen, H.; Nie, S.; Wang, X.; Chou, H.; Tan, C.;

Fallahazad, B.; Ramanarayan, H.; Magnuson, C.W.; Tutuc, E.; Yakobson, B.I.; McCarty, K.F.; Zhang, Y.W.; Kim, P.; Hone, J.; Colombo, L.; Ruoff, R.S. The role of surface oxygen in the growth of large single-crystal graphene on copper. *Science,* **2013**, *342*(6159), 720-723. [http://dx.doi.org/10.1126/science.1243879] [PMID: 24158906]

[31] Choucair, M.; Thordarson, P.; Stride, J.A. Gram-scale production of graphene based on solvothermal synthesis and sonication. *Nat. Nanotechnol.,* **2009**, *4*(1), 30-33. [http://dx.doi.org/10.1038/nnano.2008.365] [PMID: 19119279]

[32] Wang, Y.; Page, A.J.; Nishimoto, Y.; Qian, H.J.; Morokuma, K.; Irle, S. Template effect in the competition between Haeckelite and graphene growth on Ni(111): quantum chemical molecular dynamics simulations. *J. Am. Chem. Soc.,* **2011**, *133*(46), 18837-18842. [http://dx.doi.org/10.1021/ja2064654] [PMID: 21967405]

[33] Kahlert, J.U.; Austin, C.J.; Hall, A.J.; Rawal, A.; Hook, J.M.; Rendina, L.M.; Choucair, M. Site-specific synthesis of a hybrid boron-graphene salt. *Chem. Commun. (Camb.),* **2016**, *52*(6), 1290-1292. [http://dx.doi.org/10.1039/C5CC07611K] [PMID: 26627051]

[34] Kahlert, J.; Rawal, A.; Hook, J.; Rendina, L.; Choucair, M. Carborane functionalization of the aromatic network in chemically-synthesized graphene. *Chem. Commun.,* **2014**, *50*, 11332. [http://dx.doi.org/10.1039/C4CC04521A]

[35] Han, W.; Kawakami, R.K.; Gmitra, M.; Fabian, J. Graphene spintronics. *Nat. Nanotechnol.,* **2014**, *9*(10), 794-807. [http://dx.doi.org/10.1038/nnano.2014.214] [PMID: 25286274]

[36] Tombros, N.; Jozsa, C.; Popinciuc, M.; Jonkman, H.T.; van Wees, B.J. Electronic spin transport and spin precession in single graphene layers at room temperature. *Nature,* **2007**, *448*(7153), 571-574. [http://dx.doi.org/10.1038/nature06037] [PMID: 17632544]

[37] Han, W.; Pi, K.; McCreary, K.; Li, Y.; Wong, J.; Swartz, A.; Kawakami, R. Tunneling Spin Injection into Single Layer Graphene. *Phys. Rev. Lett.,* **2010**, *105*, 167202. [http://dx.doi.org/10.1103/PhysRevLett.105.167202]

[38] Krompiewski, S. *Effect of the attachment of ferromagnetic contacts on the conductivity and giant magnetoresistance of graphene nanoribbons.,* Nanotechnology, Volume 23, Number 13 **2012**. [http://dx.doi.org/10.1088/0957-4484/23/13/135203]

[39] Gmitra, M.; Konschuh, S.; Ertler, C.; Ambrosch-Draxl, C.; Fabian, J. Band-structure topologies of graphene: Spin-orbit coupling effects from first principles. Phys. Rev. B 2009, 80, 235431(1-5). [http://dx.doi.org/10.1103/PhysRevB.80.235431]

[40] Kochan, D.; Gmitra, M.; Fabian, J. Spin Relaxation Mechanism in Graphene: Resonant Scattering by Magnetic Impurities. Phys. Rev. Lett. 2014, 112, 116602(1-5). [http://dx.doi.org/10.1103/PhysRevLett.112.116602]

[41] Wang, W.; Pi, K.; Li, Y.; Chiang, Y.; Wei, P.; Shi, J.; Kawakami, R. Magnetotransport properties of mesoscopic graphite spin valves. *Phys. Rev. B,* **2008**, *77*, 020402(R). [http://dx.doi.org/10.1103/PhysRevB.77.020402]

[42] Huertas-Hernando, D.; Guinea, F.; Brataas, A. Spin-orbit coupling in curved graphene, fullerenes, nanotubes, and nanotube caps. *Phys. Rev. B,* **2006**, *74*, 155426. [http://dx.doi.org/10.1103/PhysRevB.74.155426]

[43] Dugaev, V.; Sherman, E.; Barnaś, J. Spin dephasing and pumping in graphene due to random spin-orbit interaction. *Phys. Rev. B,* **2011**, *83*, 085306. [http://dx.doi.org/10.1103/PhysRevB.83.085306]

[44] Náfrádi, B.; Forró, L.; Choucair, M. Electron spin lifetime in chemically synthesized graphene sheets. *Phys. Stat. Solidi (B),* **2014**, *251*, 2521-2524. [http://dx.doi.org/10.1002/pssb.201451224]

[45] Náfrádi, B.; Gaál, R.; Fehér, T.; Forró, L. Microwave frequency modulation in continuous-wave far-

infrared ESR utilizing a quasi-optical reflection bridge. *J. Magn. Reson.*, **2008**, *192*(2), 265-268. [http://dx.doi.org/10.1016/j.jmr.2008.03.004] [PMID: 18375160]

[46] Náfrádi, B.; Gaál, R.; Sienkiewicz, A.; Fehér, T.; Forró, L. Continuous-wave far-infrared ESR spectrometer for high-pressure measurements. *J. Magn. Reson.*, **2008**, *195*(2), 206-210. [http://dx.doi.org/10.1016/j.jmr.2008.09.014] [PMID: 18835205]

[47] Yang, T.; Balakrishnan, J.; Volmer, F.; Avsar, A.; Jaiswal, M.; Samm, J.; Ali, S.; Pachoud, A.; Zeng, M.; Popinciuc, M. Observation of Long Spin-Relaxation Times in Bilayer Graphene at Room Temperature. *Phys. Rev. Lett.*, **2011**, *107*, 047206. [http://dx.doi.org/10.1103/PhysRevLett.107.047206]

[48] Drögeler, M.; Franzen, C.; Volmer, F.; Pohlmann, T.; Banszerus, L.; Wolter, M.; Watanabe, K.; Taniguchi, T.; Stampfer, C.; Beschoten, B. Spin Lifetimes Exceeding 12 ns in Graphene Nonlocal Spin Valve Devices. *Nano Lett.*, **2016**, *16*(6), 3533-3539. [http://dx.doi.org/10.1021/acs.nanolett.6b00497] [PMID: 27210240]

[49] Han, W.; Kawakami, R. Spin Relaxation in Single-Layer and Bilayer Graphene. *Phys. Rev. Lett.*, **2011**, *107*, 047207. [http://dx.doi.org/10.1103/PhysRevLett.107.047207]

[50] Dlubak, B.; Martin, M.; Deranlot, C.; Servet, B.; Xavier, S.; Mattana, R.; Sprinkle, M.; Berger, C.; De Heer, W.; Petroff, F. Highly efficient spin transport in epitaxial graphene on SiC. *Nat. Phys.*, **2012**, *8*, 557-561. [http://dx.doi.org/10.1038/nphys2331]

[51] Guimarães, M.; Zomer, P.; Ingla-Aynés, J.; Brant, J.; Tombros, N.; van Wees, B. Controlling Spin Relaxation in Hexagonal BN-Encapsulated Graphene with a Transverse Electric Field. *Phys. Rev. Lett.*, **2011**, *113*, 086602. [http://dx.doi.org/10.1103/PhysRevLett.113.086602]

[52] Zomer, P.; Guimarães, M.; Tombros, N.; van Wees, B. Long-distance spin transport in high-mobility graphene on hexagonal boron nitride. *Phys. Rev. B.*, **2012**, *86*, 161416(R). [http://dx.doi.org/10.1103/PhysRevB.86.161416]

[53] Inagaki, M. Discussion of the formation of nanometric texture in spherical carbon bodies. *Carbon,* **1997**, *35*, 711-713. [http://dx.doi.org/10.1016/S0008-6223(97)86645-6]

[54] Serp, P.; Feurer, R.; Kalck, P.; Kihn, Y.; Faria, J.; Figueiredo, J. A chemical vapour deposition process for the production of carbon nanospheres. *Carbon,* **2001**, *39*, 621-626. [http://dx.doi.org/10.1016/S0008-6223(00)00324-9]

[55] Nieto-Márquez, A.; Valverde, J.; Keane, M. Selective low temperature synthesis of carbon nanospheres *via* the catalytic decomposition of trichloroethylene. Appl. Catal., A. 2009, 352, 159-170. [http://dx.doi.org/10.1016/j.apcata.2008.10.006]

[56] Pol, V.; Motiei, M.; Gedanken, A.; Calderon-Moreno, J.; Yoshimura, M. Carbon spherules: synthesis, properties and mechanistic elucidation. *Carbon,* **2004**, *42*, 111-116. [http://dx.doi.org/10.1016/j.carbon.2003.10.005]

[57] Ugarte, D. Formation mechanism of quasi-spherical carbon particles induced by electron bombardment. *Chem. Phys. Lett.*, **1993**, *207*, 473-479. [http://dx.doi.org/10.1016/0009-2614(93)89032-D]

[58] Choucair, M.; Stride, J. The gram-scale synthesis of carbon onions. *Carbon,* **2012**, *50*, 1109-1115. [http://dx.doi.org/10.1016/j.carbon.2011.10.023]

[59] Choucair, M.; Hill, M.; Stride, J. A low temperature reduction of CCl4 to solid and hollow carbon nanospheres using metallic sodium. *Mater. Chem. Phys.*, **2015**, *154*, 38-43. [http://dx.doi.org/10.1016/j.matchemphys.2015.01.042]

[60] Nieto-Márquez, A.; Romero, R.; Romero, A.; Valverde, J. Carbon nanospheres: synthesis,

physicochemical properties and applications. *J. Mater. Chem.*, **2011**, *21*, 1664-1672. [http://dx.doi.org/10.1039/C0JM01350A]

[61] Dong, J.; Zhang, T.; Zhang, D.; Zhang, W.; Zhang, H.; Liu, R.; Yao, M.; Liu, B. Remarkable cycle-activated capacity increasing in onion-like carbon nanospheres as lithium battery anode material. *Nanotechnology*, **2017**, *28*, 035704. [http://dx.doi.org/10.1088/1361-6528/28/3/035704]

[62] Moro, F.; Turyanska, L.; Granwehr, J.; Patanè, A. Spin manipulation and spin-lattice interaction in magnetic colloidal quantum dots. *Phys. Rev. B*, **2014**, *90*, 205428. [http://dx.doi.org/10.1103/PhysRevB.90.205428]

[63] Andersson, O.; Prasad, B.; Sato, H.; Enoki, T.; Hishiyama, Y.; Kaburagi, Y.; Yoshikawa, M.; Bandow, S. Structure and electronic properties of graphite nanoparticles. *Phys. Rev. B*, **1998**, *58*, 16387-16395. [http://dx.doi.org/10.1103/PhysRevB.58.16387]

[64] Drögeler, M.; Volmer, F.; Wolter, M.; Terrés, B.; Watanabe, K.; Taniguchi, T.; Güntherodt, G.; Stampfer, C.; Beschoten, B. Nanosecond spin lifetimes in single- and few-layer graphene-hBN heterostructures at room temperature. *Nano Lett.*, **2014**, *14*(11), 6050-6055. [http://dx.doi.org/10.1021/nl501278c] [PMID: 25291305]

[65] Halperin, W. Quantum size effects in metal particles. *Rev. Mod. Phys.*, **1986**, *58*, 533-606. [http://dx.doi.org/10.1103/RevModPhys.58.533]

[66] Dzhioev, R.; Kavokin, K.; Korenev, V.; Lazarev, M.; Meltser, B.; Stepanova, M.; Zakharchenya, B.; Gammon, D.; Katzer, D. Low-temperature spin relaxation in n-type GaAs. *Phys. Rev. B*, **2002**, *66*, 245204. [http://dx.doi.org/10.1103/PhysRevB.66.245204]

[67] Hai, P.N.; Ohya, S.; Tanaka, M. Long spin-relaxation time in a single metal nanoparticle. *Nat. Nanotechnol.*, **2010**, *5*(8), 593-596. [http://dx.doi.org/10.1038/nnano.2010.130] [PMID: 20601943]

[68] Yang, C.; Rossi, A.; Ruskov, R.; Lai, N.; Mohiyaddin, F.; Lee, S.; Tahan, C.; Klimeck, G.; Morello, A.; Dzurak, A. *Spin-valley lifetimes in a silicon quantum dot with tunable valley splitting.,* Nature Communications volume 4, Article number: 2069 **2013**. [http://dx.doi.org/10.1038/ncomms3069]

[69] Suzuki, T.; Sasaki, T.; Oikawa, T.; Shiraishi, M.; Suzuki, Y.; Noguchi, K. Room-Temperature Electron Spin Transport in a Highly Doped Si Channel. *Appl. Phys. Expr.***2011***4*, 023003(1-3). [http://dx.doi.org/10.1143/APEX.4.023003]

[70] Churchill, H.; Kuemmeth, F.; Harlow, J.; Bestwick, A.; Rashba, E.; Flensberg, K.; Stwertka, C.; Taychatanapat, T.; Watson, S.; Marcus, C. Relaxation and Dephasing in a Two-Electron ^{13}C Nanotube Double Quantum Dot. *Phys. Rev. Lett.*, **2009**, *102*, 166802. [http://dx.doi.org/10.1103/PhysRevLett.102.166802]

[71] Aromí, G.; Aguilà, D.; Gamez, P.; Luis, F.; Roubeau, O. Design of magnetic coordination complexes for quantum computing. *Chem. Soc. Rev.*, **2012**, *41*(2), 537-546. [http://dx.doi.org/10.1039/C1CS15115K] [PMID: 21818467]

[72] Cervetti, C.; Rettori, A.; Pini, M.G.; Cornia, A.; Repollés, A.; Luis, F.; Dressel, M.; Rauschenbach, S.; Kern, K.; Burghard, M.; Bogani, L. The classical and quantum dynamics of molecular spins on graphene. *Nat. Mater.*, **2016**, *15*(2), 164-168. [http://dx.doi.org/10.1038/nmat4490] [PMID: 26641019]

[73] Morton, J.; Tyryshkin, A.; Ardavan, A.; Benjamin, S.; Porfyrakis, K.; Lyon, S.; Briggs, G. Bang–bang control of fullerene qubits using ultrafast phase gates. *Nat. Phys.*, **2005**, *2*, 40-43. [http://dx.doi.org/10.1038/nphys192]

CHAPTER 4

EPR Spectroscopy on Double-Walled and Multi-Walled Carbon Nanotube Polymer Composites

Angeliki Diamantopoulou[1], Spyridon Glenis[1], Grzegorz Zolnierkiewicz[2], Anna Szymczyk[2], Nikolaos Guskos[2] and Vlassis Likodimos[1,*]

[1] *Section of Solid State Physics, Department of Physics, National and Kapodistrian University of Athens, Panepistimiopolis, 15 784, Greece*

[2] *Institute of Physics, West Pomeranian University of Technology, Al. Piastow 48, 70-311 Szczecin, Poland*

Abstract: Electron paramagnetic resonance (EPR) spectroscopy can be fruitfully applied to study the interplay of localized and itinerant spins for carbon nanomaterials, including carbon nanotubes (CNTs), and thus provides a unique spectroscopic probe of their electronic properties upon integration as active components in composite materials. In this chapter, EPR spectroscopy is exploited to investigate the magnetic properties of double-walled carbon nanotubes (DWCNTs) and composites of oxidized multi-walled carbon nanotubes (MWCNTs) embedded in an elastomeric poly(ether-ester) block copolymer. In the case of DWCNTs, an asymmetric resonance line was observed that could be accurately analyzed in terms of two independent metallic lineshapes with similar *g*-factors, a narrow and a broad one, related to the distinct contributions of defect spins located on the inner and outer DWCNTs layers, respectively. Analysis of the spin susceptibilities indicated a ferromagnetic phase transition at low temperatures, alike metallic single wall CNTs. Interlayer coupling between the DWCNT layers is accordingly suggested to enhance exchange interactions between localized spins *via* conduction electrons. Conversely, in the case of MWCNTs-polymer composites, EPR spectra in combination with static magnetization measurements revealed a drastic reduction of orbital diamagnetism and *g*-anisotropy along with a marked enhancement of spin susceptibility, with respect to the anisotropic EPR spectrum of pristine MWCNTs. These effects indicate considerable hole doping by oxygen functional groups on the MWCNTs' surface and an excessive increase of the density of paramagnetic defects, which are sensitive to the polymer relaxation and to the underlying MWCNT-polymer interfacial coupling.

Keywords: Carbon nanotubes (CNTs), Curie-Weiss law, EPR, EPR linewidth, EPR spectra deconvolution, Exchange interaction, Ferromagnetic phase transition,

* **Corresponding author Vlassis Likodimos:** Section of Solid State Physics, Department of Physics, National and Kapodistrian University of Athens, Panepistimiopolis, 15 784, Greece; Tel: +30 (210) 727-6824; E-mail: vlikodimos@phys.uoa.gr

Dariya Savchenko & Abdel Hadi Kassiba (Eds.)

Ferromagnetic resonance, Spin dynamics, Double wall carbon nanotubes (DWCNTs), Lorentzian lineshape, Magnetic properties, Multiwall carbon nanotube (MWCNTs) polymer composites, Raman spectra, Single wall carbon nanotubes (SWCNTs), Single metallic lineshape, Skin effect, Spin susceptibility, Static magnetization, Temperature dependence.

INTRODUCTION

Carbon nanotubes (CNTs) have been established as key materials in the rapidly evolving field of nanotechnology and emerging applications within diverse areas ranging from nanoelectronics and energy storage to biology and environmental engineering [1 - 3] Relying on their unique structural, chemical, mechanical, optical and electronic properties, CNTs continue to attract significant interest from both fundamental and applied perspective [4, 5]. Electron paramagnetic resonance (EPR) can be efficiently applied to investigate the interplay of localized and itinerant spins in CNTs and by this means provide a sensitive spectroscopic probe of the electronic properties of pristine or modified CNTs as well as upon their incorporation in composite materials. Being sensitive on both spin and orbital magnetism of the CNT rolled graphene layers, EPR in combination with static magnetization has been profitably exploited to investigate the magnetic properties and the effects of low dimensionality on the prototype single wall carbon nanotubes (SWCNTs) [6 - 15] as well as on multi wall carbon nanotubes (MWCNTs) [16 - 19], whose mass production has gradually evolved from the bench to the large-scale manufacturing, rendering them the material of choice for large volume applications [5].

Consisting of two concentrically rolled graphene sheets, each of which can be either metallic or semiconducting, double-walled carbon nanotubes (DWCNTs) have been attracting particular attention [20, 21], since they provide an intermediate structure between SWCNTs and multi-walled ones. This distinct structural feature provides them with enhanced functionality as they combine the properties of both SWCNTs and MWCNTs, such as high resistance under thermal, chemical and mechanical stress as well as flexibility. However, their attractiveness lies also in that they constitute an ideal system for the investigation of CNT interlayer interactions. In fact, calculations on DWCNTs have predicted that interlayer electron transfer from the outer to the inner tube can enhance the density of states at the Fermi level and, consequently, lead to a metallic state [20 - 23], while the outer shell can also serve as a shield to external modifications, like oxidation and covalent functionalization, so that the inner tube retains its structural integrity and electronic properties.

A highly promising application of CNTs is their utilization as filler materials for polymer reinforcement, endowing the polymer matrix their high strength and stiffness together with their thermal and electrical conductivity [4, 24]. Efficient dispersion and interfacial stress transfer between CNTs and the surrounding polymer are major challenges for the production of highly efficient CNT-polymer nanocomposites [24, 25]. Covalent attachment of chemical groups on the nanotube surface through oxidizing agents is a frequent processing step to improve interfacial coupling in CNT-polymer composites, which is severely impaired by the atomically smooth CNT surfaces [26]. However, despite the marked progress in the synthesis and purification of CNTs [27], controlled functionalization of their hydrophobic surface remains a major challenge even for MWCNTs [28], while such functionalization may also modify the electronic properties of CNTs through both charge transfer and defect formation [29, 30], and eventually compromise the composite's performance.

In this chapter, the application of EPR spectroscopy is exemplified on the investigation of the magnetic properties and intershell and interfacial interactions for two scientifically and technologically relevant cases of CNTs, namely DWCNTs and composites of oxidized MWCNTs embedded in elastomeric poly(ether-ester) block copolymer matrix.

DOUBLE-WALL CARBON NANOTUBES

Materials and Methods

The samples comprised DWCNTs bundles grown by catalytic chemical vapor deposition (CVD) technique with small amounts of residual catalyst particles and length reaching 50 μm (Sigma-Aldrich 637351, 50-80% DWCNT basis). Structural characterization was performed by analysis of their radial breathing modes (RBM), which correspond to the coherent vibration of carbon atoms in the radial tube direction [31]. Fig. (**1**) displays characteristic Raman spectra of the DWCNTs in comparison with those of SWCNTs bundles grown by the arc-method (519308 Sigma-Aldrich, CarboLex AP-grade, 50-70% SWCNT basis) at two excitation energies of 1.58 eV (NIR 785 nm) and 2.41 eV (Vis 514nm) using a Renishaw RM 2000 micro-Raman spectrometer. Two groups of RBM bands were observed for the DWCNTs at both laser energies, in contrast to the SWCNTs, permitting the identification of inner from the outer shells [20, 32, 33]. In particular, peak fitting analysis of the RBM spectra to the superimposition of Lorentzian lines (Fig. **2**), indicated that RBM bands at frequencies higher than 225 cm^{-1} can be attributed to the inner tubes, whereas the lower frequency modes can be mainly related to the outer ones. Relying on the fundamental relation between the RBM frequency and tube diameter $\omega_{RBM} = 227.0/d_t$ [34], the diameters of the inner shells resonantly excited at both laser energies were determined to be in the range of 0.8-0.9 nm. On the other hand, a broad distribution was obtained for the outer shells spanning the range of 1.1-2.0 nm.

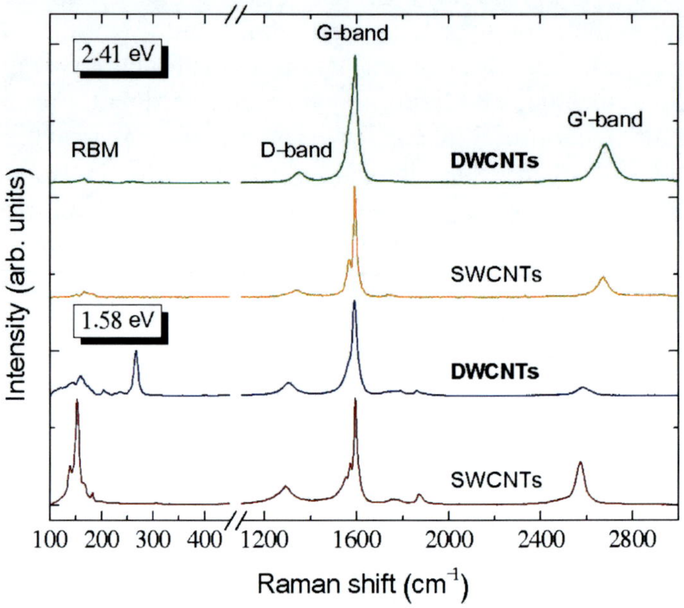

Fig. (1). Raman spectra of DWCNTs and SWCNTs at 2.41 and 1.58 eV laser excitation energies.

The tangential G band was appreciably broadened compared to that of SWCNTs (Fig. **1**), indicating a broad diameter distribution of the DWCNTs [33], while the dispersive defect-activated D band presented significant intensity, pointing to a considerable defect density that could be related to the acid treatment of DWCNTs for their purification.

Electron Paramagnetic Resonance

Fig. (**3**) shows representative EPR spectra for the SWCNTs and DWCNTs at different temperatures and magnetic field ranges. The EPR spectra were dominated by a broad resonance signal due to residual ferromagnetic catalyst particles. However, a narrow resonance line of low intensity, superimposed on the broad ferromagnetic resonance, could be clearly detected after zooming in the g = 2.0 region. Subtraction of the ferromagnetic background, using a polynomial baseline within the field scan of 150 G, revealed that the narrow EPR line was asymmetric for both SWCNTs and DWCNTs. This asymmetric line shape is typical of conducting samples when the sample thickness becomes comparable to the skin depth.

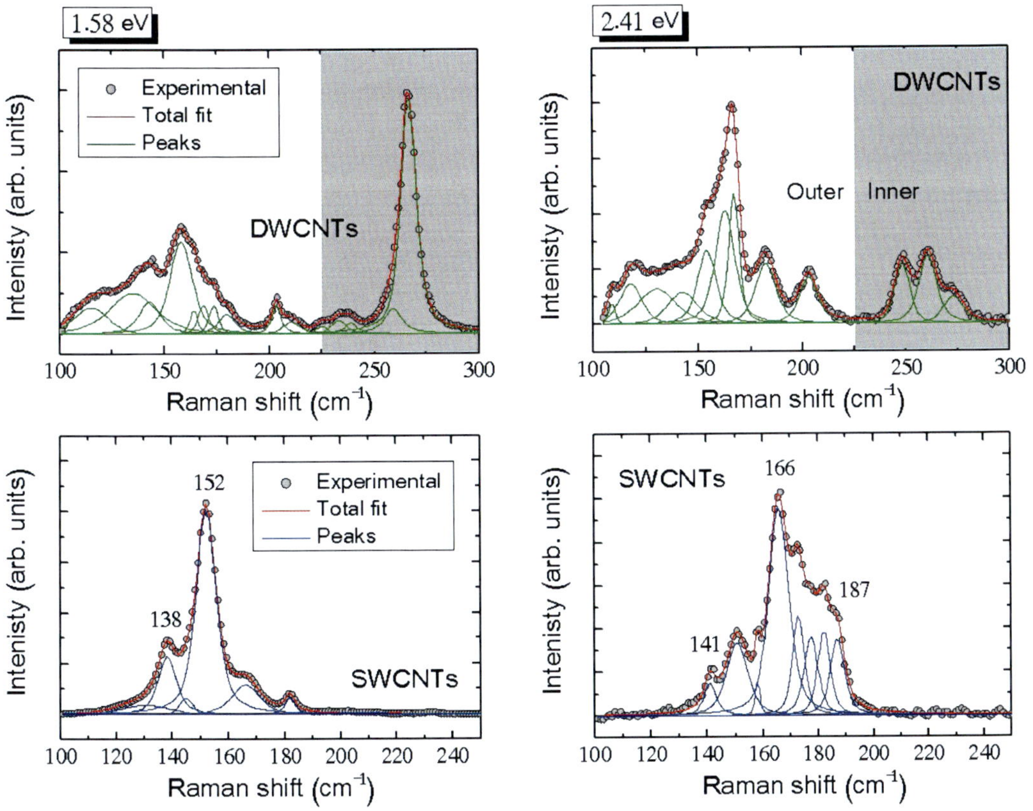

Fig. (2). Comparative RBM analysis for the DWCNTs (upper) and SWCNTs (lower) at 1.58 (left) and 2.41 eV (right) excitation energies.

In that case, the skin effect drives electric and magnetic components of the rf field out of phase, leading to the admixture of dispersion (χ') in the absorption (χ'') spectra [9]. The EPR line of SWCNTs could be well fitted to a single metallic lineshape of the form:

$$\frac{d\chi}{dH} = \left(\frac{\cos\varphi}{\Delta H^2}\right)\frac{-2y+(1-y^2)\tan\varphi}{(1+y^2)^2} \quad \textbf{(1)}$$

where y = $(H - H_0)/\Delta H$, with H_0 is the resonance field and ΔH the half-width at half-height, while $\tan\varphi = \alpha$ is the asymmetry parameter that corresponds to the ratio of dispersion to absorption of the susceptibility $\chi = \chi''\cos\phi + \chi'\sin\phi$. The asymmetry parameter varies between the values $\alpha = 0$ ($\varphi = 0$) corresponding to a symmetric Lorentzian and $\alpha = 1$ ($\varphi = \pi/4$) that produces an asymmetric resonance line with absorption and dispersion at equal weight [35, 36]. The narrow EPR line was well described by a single metallic line shape, as shown by the representative spectral fit in Fig. (**3**). However, the DWCNTs EPR signal could not be fitted with a single metallic line. Instead, satisfactory fits could be obtained only using two independent metallic lineshapes, each including the contribution of the derivatives of both absorptive and dispersive parts of a Lorentzian line [9]. These two metallic lines exhibited nearly the same resonance fields (H_0), though different half-widths at half-height (ΔH), as shown in Fig. (**3**). The narrow EPR component was attributed to the inner shells, while the broad one to the outer shells of the DWCNTs that is more susceptible to environmental effects.

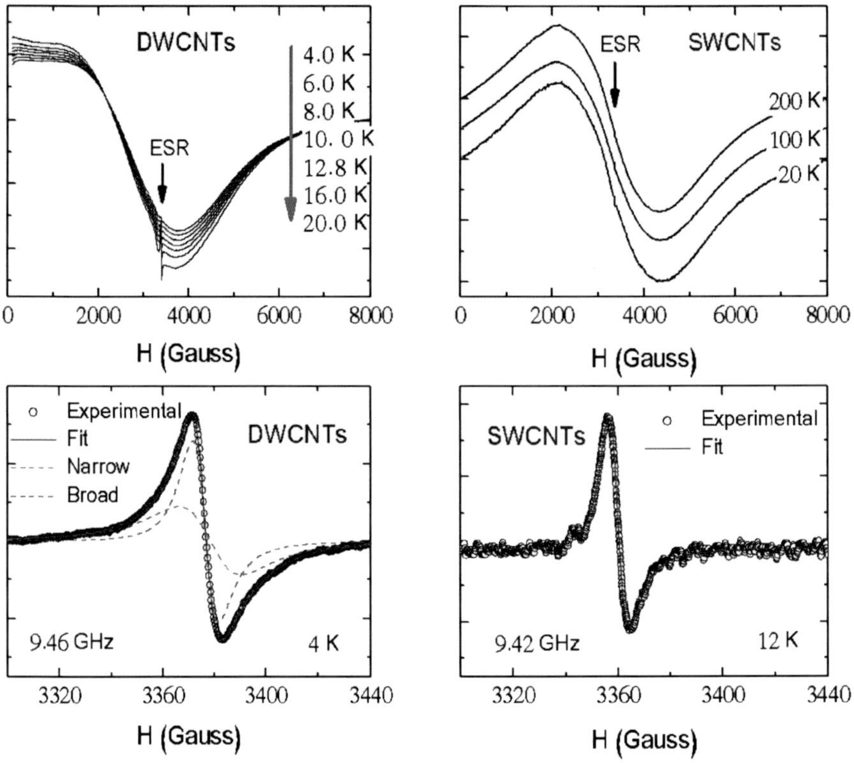

Fig. (3). (upper) EPR spectra of DWCNTs and SWCNTs at different temperatures over a wide magnetic field scan. The arrows depict the narrow EPR line superimposed on the ferromagnetic resonance of the catalyst particles. (lower) EPR spectral deconvolution using double and single metallic lineshapes after subtraction of the ferromagnetic background for the DWCNTs and SWCNTs, respectively.

The spin susceptibility (χ_s) of the constituent tubes was determined from the area under the integrated absorption component of the corresponding EPR lineshapes. Fig. (**4**) shows the temperature variation of the inverse spin susceptibility ($1/\chi_s$) of both inner and outer tubes at $T < 30$ K, where the EPR line could be clearly identified and analyzed. The inverse spin susceptibility ($1/\chi_s$) followed a linear behavior, which upon extrapolation intersects the temperature axis at positive values, indicating the presence of appreciable ferromagnetic interactions. This is further reflected in the temperature dependence of the product $\chi_s \times T$, which exhibits a continuous increase with decreasing temperature, in contrast to the variation anticipated for a simple paramagnetic system consisting of localized spins coupled with itinerant electrons (inset of Fig. **4**).

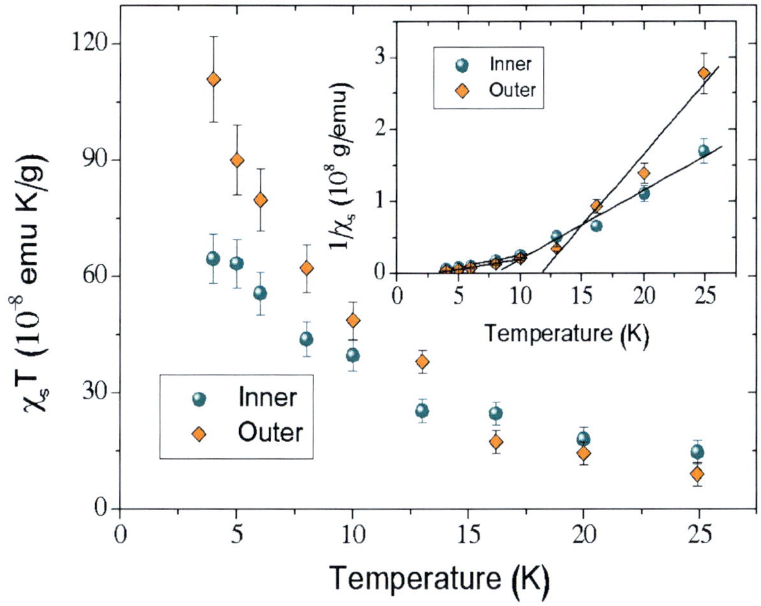

Fig. (4). Temperature dependence of the product $\chi_s \times T$ for the inner and outer DWCNT shells. The inset shows the temperature variation of the inverse spin susceptibilities ($1/\chi_s$) for the DWCNTs layers. Solid lines show the best-fit curves to the Curie-Weiss law.

The Curie-Weiss temperatures (Θ), determined by fitting to the Curie-Weiss law at $T > 10$ K, attained the positive values of 7.9(2) and 10.3(2) K for the inner and outer shells, respectively, indicative of the ferromagnetic transition at low temperature. In addition, the curvature of the inverse spin susceptibility curves $1/\chi_s = f(T)$ revealed a distribution of transition temperatures at $T < 10$ K, where the corresponding intersection tended to a lower bound of $\Theta \sim 3.0$ K for both DWCNT shells. The Pauli term (χ_0) was determined to be of the same order (10^{-9} emu/g) as those of MWCNTs [16] and SWCNTs [9]. However, the Curie constants for the inner and outer tube systems were $1.1(2)\times10^{-7}$ emuK/g and $5.6(5)\times10^{-8}$ emuK/g, respectively. These values were relatively higher than those reported for pristine SWCNTs [13, 16] while they were in qualitative agreement with the increased intensity of the defect activated D Raman band (Fig. **1**).

The ferromagnetic phase transition of the DWCNTs can be assigned to the exchange coupling between defect spins located on the shells and the edges of the constituent shells, as well as their RKKY-type coupling with conduction electrons. A similar ferromagnetic behavior has been identified by EPR in metal-semiconductor separated SWCNTs below 10 K, with the transition temperature in the metallic tubes (Θ ~ 10 K) being a factor of two higher than that in the semiconducting ones (Θ ~ 5 K) [12, 14]. This difference suggested the presence of strong exchange interactions between the localized defect spins in the metallic tubes *via* conduction electrons. In the case of DWCNTs, the Curie-Weiss temperatures for both constituent shells approached that for metallic SWCNTs, complying with the proposed tendency of DWCNTs toward metallic behavior due to interlayer electron transfer [23]. The observed difference between the Curie temperatures can be attributed to the different densities of defect spins on each tube system, which are mainly determined by the external environment and the thermal/chemical sample processing.

The low temperature ferromagnetic phase transition could also be traced in the temperature variation of the g factor and linewidths of the DWCNT EPR signals, shown in Fig. (**5**). At $T > 10$ K, the g factors of the inner and outer shells differed slightly and remained approximately constant, whereas, at $T < 10$ K, a slight decrease was detected. Similarly, ΔH did not vary significantly with temperature for both types of DWCNT layers at $T > 10$ K. However, a weak but clear broadening was observed at lower temperatures, indicative of the approach to magnetically ordered state, similar to SWCNTs [14].

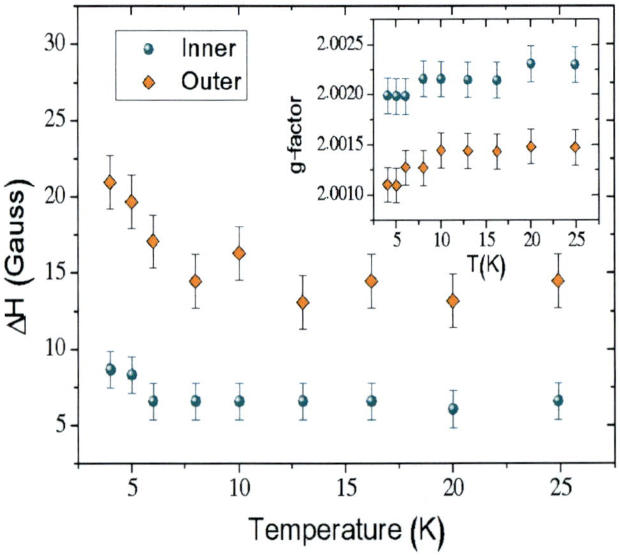

Fig. (5). Temperature dependence of the linewidth ΔH for the inner and outer DWCNT shells. The inset shows the corresponding variation of the g-factors.

In summary, EPR measurements on DWCNTs bundles revealed an asymmetric resonance line due to the contribution of a narrow and a broad resonance signal of metallic lineshapes, attributed to the inner and outer shells of the DWCNTs, respectively. At low temperatures, the spin susceptibility increased rapidly for both inner and outer tube EPR lines, indicating the presence of a ferromagnetic phase transition, similar to SWCNTs. The Curie-Weiss temperatures were determined to be approximately 8 and 10 K for the inner and outer DWCNT layers, respectively, approaching closely that reported for metallic SWCNTs. This behavior suggested significant interlayer coupling between the constituent shells of DWCNTs that would enhance the exchange interactions between localized spins on the inner and the outer DWCNT shells *via* RKKY-type coupling with conduction electrons.

MULTIWALL CARBON NANOTUBE POLY(ETHER-ESTER) NANOCOMPOSITES

Materials and Methods

Nanocomposites were prepared using the *in situ* polycondensation reaction process, where the MCWNTs are introduced as filler in the polymer during its synthesis [37]. Oxygen functional groups were introduced on the MWCNTs surface to assist grafting on the polymer matrix by heating with an excess of a mixture of sulphuric and nitric acid [37, 28, 30]. The thermoplastic poly(ether-ester) segmented block copolymer, containing 40 wt.% of poly(trimethylene terephthalate) (PTT), as rigid segment and 60 wt.% of poly(tetramethylene oxide) (PTMO) as flexible one [38], was used as polymer matrix. Composites with two different concentrations of 0.1 and 0.3 wt.% of the functionalized MWCNTs, named PTMW01 and PTMW03, respectively, were prepared by the *in situ* polycondensation process [39]. The MWCNTs were initially dispersed in N-methyl-2-pyrrolidone using an ultra-high speed stirrer and ultrasound, and then introduced into the reaction mixture in order to promote their homogenous distribution in the composite system. The solvent was then distilled off from the reaction mixture during polymerization.

Raman spectra of the PTT-PTMO/MWCNTs composites at two different energies in comparison with their pristine constituents (left panels in Fig. **6**), indicated a large increase of the photoluminescence upon incorporation of the oxidized MWCNTs in the polymer matrix, most pronounced for the lower concentration PTMW01.

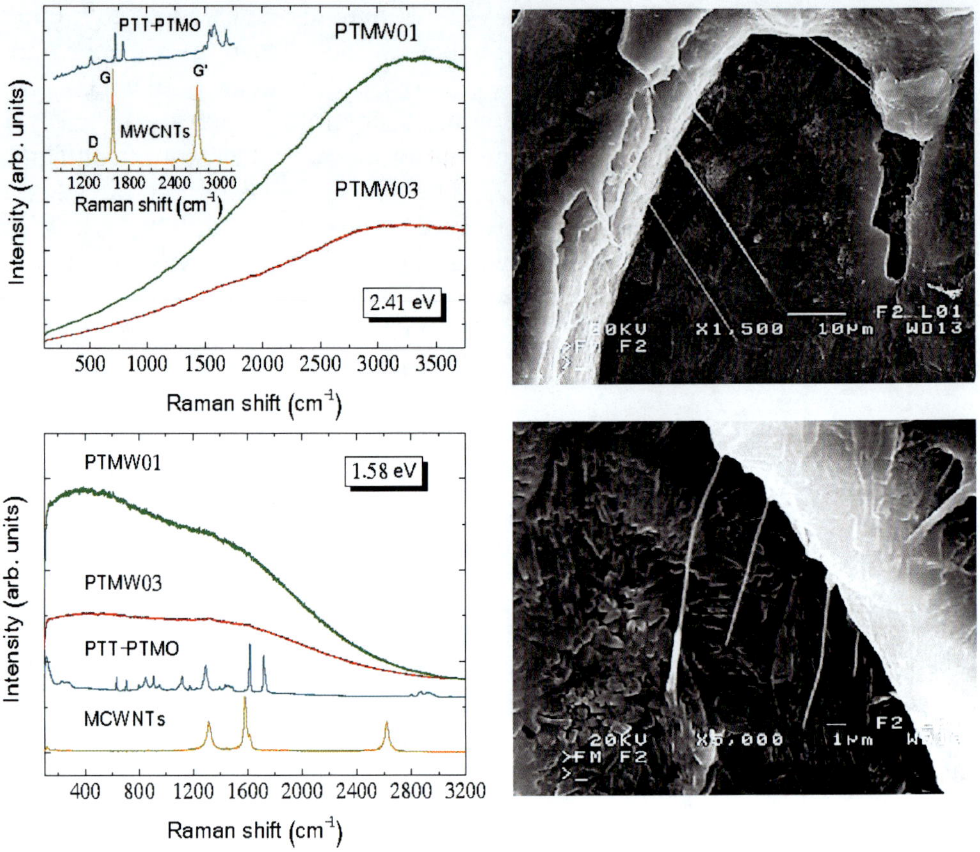

Fig. (6). (left) Raman spectra of the PTT-PTMO/MWCNT composites in comparison with the bare PTT-PTMO and MWCNTs at 2.41 and 1.58 eV excitation energies. (right) SEM images of the composites' fractured surface.

To investigate the MWCNTs' distribution in the PTT-PTMO matrix, the composite samples were fractured in liquid nitrogen and gold sputtered for imaging on a JEOL JSM-610 scanning electron microscope (SEM). Characteristic SEM images of the fractured PTT-PTMO/MWCNT nanocomposites (right panels in Fig. **6**) indicated a rather uniform distribution of the MWCNTs at 0.1 and 0.3 wt% loading in the polymer matrix without large scale agglomerates. Individual and not highly entangled nanotubes obtained from the polymer matrix after fracture were clearly observed.

Static Magnetization

Static magnetization measurements were carried out on a Quantum Design MPMS SQUID magnetometer in the temperature range of 2-300 K and magnetic fields up to 50 kOe. The diamagnetic contribution of the support cell and the PTT-PTMO copolymer matrix were independently measured and subtracted from the raw magnetization. The dc magnetization as a function of magnetic field and temperature showed a drastic variation in the magnetic response for the polymer composites compared to pristine MCWNTs, as shown in Fig. (**7**).

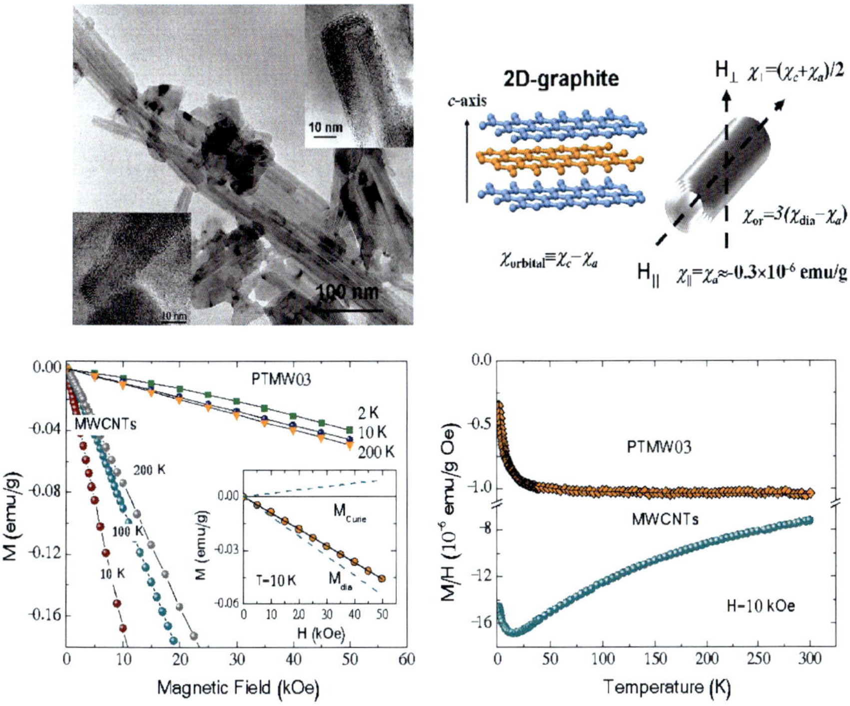

Fig. (7). (upper) TEM images of pristine MWCNTs and schematic for the graphite orbital diamagnetism adapted to the nanotube geometry [16, 40]. (lower left) Field dependence of the isothermal magnetization *M*(*H*) of PTMW03 at various temperatures. The inset shows the decomposition of the magnetization into a diamagnetic (M_{dia}) and a paramagnetic (M_{Curie}) term at 10 K. (lower right) Temperature dependence of the magnetic susceptibility *M*/*H* for PTMW03 at 10 kOe. The corresponding magnetization data for pristine MCWNTs are displayed for comparison.

Both the isothermal magnetization and the temperature dependence of the magnetization for the PTMW03 sample, where the diamagnetic contribution of the bare PTT-PTMO has been subtracted, reveal a marked reduction of the strong diamagnetic signal of the unmodified MWCNTs and its temperature variation.

The *M*(*H*) data could be consistently described at different temperatures, using the sum of a linear diamagnetic term ($M_{dia} = \chi_{dia}H$) with a paramagnetic one that is taken into account by the Brillouin function for localized spins with $S = 1/2$ and $g = 2.0$ ($M_{Curie} = N_s\mu_B \tanh(N\mu_B H/k_B T)$ (inset of Fig. 7). The magnitude of the diamagnetic susceptibility χ_{dia} was thus estimated to vary weakly about -1×10^{-7}

emu/g in the temperature range of 2-200 K. The spin density N_s was determined to be approximately 2.5×10^{18} spins/g. Likewise, a distinct paramagnetic-like contribution was observed in the temperature dependence of the static susceptibility, approximated by *M*/*H* (Fig. **7**). This variation could be described by the sum of a temperature independent term $\chi_{dia} \sim -1.0(1) \times 10^{-7}$ emu/g and a paramagnetic Curie contribution ($\chi_{Curie} = C/T$) with Curie constant of $\sim 2 \times 10^{-6}$ emu K/g, corresponding to a spin concentration of approximately 3.1×10^{18} spins/g, in agreement with the results obtained from the isothermal magnetization curves.

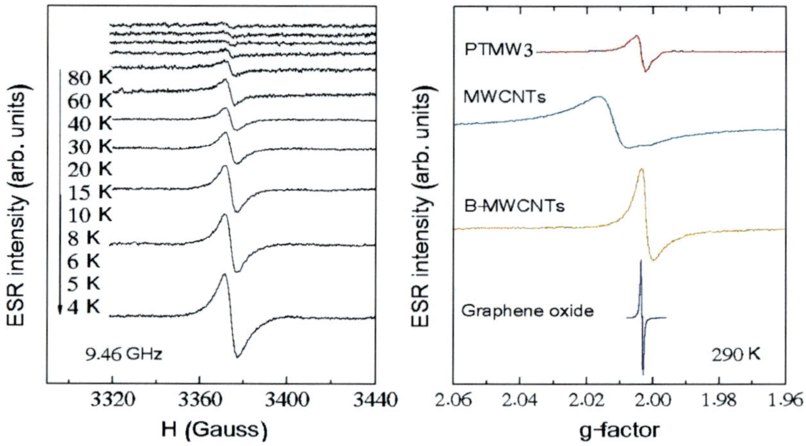

Fig. (8). (left) Temperature dependence of the EPR spectra for the PTMW3 composite at the X-band ($\nu \approx$ 9.46 GHz). (right) Comparison of the EPR spectra of PTMW03 with that of pristine MWCNTs [16], boron doped MWCNTs [19] and graphene oxide (GO) [41] at 290 K.

Electron Paramagnetic Resonance

EPR measurements were performed on an X-band Bruker E-500 spectrometer (ν = 9.46 GHz) and an Oxford flow cryostat and a standard hot air flow system for temperature dependent measurements (4-300 K). The spin susceptibility was determined from the integrated EPR intensity compared with a weighted amount of a vanadyl sulfate $VOSO_4 + 5H_2O$ calibration sample. As shown in Fig. (**8**) (left panel), PTMW03 exhibits a single EPR line at different temperatures. The *g*-factor is 2.0035(6), very close to the free electron value $g_e = 2.0023$, and the peak-to-peak linewidth $\Delta H_{pp} = 5(1)$ G at 290 K. Both EPR parameters are considerably smaller than the corresponding ones for pristine MWCNTs [16], whose EPR spectrum is dominated by the strongly anisotropic resonance of conduction electrons. This can be directly evidenced in the right panel of Fig. (**8**), where the composites EPR spectrum is compared to those of pristine MWCNTs [16], boron doped MWCNTs [19] and graphene oxide (GO) [41]. In the case of boron doped MWCNTs, an asymmetric metallic lineshape with enhanced spin susceptibility and considerably smaller g-factor and linewidth than the corresponding ones for pristine MWCNTs were observed, resulting from the increase in the density of states at the Fermi level and the concentration of localized spins upon boron doping as well as the reduction of orbital diamagnetic susceptibility and the concurrent suppression of the g-anisotropy [19]. On the other hand, a sharp EPR line with g-value close to g_e, similar, though narrower, to the PTMW composites, is observed for GO arising from the abundant localized defect spins on the strongly modified graphene planes by oxygen functional groups [41 - 43].

Fig. (**9**) compares the temperature dependence of the g-factor and linewidth ΔH_{pp} for the PTT-PTMO/MWCNT composites to the corresponding variations for the pristine MWCNTs [16]. Both parameters vary weakly with temperature, a shallow peak can be only traced at 250 K in $\Delta H_{pp}(T)$ together with a weak broadening at $T < 20$ K for the MWCNT polymer composites. This behavior deviates significantly from the strong temperature dependence of the EPR spectrum of MWCNTs, which reflects the high g-anisotropy of the CNT graphene layers. Such temperature variation of the EPR parameters can be in principle rationalized by either the doping of the MWCNTs or the generation of a high density of paramagnetic defects in the EPR spectra of the exchange coupled system of conduction and localized spins. In the former case, the Fermi-level shift and the deviation from the semimetallic 2D graphene band structure diminish the prominent g-factor anisotropy similar to the suppression of orbital diamagnetism in K and B doped MWCNTs [11, 19, 44, 45]. In the latter case, the formation of a large number of localized spins with spin susceptibility exceeding that of the conduction electron spin system, results in the effective reduction of the anisotropic conduction electron spin resonance spectrum of MWCNTs, especially at low temperatures [46].

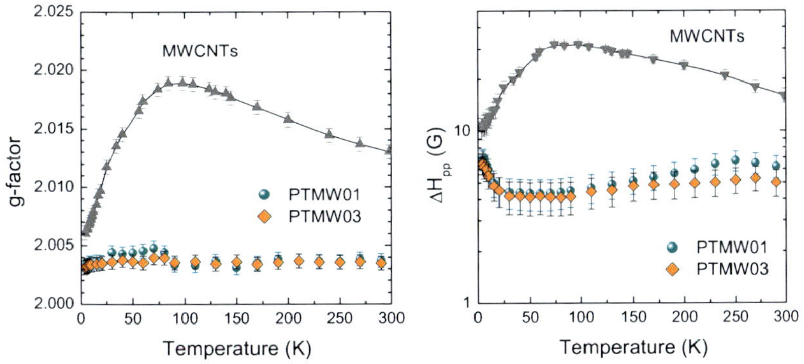

Fig. (9). Temperature dependence of the g-factor (left) and peak-to-peak linewidth ΔH_{pp} (right) for the PTT-PTMO/MWCNT composites in comparison with pristine MWCNTs.

Fig. (**10**) displays the temperature variation of the $\chi_s(T)$ that is determined from the calibrated EPR intensity, normalized to the MWCNT mass in the PTT-PTMO/MWCNT composites. The spin susceptibility is significantly enhanced compared to the pristine MWCNTs at high temperatures, followed by a marked paramagnetic upturn as temperature decreases (T < 20 K). This behavior is characteristic of the formation of localized paramagnetic centers, in contrast to pristine MWCNTs, whose $\chi_s(T)$ is dominated by the weakly temperature dependent Pauli susceptibility of conduction electrons [16]. The temperature variation of the spin susceptibility can be described by

$$\chi_S = \chi_0 + \chi_C \quad \textbf{(2)}$$

stemming from the sum of a temperature-independent Pauli term χ_0, most prominent at high temperatures, with a Curie-Weiss contribution

$$\chi_C = C/T - \Theta \quad \textbf{(3)}$$

due to localized spins. An approximate description of $\chi_s(T)$ was accordingly obtained with $\chi_0 = 1.6(2) \times 10^{-8}$ emu/g, $C = 1.4(2) \times 10^{-6}$ emu K/g and $\Theta = 2(1)$ K, for both composites. The value of the Pauli term χ_0 is higher than that of MWCNTs ($\sim 1 \times 10^{-8}$ emu/g), suggesting an increase of the density of states at the Fermi level, which, together with the reduction of the diamagnetic susceptibility, corroborates the presence of effective hole doping in the MWCNT polymer composites.

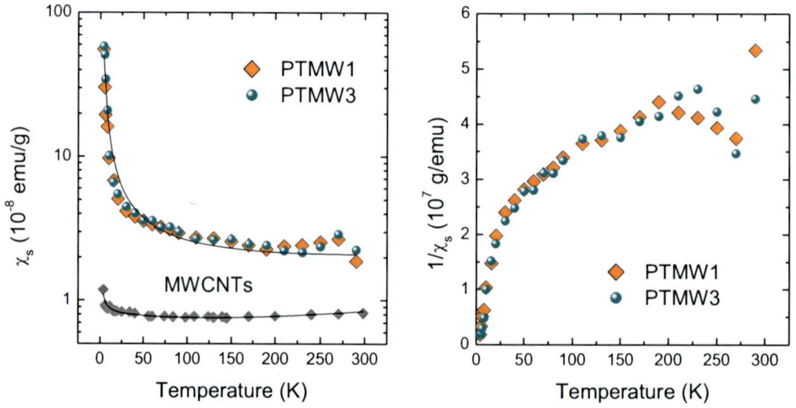

Fig. (10). (left) Temperature variation of the spin susceptibility χ_s for the PTT-PTMO/MWCNT composites in comparison with pristine MWCNTs. Solid lines correspond to the best fit curves to the sum of a temperature independent Pauli term and a Curie one $\chi_s = \chi_0 + \chi_C$ (right). The temperature variation of the inverse spin susceptibilities $1/\chi_s$ for the PTT-PTMO/MWCNT composites.

The value of the Curie-Weiss constant is close to that determined from the dc magnetization, indicating that the observed magnetism is mainly due to the formation of localized spins. The presence of a non-zero Curie-Weiss temperature implies the presence of weak exchange interactions, as well, correlating with the broadening of the resonance line at $T < 20$ K (Fig. **9**). Moreover, the localized spin density for the PTT-PTMO/MWCNT composites determined by both the EPR spin susceptibility and the dc magnetization was in the range of 2.2-3.1 x 10^{18} spins/g, which is approximately two orders of magnitude higher than that of unmodified MWCNTs [16]. This variation is similar to that previously reported for electron irradiated MWCNTs, where significant modification of the CNT electronic properties was identified by EPR [46]. In that case, both the *g*-factor and the linewidth temperature dependence were strongly suppressed when the defect spin density increased by two orders of magnitude, similar to the PTT-PTMO/MWCNTs composites. This indicates that an analogous mechanism underlies the behavior of the polymer/MWCNT composites, where both the excessive increase of the spin density and the effective doping may be brought up by the electron withdrawing nature of oxygen chemical groups on the functionalized nanotube surface and their binding with the polymer matrix.

In addition, a weak but clear anomaly is observed on $\chi_s(T)$ at high temperatures ($T \sim 250$ K) for the PTT-PTMO/MWCNT composites. This feature is most clearly resolved in the temperature variation of the inverse spin susceptibility spin (Fig. **10**), correlating with the small peak that is traced in the temperature dependence of the EPR linewidth (Fig. **9**). The anomaly of $\chi_s(T)$ and $\Delta H_{pp}(T)$ occurs within the temperature range of the dynamic mechanical relaxation of the PTT-PTMO copolymer [38]. The glass transition temperature of the soft amorphous phase and the accompanying *a*-relaxation of the 40/60 PTT-PTMO segmented block copolymer have been determined to be at 205 and 220 K, respectively. This suggests that localized spins on the PTT-PTMO/MWCNT composites are sensitive to the polymer relaxation, reflecting the interfacial coupling of the functionalized MWCNTs with the surrounding polymer. Similar effects have been reported on nanocomposites of maghemite nanoparticles embedded in elastomeric copolymer matrices [47, 48] as well as in composites of SWCNTs with high density polyethylene [49]. However, while the effective doping and the excessive defect density could be advantageous for the grafting of the CNTs with the polymer matrix, the reduction of the diamagnetic anisotropy precludes efficient alignment of CNTs by magnetic fields [50 - 52].

EPR and static magnetization measurements on composites of oxidized MWCNT embedded in PTT-PTMO segmented block copolymers reveal a drastic variation of the CNT's magnetic properties in the polymer matrix. The diamagnetic susceptibility of the oxidized MWCNTs filler and its temperature dependence was markedly reduced, indicative of the substantial shift of the Fermi level due to oxygen functional groups on the functionalized CNTs. The spin susceptibility of the composite system is considerably enhanced mainly due to the extensive formation of localized moments. The temperature dependence of the EPR spectra indicates that the underlying defects are sensitive to the mechanical relaxation of the copolymer matrix, and thus to the interfaces of the oxidized MWCNTs with the surrounding polymer.

CONCLUDING REMARKS

EPR spectroscopy is shown to be highly efficient in investigating the magnetic properties and particularly inter-shell and interfacial interactions in DWCNTs and composites comprising oxidized MWCNTs embedded in elastomeric block copolymers. In the former case, analysis of the EPR spectra provides evidence for the presence of two distinct contributions from the inner and outer DWCNTs' shells, undergoing a ferromagnetic phase transition at low temperatures. This behavior is very similar to metallic SWCNTs, pointing to significant interlayer coupling between the shells of DWCNTs that can enhance the exchange interactions between localized spins.

In the case of PTT-PTMO/MWCNT composites, the EPR spectra in combination with static magnetization measurements reveal a drastic reduction of orbital diamagnetism and *g*-anisotropy together with a marked enhancement of the spin susceptibility. These effects indicate considerable hole doping by the functional groups on the oxidized CNTs' surface and an excessive increase of the density of paramagnetic defects, which are sensitive to the dynamical polymer relaxation and thus to the MWCNT-polymer interfacial coupling.

CONSENT FOR PUBLICATION

Not applicable.

CONFLICT OF INTEREST

The author (editor) declares no conflict of interest, financial or otherwise.

ACKNOWLEDGEMENT

Declare none.

REFERENCES

[1] Tománek, D.; Jorio, A.; Dresselhaus, M.; Dresselhaus, G. Introduction to the Important and Exciting Aspects of Carbon-Nanotube Science and Technology.*Carbon Nanotubes*; Jorio, A.; Dresselhaus, G.; Dresselhaus, M., Eds.; Springer Berlin Heidelberg: Berlin, **2008**, pp. 1-12.

[2] Eklund, P.; Ajayan, P.; Blackmon, R.; Hart, A.; Kong, J.; Pradhan, B.; Rao, A.; Rinzler, A. *On International Assessment of Research and Development of Carbon Nanotube Manufacturing and Applications,* **2007**.http://www.wtec.org/cnm/CNM_final_report.pdf

[3] Schnorr, J.; Swager, T. Emerging Applications of Carbon Nanotubes. *Chem. Mater.,* **2011**, *23*, 646-657.
[http://dx.doi.org/10.1021/cm102406h]

[4] Endo, M.; Strano, M.S.; Ajayan, P.M. Potential Applications of Carbon Nanotubes.*Carbon Nanotubes*; Jorio, A.; Dresselhaus, G.; Dresselhaus, M., Eds.; Springer Berlin Heidelberg: Berlin, **2008**, pp. 13-62.

[5] Lehman, J.; Terrones, M.; Mansfield, E.; Hurst, K.; Meunier, V. Evaluating the characteristics of multiwall carbon nanotubes. *Carbon,* **2011**, *49*, 2581-2602.
[http://dx.doi.org/10.1016/j.carbon.2011.03.028]

[6] Petit, P.; Jouguelet, E.; Fischer, J.; Rinzler, A.; Smalley, R. Electron spin resonance and microwave resistivity of single-wall carbon nanotubes. *Phys. Rev. B,* **1997**, *56*, 9275-9278.
[http://dx.doi.org/10.1103/PhysRevB.56.9275]

[7] Salvetat, J.; Fehér, T.; L'Huillier, C.; Beuneu, F.; Forró, L. Anomalous electron spin resonance behavior of single-walled carbon nanotubes. *Phys. Rev. B,* **2005**, *72*, 075440.
[http://dx.doi.org/10.1103/PhysRevB.72.075440]

[8] Corzilius, B.; Dinse, K.; van Slageren, J.; Hata, K. Low-temperature anomaly of microwave absorption and ac susceptibility of single-wall carbon nanotubes: Bulk superconductivity and weak ferromagnetism. Phys. Rev. B 2007, 75, 235416(1-7).
[http://dx.doi.org/10.1103/PhysRevB.75.235416]

[9] Likodimos, V.; Glenis, S.; Guskos, N.; Lin, C. Antiferromagnetic behavior in single-wall carbon nanotubes. *Phys. Rev. B,* **2007**, *76*, 075420(1-5).
[http://dx.doi.org/10.1103/PhysRevB.76.075420]

[10] Corzilius, B.; Dinse, K.; Hata, K.; Haluška, M.; Skákalová, V.; Roth, S. SWNT probed by multi-frequency EPR and nonresonant microwave absorption. *Phys. Stat. Solidi B,* **2008**, *245*, 2251-2254.
[http://dx.doi.org/10.1002/pssb.200879574]

[11] Galambos, M.; Fábián, G.; Simon, F.; Ćirić, L.; Forró, L.; Korecz, L.; Rockenbauer, A.; Koltai, J.; Zólyomi, V.; Rusznyák, Á. Identifying the electron spin resonance of conduction electrons in alkali doped SWCNTs. *Phys. Stat. Solidi B,* **2009**, *246*, 2760-2763.
[http://dx.doi.org/10.1002/pssb.200982330]

[12] Havlicek, M.; Jantsch, W.; Rümmeli, M.; Schönfelder, R.; Yanagi, K.; Miyata, Y.; Kataura, H.; Simon, F.; Peterlik, H.; Kuzmany, H. Electron spin resonance from semiconductor-metal separated SWCNTs. *Phys. Stat. Solidi B,* **2010**, *247*, 2851-2854.
[http://dx.doi.org/10.1002/pssb.201000317]

[13] Rice, W.D.; Weber, R.T.; Leonard, A.D.; Tour, J.M.; Nikolaev, P.; Arepalli, S.; Berka, V.; Tsai, A.L.; Kono, J. Enhancement of the electron spin resonance of single-walled carbon nanotubes by oxygen removal. *ACS Nano,* **2012**, *6*(3), 2165-2173.
[http://dx.doi.org/10.1021/nn204094s] [PMID: 22324937]

[14] Havlicek, M.; Jantsch, W.; Wilamowski, Z.; Yanagi, K.; Kataura, H.; Rümmeli, M.; Malissa, H.; Tyryshkin, A.; Lyon, S.; Chernov, A. Indirect exchange interaction in fully metal-semiconductor separated single-walled carbon nanotubes revealed by electron spin resonance. *Phys. Rev. B,* **2012**, *86*, 045402(1-5).
[http://dx.doi.org/10.1103/PhysRevB.86.045402]

[15] Rice, W.; Weber, R.; Nikolaev, P.; Arepalli, S.; Berka, V.; Tsai, A.; Kono, J. Spin relaxation times of single-wall carbon nanotubes. *Phys. Rev. B,* **2013**, *88*, 041401(R)(1-5).
[http://dx.doi.org/10.1103/PhysRevB.88.041401]

[16] Likodimos, V.; Glenis, S.; Guskos, N.; Lin, C. Magnetic and electronic properties of multiwall carbon

nanotubes. *Phys. Rev. B,* **2003**, *68*, 045417(1-6).
[http://dx.doi.org/10.1103/PhysRevB.68.045417]

[17] Kotosonov, A.; Shilo, D. Electron spin resonance study of carbon nanotubes. *Carbon,* **1998**, *36*, 1649-1651.
[http://dx.doi.org/10.1016/S0008-6223(98)00159-6]

[18] Chauvet, O.; Baumgartner, G.; Carrard, M.; Bacsa, W.; Ugarte, D.; Forró, L.; Forró, L. ESR study of potassium-doped aligned carbon nanotubes. *Phys. Rev. B Condens. Matter,* **1996**, *53*(21), 13996-13999.
[http://dx.doi.org/10.1103/PhysRevB.53.13996] [PMID: 9983182]

[19] Likodimos, V.; Glenis, S.; Lin, C. Electronic properties of boron-doped multiwall carbon nanotubes studied by ESR and static magnetization. *Phys. Rev. B,* **2005**, *72*, 045436(1-7).
[http://dx.doi.org/10.1103/PhysRevB.72.045436]

[20] Pfeiffer, R.; Pichler, T.; Kim, Y.A.; Kuzmany, H. Double-Wall Carbon Nanotubes.*Carbon Nanotubes*; Jorio, A.; Dresselhaus, G.; Dresselhaus, M., Eds.; Springer Berlin Heidelberg: Berlin, **2008**, pp. 495-530.

[21] Shen, C.; Brozena, A.H.; Wang, Y. Double-walled carbon nanotubes: challenges and opportunities. *Nanoscale,* **2011**, *3*(2), 503-518.
[http://dx.doi.org/10.1039/C0NR00620C] [PMID: 21042608]

[22] Zólyomi, V.; Koltai, J.; Rusznyák, Á.; Kürti, J.; Gali, Á.; Simon, F.; Kuzmany, H.; Szabados, Á.; Surján, P. Intershell interaction in double walled carbon nanotubes: Charge transfer and orbital mixing. *Phys. Rev. B.***2008**, *77*, 245403(1-10).
[http://dx.doi.org/10.1103/PhysRevB.77.245403]

[23] Singer, P.; Wzietek, P.; Alloul, H.; Simon, F.; Kuzmany, H. NMR Evidence for Gapped Spin Excitations in Metallic Carbon Nanotubes. *Phys. Rev. Lett.,* **2005**, *95*, 236403(1-4).
[http://dx.doi.org/10.1103/PhysRevLett.95.236403]

[24] Coleman, J.; Khan, U.; Blau, W.; Gun'ko, Y. Small but strong: A review of the mechanical properties of carbon nanotube–polymer composites. *Carbon,* **2006**, *44*, 1624-1652.
[http://dx.doi.org/10.1016/j.carbon.2006.02.038]

[25] Miranda, S.; Romanos, G.; Likodimos, V.; Marques, R.; Favvas, E.; Katsaros, F.; Stefanopoulos, K.; Vilar, V.; Faria, J.; Falaras, P. Pore structure, interface properties and photocatalytic efficiency of hydration/dehydration derived TiO2/CNT composites. *Appl. Catal. B,* **2014**, *147*, 65-81.
[http://dx.doi.org/10.1016/j.apcatb.2013.08.013]

[26] Barber, A.; Cohen, S.; Eitan, A.; Schadler, L.; Wagner, H. Fracture Transitions at a Carbon-Nanotube/Polymer Interface. *Adv. Mater.,* **2006**, *18*, 83-87.
[http://dx.doi.org/10.1002/adma.200501033]

[27] Hou, P.; Liu, C.; Cheng, H. Purification of carbon nanotubes. *Carbon,* **2008**, *46*, 2003-2025.
[http://dx.doi.org/10.1016/j.carbon.2008.09.009]

[28] Likodimos, V.; Steriotis, T.; Papageorgiou, S.; Romanos, G.; Marques, R.; Rocha, R.; Faria, J.; Pereira, M.; Figueiredo, J.; Silva, A. Controlled surface functionalization of multiwall carbon nanotubes by HNO3 hydrothermal oxidation. *Carbon,* **2014**, *69*, 311-326.
[http://dx.doi.org/10.1016/j.carbon.2013.12.030]

[29] Zhou, W.; Vavro, J.; Nemes, N.; Fischer, J.; Borondics, F.; Kamarás, K.; Tanner, D. Charge transfer and Fermi level shift inp-doped single-walled carbon nanotubes. *Phys. Rev. Lett.,* **2005**, *71*, 205423(1-7).
[http://dx.doi.org/10.1103/PhysRevB.71.205423]

[30] Romanos, G.; Likodimos, V.; Marques, R.; Steriotis, T.; Papageorgiou, S.; Faria, J.; Figueiredo, J.; Silva, A., Falaras, P. Controlling and Quantifying Oxygen Functionalities on Hydrothermally and Thermally Treated Single-Wall Carbon Nanotubes. *J. Phys. Chem. C,* **2011**, *115*, 8534-8546.

[http://dx.doi.org/10.1021/jp200464d]

[31] Dresselhaus, M.; Dresselhaus, G.; Saito, R.; Jorio, A. Raman spectroscopy of carbon nanotubes. *Phys. Rep.*, **2005**, *409*, 47-99.
[http://dx.doi.org/10.1016/j.physrep.2004.10.006]

[32] Pfeiffer, R.; Simon, F.; Kuzmany, H.; Popov, V. Fine structure of the radial breathing mode of double-wall carbon nanotubes. *Phys. Rev. B.*, **2005**, *72*, 161404(R)(1-4).
[http://dx.doi.org/10.1103/PhysRevB.72.161404]

[33] Grüneis, A.; Rümmeli, M.; Kramberger, C.; Barreiro, A.; Pichler, T.; Pfeiffer, R.; Kuzmany, H.; Gemming, T.; Büchner, B. High quality double wall carbon nanotubes with a defined diameter distribution by chemical vapor deposition from alcohol. *Carbon*, **2006**, *44*, 3177-3182.
[http://dx.doi.org/10.1016/j.carbon.2006.07.003]

[34] Araujo, P.; Maciel, I.; Pesce, P.; Pimenta, M.; Doorn, S.; Qian, H.; Hartschuh, A.; Steiner, M.; Grigorian, L.; Hata, K. Nature of the constant factor in the relation between radial breathing mode frequency and tube diameter for single-wall carbon nanotubes. *Phys. Rev. B.*, **2008**, *77*, 241403(R)(1-4).
[http://dx.doi.org/10.1103/PhysRevB.77.241403]

[35] Feher, G.; Kip, A. Electron Spin Resonance Absorption in Metals. I. Experimental. *Phys. Rev.*, **1955**, *98*, 337-348.
[http://dx.doi.org/10.1103/PhysRev.98.337]

[36] Pifer, J. Magno, R. Conduction-Electron Spin Resonance in a Lithium Film. *Phys. Rev. B*, **1971**, *3*, 663-673.
[http://dx.doi.org/10.1103/PhysRevB.3.663]

[37] Gojny, F.; Nastalczyk, J.; Roslaniec, Z.; Schulte, K. Surface modified multi-walled carbon nanotubes in CNT/epoxy-composites. *Chem. Phys. Lett.*, **2003**, *370*, 820-824.
[http://dx.doi.org/10.1016/S0009-2614(03)00187-8]

[38] Szymczyk, A.; Senderek, E.; Nastalczyk, J.; Roslaniec, Z. New multiblock poly(ether-ester)s based on poly(trimethylene terephthalate) as rigid segments. *Eur. Polym. J.*, **2008**, *44*, 436-443.
[http://dx.doi.org/10.1016/j.eurpolymj.2007.11.005]

[39] Broza, G.; Kwiatkowska, M.; Rosłaniec, Z.; Schulte, K. Processing and assessment of poly(butylene terephthalate) nanocomposites reinforced with oxidized single wall carbon nanotubes. *Polymer (Guildf.)*, **2005**, *46*, 5860-5867.
[http://dx.doi.org/10.1016/j.polymer.2005.05.073]

[40] Kotosonov, A. Texture and magnetic anisotropy of carbon nanotubes in cathode deposits obtained by the electric-arc method. *JETP Lett.*, **1999**, *70*, 476-480.
[http://dx.doi.org/10.1134/1.568199]

[41] Diamantopoulou, A.; Glenis, S.; Zolnierkiwicz, G.; Guskos, N.; Likodimos, V. Magnetism in pristine and chemically reduced graphene oxide. *J. Appl. Phys.*, **2017**, *121*, 043906(1-11).
[http://dx.doi.org/10.1063/1.4974364]

[42] Ćirić, L.; Sienkiewicz, A.; Gaál, R.; Jaćimović, J.; Vâju, C.; Magrez, A.; Forró, L. Defects and localization in chemically-derived graphene. *Phys. Rev. B.*, **2012**, *86*, 195139(1-6).
[http://dx.doi.org/10.1103/PhysRevB.86.195139]

[43] Panich, A.; Shames, A.; Sergeev, N. Paramagnetic Impurities in Graphene Oxide. *Appl. Magn. Reson.*, **2012**, *44*, 107-116.
[http://dx.doi.org/10.1007/s00723-012-0392-z]

[44] Claye, A.; Nemes, N.; Jánossy, A.; Fischer, J. Structure and electronic properties of potassium-doped single-wall carbon nanotubes. *Phys. Rev. B*, **2000**, *62*, R4845-R4848.
[http://dx.doi.org/10.1103/PhysRevB.62.R4845]

[45] Numao, S.; Bandow, S.; Iijima, S. Control of the Innermost Tube Diameters in Multiwalled Carbon

Nanotubes by the Vaporization of Boron-Containing Carbon Rod in RF Plasma. *J. Phys. Chem. C,* **2007**, *111*, 4543-4548.
[http://dx.doi.org/10.1021/jp0682558]

[46] Beuneu, F.; l'Huillier, C.; Salvetat, J.; Bonard, J.; Forró, L. Modification of multiwall carbon nanotubes by electron irradiation: An ESR study. *Phys. Rev. B,* **1999**, *59*, 5945-5949.
[http://dx.doi.org/10.1103/PhysRevB.59.5945]

[47] Guskos, N.; Glenis, S.; Likodimos, V.; Typek, J.; Maryniak, M.; Roslaniec, Z.; Kwiatkowska, M.; Baran, M.; Szymczak, R.; Petridis, D. Matrix effects on the magnetic properties of γ-Fe2O3 nanoparticles dispersed in a multiblock copolymer. *J. Appl. Phys.,* **2006**, *99*, 084307(1-7).
[http://dx.doi.org/10.1063/1.2189216]

[48] Guskos, N.; Likodimos, V.; Glenis, S.; Maryniak, M.; Baran, M.; Szymczak, R.; Roslaniec, Z.; Kwiatkowska, M.; Petridis, D. Magnetic properties of γ-Fe_2O_3/poly(ether-ester) nanocomposites. *J. Nanosci. Nanotechnol.,* **2008**, *8*(4), 2127-2134.
[http://dx.doi.org/10.1166/jnn.2008.063] [PMID: 18572623]

[49] Chipara, M.; Lozano, K.; Wilkins, R.; Barrera, E.; Pulikkathara, M.; Penia-Para, L.; Chipara, M. ESR investigations on polyethylene-single wall carbon nanotube composites. *J. Mater. Sci.,* **2007**, *43*, 1228-1233.
[http://dx.doi.org/10.1007/s10853-007-2264-8]

[50] Fischer, J.; Zhou, W.; Vavro, J.; Llaguno, M.; Guthy, C.; Haggenmueller, R.; Casavant, M.; Walters, D.; Smalley, R. Magnetically aligned single wall carbon nanotube films: Preferred orientation and anisotropic transport properties. *J. Appl. Phys.,* **2003**, *93*, 2157-2163.
[http://dx.doi.org/10.1063/1.1536733]

[51] Choi, E.; Brooks, J.; Eaton, D.; Al-Haik, M.; Hussaini, M.; Garmestani, H.; Li, D.; Dahmen, K. Enhancement of thermal and electrical properties of carbon nanotube polymer composites by magnetic field processing. *J. Appl. Phys.,* **2003**, *94*, 6034-6039.
[http://dx.doi.org/10.1063/1.1616638]

[52] Steinert, B.; Dean, D. Magnetic field alignment and electrical properties of solution cast PET–carbon nanotube composite films. *Polymer (Guildf.),* **2009**, *50*, 898-904.
[http://dx.doi.org/10.1016/j.polymer.2008.11.053]

CHAPTER 5

Impact of Point Defects on Graphene Oxide and Carbon Nanotubes: Study of Electron Paramagnetic Resonance Spectroscopy

Chuyen V. Pham[1,*], **Sergej Repp**[2], **Michael Krueger**[3] and **Emre Erdem**[2]

[1] *Laboratory for MEMS Applications, Department of Microsystems Engineering - IMTEK, University of Freiburg, Georges-Koehler-Allee 103, D79110 Freiburg, Germany*

[2] *Universite Francois Rabelais, GREMAN, Faculté de Sciences et Techniques, 37200 Tours, France*

[3] *Carl-von-Ossietzky University Oldenburg, Institute of Physics, Carl-von-Ossietzky Str. 9-11, D-26129 Oldenburg, Germany*

Abstract: The electron paramagnetic resonance (EPR) spectroscopy is a powerful and sensitive method to detect intrinsic and extrinsic paramagnetic point defects in a material system. EPR has recently been proven an effective tool for studying the lattice defect of nanostructured carbon materials. In particular, EPR can be used to elucidate the spin properties, including unpaired spins, conduction electrons, and dangling bonds as well as the electronic states of different carbon nanostructures. EPR studies on point-defects of carbon materials such as graphene and carbon nanotubes help to unearth several electronic and optical features of the materials. Though the magnetic feature of graphene has been studied intensively, EPR research on graphene and graphene-like structures is still a new field. This chapter focuses on discussing EPR investigations on graphene oxide, functional reduced graphene oxide, and carbon nanotubes. In that, EPR has demonstrated as a suitable tool to detect spin density changes in different functionalized nanocarbon materials. A novel approach to studying the charge transfer within quantum dots-graphene hybrids, using continuous wave EPR, will be discussed. It also enables the study of the change in the electronic properties of graphene before and after attaching of quantum dots. This contributes to improved understanding of electronic coupling effects in nanocarbon-nanoparticle hybrid materials which are promising for various electronic and optoelectronic applications.

Keywords: Carbon-centered radicals, Carbon nanotubes (CNTs), Charge transfer, Curie's law, Defects, Defect concentration, Electron spin delocalization, EPR spectroscopy, Functionalized grapheme, Graphene oxide (GO), Number of spins,

* **Corresponding authors Chuyen V. Pham:** Laboratory for MEMS Applications, Department of Microsystems Engineering - IMTEK, University of Freiburg, Georges-Koehler-Allee 103, D79110 Freiburg, Germany; Tel: 0049 761 203 950 81; E-mail: chuyen.pham@imtek.uni-freiburg.de; **Emre Erdem:** Universite Francois Rabelais, GREMAN, Faculté de Sciences et Techniques, 37200 Tours, France; Tel: 0033247367912; Email: emre.erdem@univ-tours.fr

Dariya Savchenko & Abdel Hadi Kassiba (Eds.)

Oxidized CNT, Pauli-type paramagnetism, Photoluminescence, Quantum dots (QDs), Reduced graphene oxide (rGO), Temperature dependence, Thiol-functionalized rGO (TrGO), ZnO NP decorated rGO hybrid material (rGO-ZnO), ZnO NP decorated TrGO (TrGO-ZnO).

INTRODUCTION

Graphene represents a promising material for the synergetic application as the second compound in semiconductor-based advanced materials. It possesses a single atomic layer where sp^2 carbon atoms are arranged in a honeycomb-like a 2-dimensional network. These unique structural properties result in a high specific surface area, high thermal conductivity as well as an ultrahigh intrinsic mobility of charge carriers [1, 2]. The synthesis of advanced materials by using pristine graphene is too complicated. Therefore, the usage of graphene oxide (GO) is very popular for the production of advanced materials. The highly oxidized monolayer GO thereby is usually synthesized by treating natural graphite with the well-known Hummers method [3]. Subsequent to the oxidation process, the chemical reduction of the synthesized GO delivers so-called reduced graphene oxide (rGO), which shows similar characteristics to pristine graphene. The basic concept of this reaction route consists in the extension of the multilayer structure of graphite by the introduction of oxygen functionalities during the oxidation process. The introduced oxygen functionalities cannot be completely eliminated during the following reduction process and remain as defects in the obtained monolayer structure. The quality of the produced rGO is therefore determined by the number of oxygenated functional groups left. The sp^3 defects interrupt the aromatic honeycomb structure of rGO and lead to slight decreases of the electrical properties in comparison to the ones of pristine graphene. In summary, it can be said that the chemical reduction of GO represents a versatile, cost-effective method to produce graphene in an upscalable way.

Electron paramagnetic resonance (EPR) spectroscopy, a powerful and sensitive method to detect active intrinsic and extrinsic paramagnetic defects in a material system [4, 5], has demonstrated an effective method to study point defects within carbon nanostructures [6]. In this, EPR is employed not only to elucidate the paramagnetic characters, induced by itinerant electrons, unpaired electrons and carbon-dangling bonds, also to allow the study of the structure of carbon materials in different functional forms [7]. The carbon-related EPR-induced defects are influencing the optical and electronic features of carbon material. Though the magnetic features of graphene have been studied intensively, EPR research on graphene and graphene-like structures is still a new field. Due to the mostly paramagnetic nature of the defects, this high sensitive method provides valuable information about the rGO defect structure and concentration based on the measured g-values. A high g-value (> 2.004) would, for example, mean that the analysed rGO sample has an inferior quality originated in an elevated number of defects in the aromatic honeycomb network.

This book chapter focuses on discussing EPR investigations on graphene oxide, and functional reduced graphene oxide and carbon nanotubes. In that, EPR has demonstrated as a suitable tool to detect spin density changes in different functionalized nanocarbon materials. The last part will discuss a novel approach to studying the charge transfer within quantum dots-graphene hybrid materials using EPR, which examines the change in the electronic properties of graphene before and after being attached with quantum dots. This contributes to improved understanding of electronic coupling effects in nanocarbon-nanoparticle hybrid materials promising for various electronic and optoelectronic applications.

EPR INVESTIGATIONS OF FUNCTIONALIZED GRAPHENE AND CARBON NANOTUBES

EPR has recently proven to be an effective method for studying the defect structure of nanostructured carbon materials [6] as it is sensitive to both extrinsic and intrinsic lattice defects within the carbon nanostructures. The EPR signals are induced by itinerant electrons, and unpaired electrons caused by dangling bonds [6], which provide detailed information about different forms of carbon [8]. Although pure and perfect carbon nanotubes (CNTs) are inactive to EPR, experimental data often show the EPR signals with a Lorentzian lineshape at $g = 2.001$ [6]. The signals are believed to result from the impurities within the CNTs that are the remained catalysts from the CNT synthesis. In contrast, when graphite particles are broken into nano sizes, abundant edge-structures are created, facilitating the localization of electrons due to edge-defective dangling bonds. This thus results in a shape signals for the nanosized graphite upon EPR experiments [8].

The magnetic features of graphene are considered promising for application in spintronic devices. This motivates the use of EPR to study magnetic properties of graphene. For example, in Ref [9], EPR was employed to study lattice structures of rGO, two EPR signals were observed: one broad signal at $g = 2.0027$ attributed to graphitic carbon structures; another narrow signal at $g = 2.0028$ assignable to carbon-dangling bond radicals. In Ref [10], EPR experiment was *in situ* performed to observe the grafting of polymers onto functional graphene upon polymerization. Based on EPR investigations on GO and rGO, Mn^{2+} impurities are detected in the samples. These impurities result from $KMnO_4$ precursor used for the preparation graphene oxide (GO), and remain even after washing. This indicates that Mn^{2+} binds to the graphene lattice and is not easy to be removed *via* conventional washing procedures [11]. Su *et al.* [12] shown that the catalytic activity of GO results from the localized electron spins at the edge of the GO lattice *via* EPR quenching experiments.

In this chapter, the conversion of oxygen-functional groups, such as hydroxyls and carboxylic, of the GO to thiol-functional groups of thiol-functionalized rGO (TrGO) and carbon nanotubes (CNT-SH) is investigated by EPR.

MATERIALS AND METHODS

Synthesis

GO was synthesized using a modified Hummer's method [13]. Thionation of GO to TrGO was performed as follows. First, 100 mg GO was dispersed in 100 ml dimethylformamide (DMF) and sonicated for 1 h. After removing undispersed GO by centrifugation at 1000 rpm for 10 min, a homogeneous solution of GO in DMF was obtained. Then, 300 mg P_4S_{10} was added to the solution and the reaction flask was evacuated to $5 \cdot 10^{-3}$ mbar at 100°C for 2 min to remove humidity in the flask atmosphere. The thionation was performed for 12 h in vacuum under continuous stirring at 120°C. Finally, the reaction product was collected by filtering the solution through a 0.45 μm polyamide membrane filter and was extensively washed in succession with 100 ml of water, ethanol and acetone, respectively.

In addition, oxidized carbon nanotube (O-CNT1) and thiol-functionalized carbon nanotube (CNT-SH) were also synthesized by the identical procedures to GO and TrGO, respectively, for comparative EPR investigations.

A Mn-free oxidation method was used to exclude the influence of Mn EPR signal on the studied materials. 200 mg CNT with inner and outer diameter: ~4 and 13 nm, and tube-length of >1 μm (Bayer Material Science, purity >99%) was put in 40 ml strong acid mixture of 1:3 HNO_3 (Merck, 63%) and H_2SO_4 (Merck, 98%). The dispersion was then stirred for 4 h at 100 °C. Next, the dispersion mixture was diluted into 150 ml DI-H_2O. The final Mn-free oxidized CNT (O-CNT2) product was removed from suspension by centrifugation for 2 h at 4400 rpm and then washed 3 times with 100 ml DI-H_2O. Finally, the O-CNT2 powder was dried for 12h in a vacuum oven at 50°C.

RGO powder was obtained *via* chemical reduction of GO by hydrazine (N_2H_4, Sigma, 64%) in DMF. Briefly, 20 mg GO was ultrasonically dispersed in 50 ml DMF. After that, 20 μl of the N_2H_4 solution was added and the solution was stirred for 12 h at 90°C. The received rGO was purified by washing with DMF three times.

EPR EXPERIMENTS

EPR investigations were carried out at X-band (9.86 GHz) with a Bruker EMX spectrometer and an NMR gaussmeter (ER 035M, Bruker) for measuring the magnetic field. Polycrystalline diphenylpicrylhydrazyl (DPPH) ($g = 2.0036$) was used as a reference for accurate calibration of the g-factor and the resonance magnetic-field values. EPR experimental conditions were as follows: time constant: 20.48 ms and receiver gain: 2×10^3, microwave power: 1 mW, modulation amplitude: 0.5 G (0.01 mT).

Spin-counting Procedure

For accurately determining the spins density, following important parameters have to be carefully taken into account before the EPR experiment, and also for data analyses after experiments [14].

i. The weight of samples has to be accurately measured prior to the experiments for weight-intensity normalization. If the amount of sample in an EPR tube is not the same, the obtained spectrum should be normalized by multiplying by a filling factor [14, 15] (derived from the sample weight in the EPR tube).
ii. The samples must be placed in the middle of the microwave resonator.
iii. The background signal (*e.g.*, from impurities in the cavity) should be subtracted out from the EPR spectrum of the measured sample.
iv. The EPR signal should not be saturated with high microwave power applied. The microwave phase must be cautiously tuned when critical coupling (tuning) of the cavity.
v. If an offset exists in a magnetic field, it should be calibrated.
vi. The Q value of the cavity must be determined, and EPR spectra should use the received Q value.

The defect density can be quantitatively determined by doubly integrating each first-derivative EPR spectrum, which is independent of the microwave frequency. Based on the integral ratios of the measured samples to a standard reference sample (here, MnO powder), the spin density, and as a result, the spin density of defects can be calculated. For an accurate calculation of defect density, a normalized-corrected value shown below, which includes experimental parameters of the reference and the investigated samples has to be taken into account:

$$N_S^* \cdot \frac{RG^*}{RG} \cdot \frac{MF^*}{MF} \cdot \frac{MA^*}{MA} \frac{CT^*}{CT} \cdot \sqrt{\frac{P^*}{P}} \times$$

$$\times \left(\frac{Scans^*}{Scans}\right)^2 \cdot \left(\frac{SW}{SW^*}\right)^2 \cdot \frac{S^*(S^*+1)}{S(S+1)} = corrected\ value \tag{1}$$

where N_S* is the number of spins in the reference sample, *RG* is the receiver-gain, *MF* is the modulation frequency (in kHz), *MA* is the modulation amplitude (in G), *CT* is the conversion time (in ms), *P* is the microwave power (in mW), *SW* is the magnetic field sweep (in G), *Scans* is the number of scans, *S* is the spin quantum number. The note (*) refers to the measured parameters for the standard samples. After N_S* is acquired, the defect density N_D of the studied sample is calculated as follows:

$$N_D = \frac{(Area)_D \cdot N_S^*}{(Area)^*} \tag{2}$$

where $(Area)_D$ is the area of the EPR spectrum under integration for the related defect sites, N_D is the number of spins of the sample. Here, as a standard reference sample the MnO powder (3.34×10^{19} spins/g) was used.

RESULTS AND DISCUSSIONS

Elimination of Mn^{2+} Impurities Within GO and O-CNT Samples

In Fig. (**1**), the comparative EPR spectra of GO, rGO, and TrGO are shown. EPR spectra (Fig. **1a**) of CNT with different functionalities were recorded under identical experimental conditions for direct comparison.

EPR investigations of GO and O-CNT samples revealed characteristic Mn^{2+} (S = 5/2, I = 5/2) hyperfine patterns centered at g = 2.0019, both arising from the electronic transition $M_S = -1/2 \leftrightarrow M_S = +1/2$. The characteristic sextet results from Mn impurities remained from the preparation step in which $KMnO_4$ was employed as an oxidant [11]. Besides the Mn^{2+} lines, there are additional EPR signals which are attributed to the defects within GO lattices induced by oxygen-functionalities such as −COOH and −OH. The hyperfine patterns for Mn^{2+} were proven in the previous study on complexes containing Mn^{2+} [16].

To eliminate the influence of the EPR signal of Mn^{2+} traces on the investigation of carbon-related defects signal, a Mn-free oxidation route using an HNO_3/H_2SO_4

mixture oxidant was applied on CNT (O-CNT2). As a result, the EPR spectrum of O-CNT2 (Fig. **1a**) reveals only one sharp line arose from C-related spin(s), and the Mn^{2+} hyperfine lines are therefore invisible in the sample. In addition, the Mn^{2+} hyperfine lines are also no longer visible in CNT-SH, rGO, and TrGO. This indicates the removal of Mn^{2+} ions from the samples after the reduction or thionation steps of both CNT and graphene.

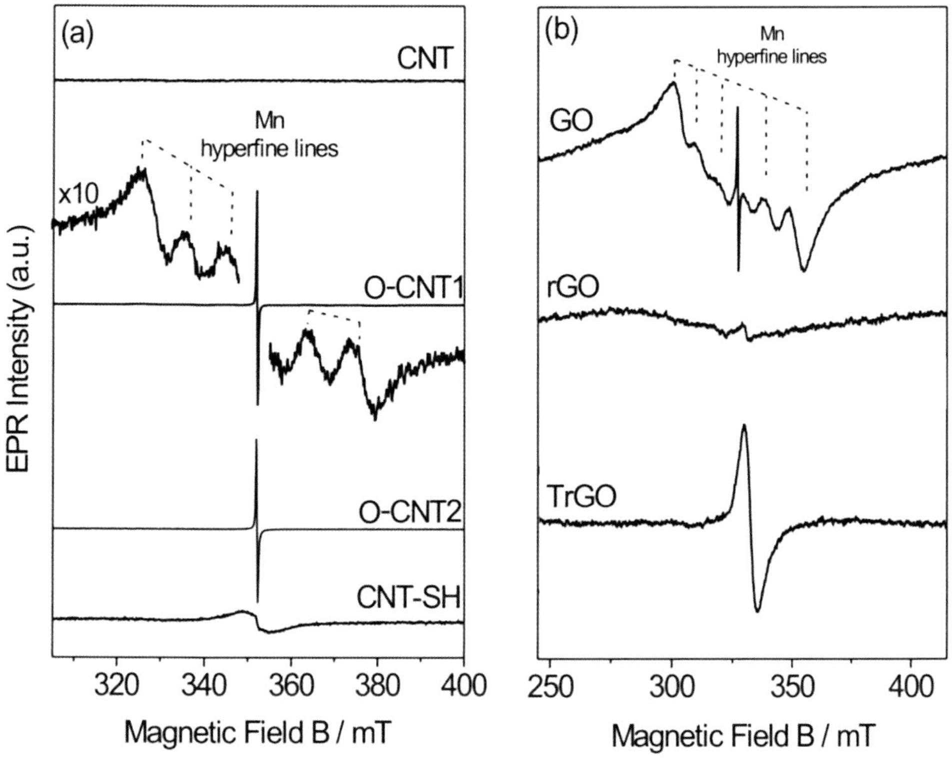

Fig. (1). X-band EPR spectra of CNT (a) with different functional groups and graphene (b) with different functionalities. All spectra were recorded at room temperature. The spectra with and without Mn^{2+} hyperfine lines indicate the elimination of Mn impurities in O-CNT2, CNT-SH, rGO, and TrGO samples (reused from Ref [17]. with permission from AIP publishing).

EPR investigation of Graphene and CNT with Different Functional Groups

In both O-CNT and GO, besides the Mn^{2+} signal, single EPR lines at $g \sim 2.0029$ are visible, such signals are typically found for C-related dangling bonds (2.0022-2.0035) [10]. The dangling bonds are induced by oxygen-containing groups such as carboxylic, carbonyl, hydroxyl, and epoxide in O-CNT and GO. Subsequently, by chemical reduction on GO, the oxygen-groups are significantly reduced. The hexagonal carbon lattice of graphene is mostly restored. As a result, its electron mobility increases. The localized spin density in rGO is therefore considerably reduced, leading to the quenched EPR signal on the rGO sample as shown in Fig. (**1b**). Pristine graphene is still EPR-active with a small EPR signal intensity due to defects arising from the zigzag and armchair edges of graphene lattices [9].

Upon thiol-functionalization and simultaneous reduction on GO, TrGO is obtained with restored the sp^2-carbon structure of graphene and thiol functional groups being attached to the graphene lattice. While GO is an electrical insulator, TrGO exhibits a conductivity of 5300 S/m [13]. Fig. (**1b**) reveals that TrGO has an EPR signal with a broader peak (larger peak-to-peak linewidth (ΔB_{pp}), see Table **1**), in comparison with that of GO. This indicates the lower spin density of TrGO compared with GO. This is explained by the partially restored sp^2-carbon structure of TrGO by the chemical reduction. A part of oxygen-containing chemical groups is transformed into thiol-containing groups such as -CSSH, and -SH. These thiol groups play a role to remain localized radicals (*e.g.*, trapped electrons), leading to the EPR signal with g-value close to that of free electrons (g = 2.0023) [15].

Table 1. Peak-to-peak linewidth (ΔBpp) and spin densities of functional graphene and CNT (reused from Ref [17]. with permission from AIP publishing).

Sample	ΔB_{pp} (mT)	Spin density (# of spins/g)	Sample	ΔB_{pp} (mT)	Spin density (# of spins/g)
GO	0.36	2.4×10^{17}	**CNT**	-	-
rGO	-	-	**O-CNT1**	0.23	4.3×10^{17}
TrGO	5.5	1.1×10^{17}	**O-CNT2**	0.37	0.9×10^{17}
QD-TrGO	22.6 (weak)	2×10^{15}	**CNT-SH**	6.42	2×10^{16}

The peak-to-peak widths (ΔB_{pp}) and the spin densities of the studied materials are summarized In Table **1**. The spin densities were then calculated by employing the spin-counting method described above. TrGO exhibits a ΔB_{pp} of 5.5 mT, which is ~15 times the value of the GO (0.36 mT). As a result of calculation, a spin-density of 2.4×10^{17} spins/g for GO is larger than that of 1.1×10^{17} spins/g for TrGO. This indicates fewer defects in TrGO than in GO as a result of the chemical reduction.

In addition, while pristine CNTs does not contain any spins, the oxidation of the CNTs creates lattice defects into the CNTs, leading to spin densities of 4.3×10^{17} spins/g and 0.9×10^{17} spins/g detected for O-CNT1 and O-CNT2, respectively. These results show that the oxidation of CNT using the mixture of $KMnO_4$, $NaNO_3$ and H_2SO_4 is more effective than using the mixture of HNO_3 and H_2SO_4. After the transformation of O-CNT1 into CNT-SH, the spin density decreases from 4.3×10^{17} spins/g to 6.42×10^{16} spins/g, respectively, as a consequence of the chemical reduction during the chemical transformation.

Comparison of EPR Signals of TrGO and CNT-SH

CNT-SH is synthesized using the same method as for TrGO, the defect densities of the received samples are investigated using EPR. CNT-SH shows smaller intensity (Fig. **2**) as compared to TrGO. As a result, a spin density of 2×10^{16} spins/g is determined for CNT-SH. This is one order of magnitude lower than that of 1.1×10^{17} spins/g for TrGO, as shown in Table **1**. This can be interpreted in the following way. GO is synthesized from graphene nanoplatelets, which have significantly more edge-structures than CNT with nanotube structures. It is known that the edges are more energy-favorably functionalized than the planes of graphene structures. Thus GO is more intensively oxidized than the O-CNT. Moreover, only the outer walls of multi-walled CNT (MWCNT) are chemically functionalized leading to fewer defects per weight [18, 19]. As a result, after the simultaneous reduction and thionation steps, TrGO is supposed to contain more thiol groups and defects than CNT-SH. Consequently, TrGO exhibits an EPR line with substantially higher peak intensity than that of the CNT-SH with the equal weight of measured samples, as shown in Fig. (**2**).

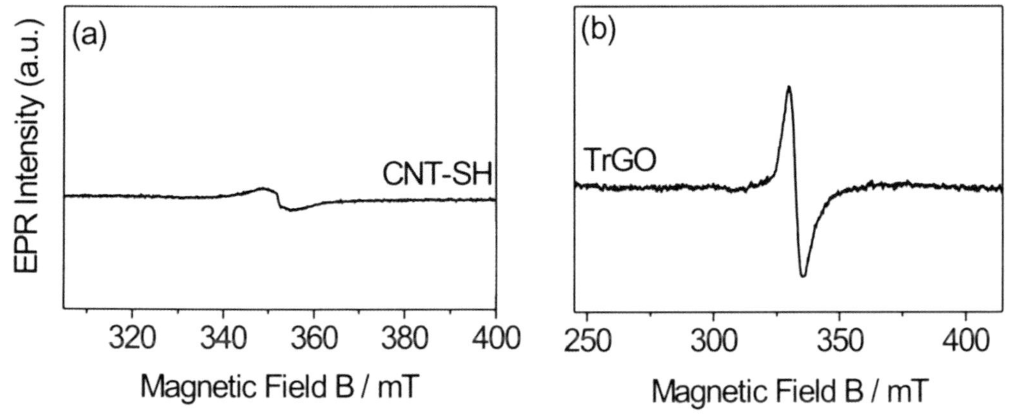

Fig. (2). X-band EPR spectra of CNT-SH (a) and TrGO (b) measured at the same conditions.

Dependence of EPR Signals of O-CNT and TrGO on Temperatures

Generally, the EPR sensitivity is enhanced by lowering the temperatures. In this section, to study the temperature-dependence of EPR signals of O-CNT and TrGO, EPR experiments are performed at different temperatures in a range of 100 K – 300 K.

Fig. (**3a,b**) presents the temperature variation of the EPR spectra of O-CNT1 and TrGO at different temperatures ranging from 100 to 300 K. The EPR spectrum of O-CNT1 shows typical temperature dependence described by Curie law, whereas the EPR spectrum of TrGO does not vary considerably with the temperature, hence exhibiting a Pauli-type behavior. This Pauli-type paramagnetism results from itinerant electrons in TrGO lattice and is thus temperature-independent.

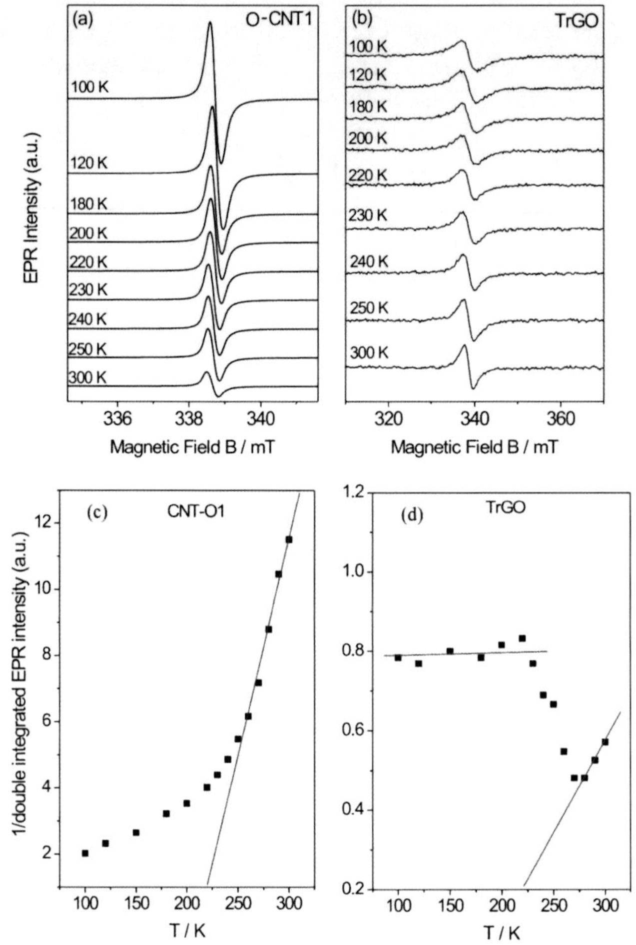

Fig. (3). The temperature dependence of the X-band EPR spectra (first-derivative) measured for O-CNT1 (a) and TrGO (b). The temperature variation of the EPR-susceptibility of O-CNT1 (c) and TrGO (d), acquired after doubly integrating the EPR spectra in (a) and (b), correspondingly. (reused from Ref [17]. with permission from AIP publishing).

The magnetization of OCNT1 exhibits typical Curie-law indicates that the number of spins in the carbon matrix decreases with treated temperatures. In contrast, in TrGO, its EPR Pauli-type behavior is attributed to the fact that thiol groups are spin trapping centers, holding them more localized even with varying temperature. As a result, EPR spectra of TrGO are temperature-independent below 222 K (Curie temperature).

The EPR spectra of O-CNT1 and TrGO are further analyzed by doubly integrating the dispersive-type continuous-wave EPR, the received result known as EPR susceptibility. The received data referred as Curie-plots are shown in Fig. (**3c** and **3d**). At $T < 222$ K, the EPR-susceptibility of O-CNT1 increases exponentially with temperatures, whereas, in a range of 222–300 K, it increases proportionally with temperatures. Therefore, the EPR signal of O-CNT1 displays a ferromagnetic behavior, with the exact Curie temperature of $T_C = 222$ K (Fig. **3a**).

In contrary, TrGO exhibits a complex behavior (Fig. **3b,c**). The EPR lines of TrGO are temperature-independent below T_C (222 K), which is the property of Pauli paramagnetism. From 222 K to 275 K, the EPR signal of TrGO begins to tumble deteriorating the magnetic interaction of the electron spins, which is also observed in the EPR study of fullerene anions [20].

CONCLUSIONS ON THE EPR BEHAVIORS OF FUNCTIONALIZED MWCNTS AND GRAPHENE

The EPR investigations on functionalized graphene and MWCNTs show that Mn^{2+} impurities remain chemically bound in GO lattices, and cannot be removed by conventional washing. Mn^{2+} impurities are eliminated from TrGO during the functionalization and simultaneous reduction steps. The EPR signal of TrGO induced by spins localized at the functional groups and carbon defects. Whereas the temperature variation of EPR spectra for O-CNT1 obeys the typical Curie law, the variation of TrGO reveals characteristics of Pauli paramagnetism at $T < 222$ K.

INVESTIGATION OF CHARGE TRANSFER WITHIN QUANTUM DOTS – GRAPHENE HYBRID MATERIALS USING EPR QUENCHING EXPERIMENTS

Charge Transfer Within CdSe QD-TrGO Hybrid Materials

With astonishing electrical and optical characters, semiconducting CdSe quantum dots (QDs) have been proven promising for applications in hybrid solar cells and photodetector devices [21, 22]. However, the quantum dots are often synthesized while encapsulating with organic ligands to keep the particles separated. The organic ligands decrease their activity in catalysis or charge transfer in photovoltaic applications [23]. If the synthesis ligands are removed or exchanged by more conductive ligands, the quantum dots tend to agglomerate, leading to a significant reduction of their active surface area. This problem can be resolved by using conductive support materials to keep them from aggregation [24, 25].

Graphene is demonstrated as an excellent framework and support for QD nanoparticles [26, 27]. This is because with the ultra-large specific area and extremely high electron mobility (15000 cm^2 $V^{-1}s^{-1})^2$ of graphene (2630 m^2/g) [31] graphene facilitates charge transfer processes at the QD-graphene interfaces [28] and charge transport processes through the graphene phases [29, 30]. The obtained QD-graphene hybrids possess synergetic properties of both QDs and graphene and therefore are very promising for photo energy harvesting applications. For example, CdSe QDs decorated graphene have been realized for applications in photovoltaics and OLEDs [28, 32 - 34].

QD-graphene hybrid materials can be synthesized by an in-situ growth of QDs on graphene oxide sheets (GO). Another approach evolves the functionalization of graphene with dedicated chemical groups which have a high affinity to QDs to attach QDs on graphene by self-assembly. For example, graphene was first functionalized with poly-(diallyldimethylammonium chloride) (PDDA), then the positively charged PDDA functionals attach negatively charged CdSe QDs to form CdSe-graphene hybrids [32]. Benzyl mercaptan can also be used as interlinkers on the one hand binding to CdS QDs *via* thiol group (–SH), and on the other hand attaching to graphene *via* non-covalent π-π stacking [35]. The non-covalent interlinker molecules approach might overcome the solubility problem of graphene in various solvents. However, the interlinkers might hinder the effective charge transfer between QDs and graphene when the QD-graphene hybrids are used in photoenergy harvesting applications.

This chapter focuses on a scalable synthesis route towards QD-rGO hybrids *via* thiol-functionalization directly on GO to obtain TrGO as described above, followed by the decoration of QDs on TrGO *via* a self-assembly process. This self-assembly process is realized due to the high affinity of thiol functionalities to transition metal chalcogenides such as CdS and CdSe as proven by Mann et al. [36] and Colvin et al. [37] in previous reports. In this section, the electronic features of TrGO before and after attaching with CdSe QDs are investigated by EPR. By this, charge transfer between CdSe QDs and TrGO might be elucidated.

In fact, it is observed that after decoration with CdSe QDs, forming the CdSe QD-TrGO hybrids, the EPR signal of TrGO is almost quenched (see Fig. **4**). This phenomenon can be explained as follows: the chemical binding of the QDs to the thiol groups on the surface of TrGO leads to an efficient charge or/and energy transfer between QDs and TrGO; the transferred electrons couple with electron spins within TrGO, and therefore quench the EPR signal of TrGO. This suggestion is consistent with the previous study by photoluminescence quenching experiments [13]. The spin density of CdSe QD-TrGO was dramatically reduced to the low value of 2×10^{15} spins/g from the original value of 1.1×10^{17} spins/g

(Table **1**) due to a charge transfer and coupling between the two components.

Moreover, the quenching of EPR signal CdSe QD-TrGO is even observed when the experiments are conducted in the dark. Therefore, it can be concluded that the chemical bondings between the QDs and TrGO occur during the synthesis of the CdSe QD-TrGO hybrid.

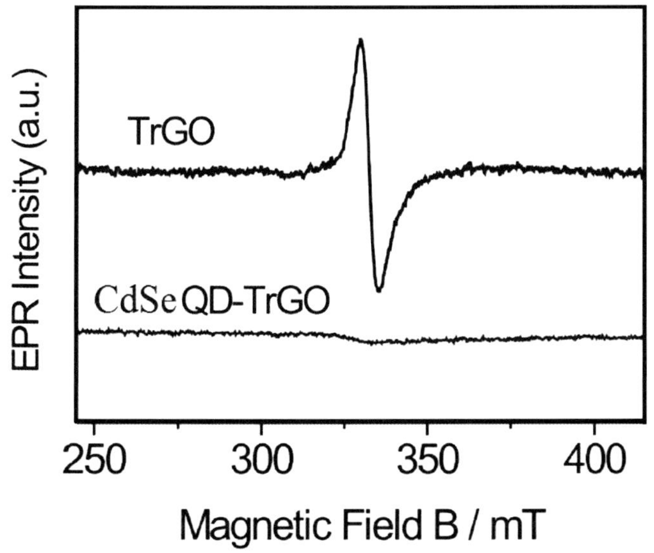

Fig. (4). Quenching of EPR signal of CdSe QD-TrGO hybrid (lower line) in comparison with the EPR signal of pristine TrGO (upper line) before QD decoration [38]. Reproduced from Ref [38]. with permission from the Royal Society of Chemistry.

The quenching of TrGO EPR signal in CdSe QD-TrGO hybrid demonstrates a direct chemical binding of CdSe QDs to TrGO that leads to the charge transfer between CdSe QDs and TrGO. The charge transfer and electric transport properties are necessary for various electronic applications. Therefore, CdSe QD-TrGO hybrid is promising for optoelectronic applications such as photodetectors and solar cells.

Charge Transfer Within GO-ZnO and TrGO-ZnO Hybrid Materials

Zinc oxide (ZnO) nanoparticles have recently attracted great interest of the research community due to their astonishing characters such as magnetic, electronic, optical, and electromechanical properties [39, 40]. In addition, with a large exciton binding energy (60 meV), ZnO nanoparticles enable an efficient excitonic lasing at room temperature [41]. Therefore, ZnO nanomaterials have been demonstrated promising for various applications in light-emitting diodes [42], photovoltaic cells [43], laser diodes, spintronic devices, ultra-violet photodetectors, and field effect transistors [44].

Meanwhile, graphene has been demonstrated to be highly potential for various applications due to its attractive optical, electrical and mechanical properties [45]. For example, its high mechanical stability and flexibility in combination with electrical conductivity are suitable for flexible electronic devices [46]. With the high conductivity and optical transparency, graphene is a great candidate as transparent electrodes for optoelectronics [47]. Also, its extremely high specific surface area is promising for sensor applications or as a catalyst support [48].

To utilize the attractive properties of both ZnO and graphene, ZnO-graphene hybrid materials have recently synthesized to create the synergetic ability for novel applications [40, 49]. The hybrids have interesting properties, which are not a simple addition of those of the two components, but novel ones might arise from the electronic coupling of the two materials. To this end, ZnO QDs were synthesized and encapsulated by GO, thereby an additional energy band is created for within the ZnO QDs, therefore resulting in the additional blue-light emissions [40]. Subsequently, the ZnO-graphene hybrids were used as active material for a white-light-emitting diode device.

In another work [50], ZnO-graphene and ZnO-GO nanocomposite films were fabricated using ultrasonic-assisted spray pyrolysis. The photoluminescence (PL) property of the hybrid films exhibited a high intensity and broad green light emission, which was not shown by a pure comparable ZnO film. The new emission was originated by the decreased crystallinity of the ZnO hybrid films induced by increased oxygen vacancies in the presence of graphene. However, Kavitha *et al.* [51] recently observed that oxygen vacancy defect states can be healed by oxygen diffusion from GO to ZnO while GO was being reduced during the pyrolysis synthesis. Also, the PL of rGO-ZnO hybrid material was significantly quenched due to the photoexcited charge transfer between ZnO and graphene sheets.

This chapter discusses on charge transfer within TrGO-ZnO and GO-ZnO hybrid materials observed by EPR. The (V^+_O) surface defects healing effect of both oxygen-containing groups of GO and thiol functionals of TrGO were investigated and compared by EPR. GO was synthesized by a modified Hummer's method as presented in our publications [13]. The oxygen-containing groups, such as hydroxyl (OH) and carboxylic (COOH), of GO are used to attach ZnO NP to form ZnO NP decorated rGO hybrid material (rGO-ZnO). Moreover, oxygen-functionalities was then transformed into thiol groups to obtain thiol-functionalized reduced graphene oxide (TrGO) by refluxing GO with phosphorus pentasulfide (P_4S_{10}) in dimethylformamide (DMF) at 154°C for 24 h according to a procedure reported elsewhere [13]. Because the thiol groups have a strong affinity to the transition metal [13], they attach well ZnO NPs to form ZnO NP decorated TrGO (TrGO-ZnO) *via* a self-assembly process. Further details about the syntheses of ZnO NPs, GO-ZnO, and TrGO-ZnO hybrid materials are referred to the literature [52].

Fig. (**5**) presents transmission electron microscopy (TEM) images of GO-ZnO and TrGO-ZnO hybrid. ZnO NPs are observed attached to the graphene surface, and no ZnO NPs are visible outside of the graphene sheets. This proves the successful syntheses of the hybrid materials based on a self-assembly decoration.

EPR was then used to investigate the electronic interaction between components of the hybrids. As shown in Fig. (**6a**) (upper line), the EPR spectrum of ZnO NPs reveals two signals at a *g*-factor of $g1 = 2.0037$ close to that of the free-electron ($g_e = 2.0023$), and at $g2 = 1.9600$. The signal at $g1$ is attributed to singly charged oxygen vacancies (V^+_{OS}), and the signal centered at $g2$ is associated to the Zn-core defects (V^-_{Zn}) of the ZnO NPs [53].

The core signal was vanished upon the attachment of ZnO NPs to GO sheets, as presented in Fig. (**6a**) (lower line). The signal at $g3 = 2.0024$ is the composition of a surface defect signal (at $g1 = 2.0037$) from ZnO NPs and a carbon dangling bond signal (at $g = 2.0023$) from GO lattices, which is also observed in the literature [17, 54].

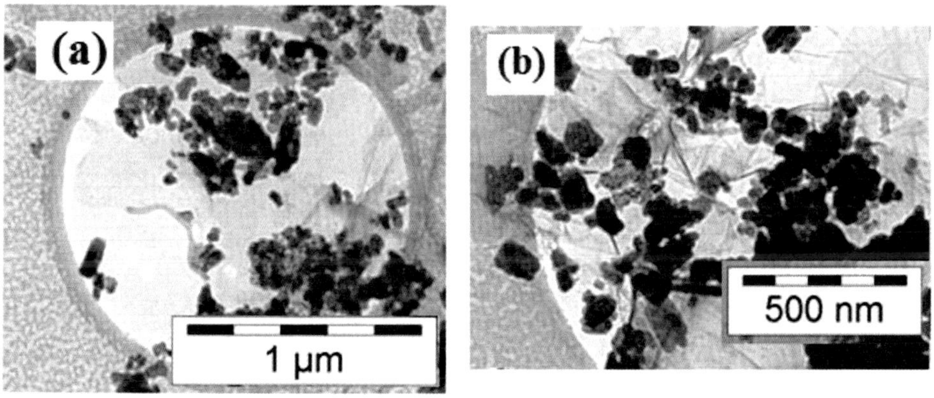

Fig. (5). Morphology investigation: TEM images of (a) ZnO NPs decorated on GO nanosheets and (b) ZnO NPs decorated on TrGO nanosheets. Reproduced from Ref [52]. with permission from the Royal Society of Chemistry.

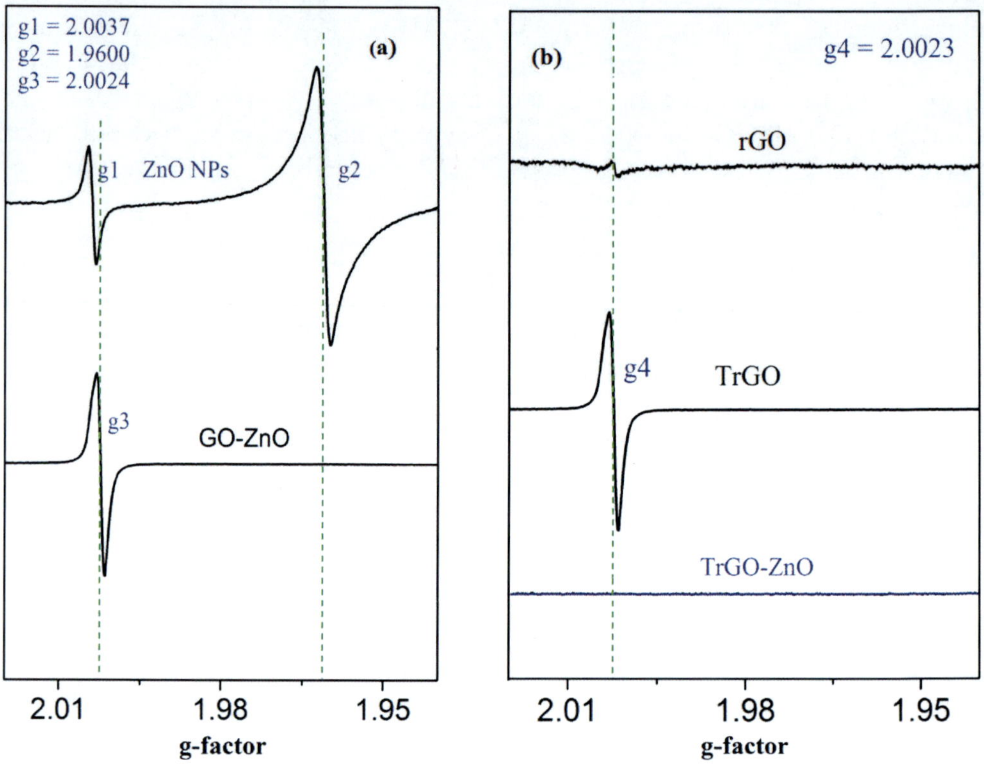

Fig. (6). EPR investigation: EPR spectra of ZnO NPs and GO-ZnO hybrid materials (a), and rGO, TrGO, and ZnO-TrGO (b). While rGO has low EPR signal intensity, TrGO exhibits high EPR intensity. Defect healing and electron transfer quench the EPR signals of both ZnO NPs and TrGO within the ZnO-TrGO hybrid. Reproduced from Ref [52]. with permission from the Royal Society of Chemistry.

The vanishing of the core signals might be originated by the charge transfer from ZnO NPs to either GO or TrGO. This can be explained based on a reported core-shell model for ZnO nanocrystals [54, 55]. In that model, the core of ZnO NPs is negatively charged by electrons trapped in Zn vacancy defects (V^{-}_{Zn}), whereas the shell has the surface defects due to singly charged oxygen vacancies (Fig. **7**).

The trapped electrons within surface defects are well restrained due to the dielectric property of the ZnO NP surfaces. Thus, they cannot be transferred to the GO sheets, and stay in the surface defects conserving the EPR signal at *g*3 = 2.0024, as revealed in Fig. (**6a**) lower line. The trapped electrons in the cores can delocalize because of the electrical mobility of the cores [55]. This results in an intriguing physical phenomenon happening in the GO-ZnO and TrGO-ZnO hybrid that the core electrons move to the surfaces, and then are transferred to GO and TrGO sheets, as depicted in Fig. (**7**).

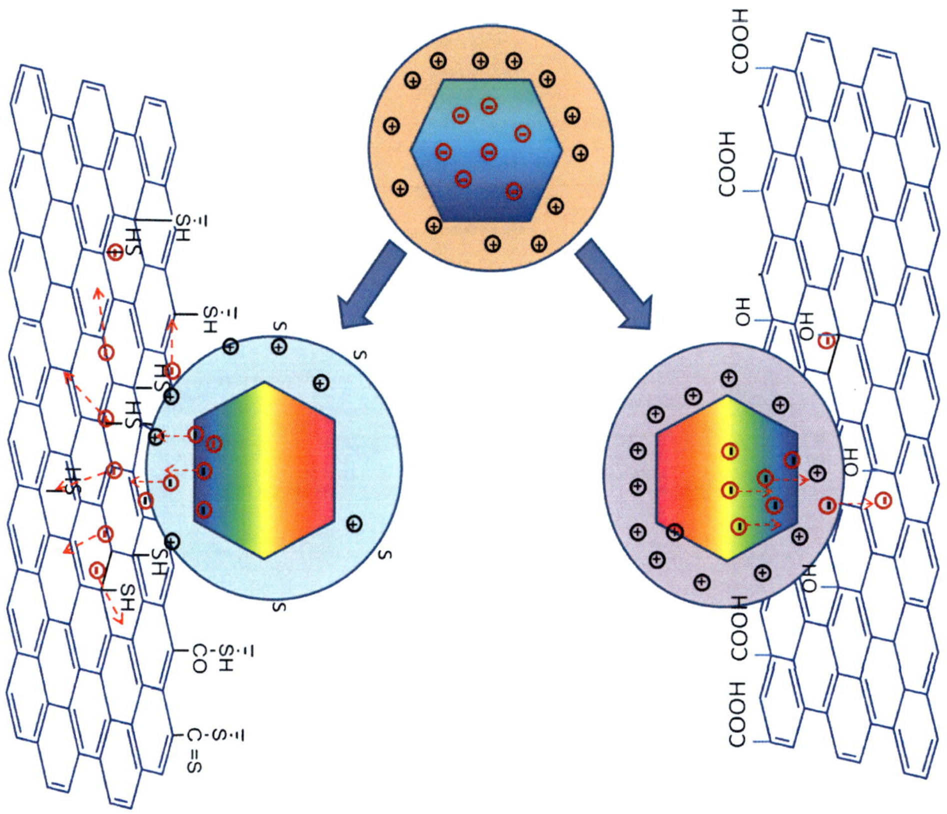

Fig. (7). Schematic demonstration for the delocalization of electron spins within GO-ZnO and TrGO-ZnO hybrid materials resulting in the quenching of the EPR core signals of ZnO NPs. Reproduced from Ref [52]. with permission from the Royal Society of Chemistry.

The phenomenon can be explained by the energy-levels matching of the Fermi levels of either GO or TrGO with the conduction band (CB) of ZnO NPs, which facilitates the electron transfer. As schematically illustrated in Fig. (**8**), the CB of ZnO NPs is at around –4.19 eV a little higher than the Fermi level of GO and TrGO, which were reported in a range from –4.5 to –4.7 eV [56].

Therefore, electrons from CB of ZnO are preferably transferred to GO or TrGO. As a result, the trapped electrons in the ZnO core vanish, thus leading to the absence of the core signals in both GO-ZnO and TrGO-ZnO hybrids.

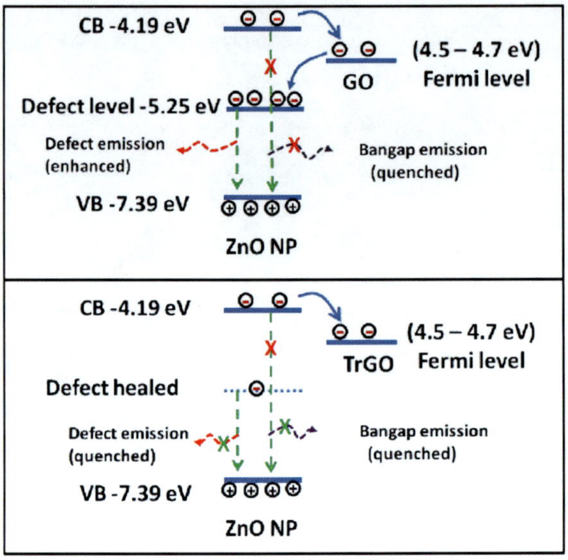

Fig. (8). Schematic illustration of energy level diagram and electron transfer leads to PL quenching and enhanced defect emission in the GO-ZnO (upper image) and TrGO-ZnO (lower image) hybrid. Reproduced from Ref [52]. with permission from the Royal Society of Chemistry.

EPR signals of the rGO, TrGO, and TrGO-ZnO are shown in Fig. (**6**). RGO shows low signal intensity. TrGO exhibits a pronounced EPR signal at $g4 = 2.0023$ due to the sulfur-doping of rGO creating dangling bonds within the lattice of graphene. Interestingly, after attachment of TrGO with ZnO NPs *via* thiol functionalities, both original EPR signals of TrGO and ZnO are completely quenched, as revealed in Fig. (**6b**). In contrast, the EPR signal at $g3 = 2.0024$ of GO-ZnO still remains. Here, the quenching of the ZnO NP's core signal of TrGO-ZnO can be explained the same as for GO-ZnO.

The surface defect signal quenching is ascribed to the healing property of thiol groups to oxygen vacancies in the surface defects. In addition, traces of the molecular sulfur byproduct during the TrGO synthesis will also fill up the oxygen vacancies on the ZnO NPs surface, which are not covered by TrGO sheet, as depicted in Fig. (**7**). Therefore, it can be concluded that the oxygen vacancy defects within original ZnO NPs are completely healed when the NPs are integrated into TrGO-ZnO hybrid materials. Such effect can also be observed in photoluminescence (PL) spectroscopy investigations, as discussed in the following. Hereby one interesting question arises as to why is the EPR signal of TrGO quenched, but not that of GO when they are decorated with ZnO NPs? This can be explained as follows: The transferred electrons within the TrGO lattice can move within the graphene lattices of TrGO due to their high electron mobility as demonstrated by our previous report [13]. The transferred electron spins then coupled with localized electron spins within TrGO, leading to the quenching of the primary EPR signal of TrGO. In contrary, due to the dielectric property of GO, the transferred electrons cannot move to where the localized electron spins located, and thus no electron coupling occurs, as a result, the EPR signal of GO persists in the GO-ZnO hybrid materials (Fig. **7**).

Photoluminescence investigation further corroborates the results and conclusion drawn from above EPR study [52]. As schematically demonstrated in Fig. (**8**), the free electrons in the CB (–4.19 eV) of ZnO NPs are energetic-favorably transferred to GO and TrGO (Fermi level: –4.5 to –4.7 eV) [56] due to the energy match. Therefore the PL intensity of the bandgap emission of ZnO NPs of the hybrid materials is decreased [52]. In the opposite way, electrons in graphene are also transferred back to the defect states (defect level: –5.25 eV) of ZnO NPs, and thus enhances the electron density in defect states. This explains for the enhancement of the defect emission band of the PL spectrum of GO-ZnO. In contrary, the surface defects of ZnO NPs are healed in the TrGO-ZnO hybrid, leading to the defect emission quenching [52].

CONCLUSION

EPR spectroscopy demonstrates as a suitable tool to detect spin density changes in different functionalized graphene and carbon nanotubes. The spin density changes result from different functionalities, lattice defects, and impurities within the carbon materials. In contrast, pristine carbon nanotubes and reduced graphene oxide exhibit negligible spin density and therefore EPR signal. Continuous wave EPR also demonstrates as an extensive approach to studying the charge transfer within quantum dots-graphene hybrid materials, revealing the change in the electronic properties of graphene before and after attaching of quantum dots such as CdSe and ZnO.

The quenching of the EPR signal of TrGO in CdSe QD-TrGO hybrid owning to a chemical binding of CdSe QDs to the thiol groups on the surface of TrGO that leads to the charge transfer between them. EPR investigations show that the electron spins trapped in Zn vacancy defects (V^-_{Zn}) within the core of ZnO NPs move to the surface and are then transferred to GO and TrGO in the hybrid materials thus leading to the disappearance of the core signal from EPR spectra for ZnO-TrGO. On the other hand, thiol groups of TrGO and sulfur can effectively heal the oxygen vacancy surface defects of ZnO NPs while oxygen-containing functionalities have low healing ability at low temperatures. The electrons transferred from ZnO NPs can hop in the TrGO lattice and its spins coupled with the localized electron spins, thus leading to the quenching of the

EPR signal of TrGO. In contrast, this does not occur in GO due to its low charge mobility. Charge transfer from graphene to defect states of ZnO NPs is the origin of the enhanced green defect emission of GO-ZnO. This book chapter contributes to a deeper understanding of electronic coupling effects in nanocarbon-nanoparticle hybrid materials which are promising for various electronic and optoelectronic applications.

CONSENT FOR PUBLICATION

Not applicable.

CONFLICT OF INTEREST

The author (editor) declares no conflict of interest, financial or otherwise.

ACKNOWLEDGEMENT

E. Erdem acknowledges financial support from Deutsche Forschungsgemeinschaft (DFG; Grant No.: Er662/1-2) and Smart Loire Valley General Programme for LE STUDIUM/Marie Skłodowska-Curie Research Fellowship.

REFERENCES

[1] Geim, A.; Novoselov, K. The rise of graphene. *Nat. Mater.*, **2007**, *6*, 183-191.

[2] Novoselov, K. Electric Field Effect in Atomically Thin Carbon Films. *Science,* **2004**, *306*, 666-669.

[3] Hummers, W.; Offeman, R. Preparation of Graphitic Oxide. *J. Am. Chem. Soc.,* **1958**, *80*, 1339-1339.

[4] Erdem, E. Microwave power, temperature, atmospheric and light dependence of intrinsic defects in ZnO nanoparticles: A study of electron paramagnetic resonance (EPR) spectroscopy. *J. Alloys Compd.,* **2014**, *605*, 34-44.

[5] Repp, S.; Weber, S.; Erdem, E. Defect Evolution of Nonstoichiometric ZnO Quantum Dots. *J. Phys. Chem. C,* **2016**, *120*, 25124-25130.

[6] Zaka, M.; Ito, Y.; Wang, H.; Yan, W.; Robertson, A.; Wu, Y.; Rümmeli, M.; Staunton, D.; Hashimoto, T.; Morton, J. Electron Paramagnetic Resonance Investigation of Purified Catalyst-free Single-Walled Carbon Nanotubes. *ACS Nano,* **2010**, *4*, 7708-7716.

[7] Bateni, A.; Erdem, E.; Repp, S.; Acar, S.; Kokal, I.; Häßler, W.; Weber, S.; Somer, M. *Electron paramagnetic resonance and Raman spectroscopy studies on carbon-doped MgB2 superconductor nanomaterials.,* **2015**.

[8] Kausteklis, J.; Cevc, P.; Arčon, D.; Nasi, L.; Pontiroli, D.; Mazzani, M.; Riccò, M. *Electron paramagnetic resonance study of nanostructured graphite.,* **2011**.

[9] Rao, S.; Stesmans, A.; Kosynkin, D.; Higginbotham, A.; Tour, J. *Paramagnetic centers in graphene nanoribbons prepared from longitudinal unzipping of carbon nanotubes.,* **2011**.

[10] Beckert, F.; Rostas, A.; Thomann, R.; Weber, S.; Schleicher, E.; Friedrich, C.; Mülhaupt, R. Self-Initiated Free Radical Grafting of Styrene Homo- and Copolymers onto Functionalized Graphene. *Macromolecules,* **2013**, *46*, 5488-5496.

[11] Panich, A.; Shames, A.; Sergeev, N. Paramagnetic Impurities in Graphene Oxide. *Appl. Magn. Reson.*, **2012**, *44*, 107-116.

[12] Su, C.; Acik, M.; Takai, K.; Lu, J.; Hao, S.; Zheng, Y.; Wu, P.; Bao, Q.; Enoki, T.; Chabal, Y.; Loh, K. *Probing the catalytic activity of porous graphene oxide and the origin of this behaviour.*, **2012**.

[13] Pham, C.; Eck, M.; Krueger, M. Thiol functionalized reduced graphene oxide as a base material for novel graphene-nanoparticle hybrid composites. *Chem. Eng. J.*, **2013**, *231*, 146-154.

[14] Eaton, G.; Eaton, S.; Barr, D.; Weber, R. Basics of Continuous Wave EPR.*Quantitative EPR*; Eaton, G.; Eaton, S.; Barr, D.; Weber, R., Eds.; Springer: Vienna, **2010**, pp. 1-14.

[15] Eaton, G.; Eaton, S.; Barr, D.; Weber, R. Filling Factor.*Quantitative EPR*; Eaton, G.; Eaton, S.; Barr, D.; Weber, R., Eds.; Springer: Vienna, **2010**, pp. 89-90.

[16] Panich, A.; Shames, A.; Aleksenskii, A.; Dideikin, A. Magnetic resonance evidence of manganese–graphene complexes in reduced graphene oxide. *Solid State Commun.*, **2012**, *152*, 466-468.

[17] Pham, C.; Krueger, M.; Eck, M.; Weber, S.; Erdem, E. Comparative electron paramagnetic resonance investigation of reduced graphene oxide and carbon nanotubes with different chemical functionalities for quantum dot attachment. *Appl. Phys. Lett.*, **2014**, *104*, 132102.

[18] Bagri, A.; Mattevi, C.; Acik, M.; Chabal, Y.; Chhowalla, M.; Shenoy, V. Structural evolution during the reduction of chemically derived graphene oxide. *Nat. Chem.*, **2010**, *2*, 581-587.

[19] Park, S.; Ruoff, R. Chemical methods for the production of graphenes. *Nat. Nanotechnol.*, **2009**, *4*, 217-224.

[20] Boeddinghaus, M.; Klein, W.; Wahl, B.; Jakes, P.; Eichel, R.; Fässler, T.C. 60 3- versus C 60 4- /C 60 2- - Synthesis and Characterization of Five Salts Containing Discrete Fullerene Anions. *Z. Anorg. Allg. Chem.*, **2014**, *640*, 701-712.

[21] Zhou, Y.; Eck, M.; Men, C.; Rauscher, F.; Niyamakom, P.; Yilmaz, S.; Dumsch, I.; Allard, S.; Scherf, U.; Krüger, M. Efficient polymer nanocrystal hybrid solar cells by improved nanocrystal composition. *Sol. Energy Mater. Sol. Cells,* **2011**, *95*, 3227-3232.

[22] Ren, S.; Chang, L.; Lim, S.; Zhao, J.; Smith, M.; Zhao, N.; Bulović, V.; Bawendi, M.; Gradečak, S. Inorganic–Organic Hybrid Solar Cell: Bridging Quantum Dots to Conjugated Polymer Nanowires. *Nano Lett.*, **2011**, *11*, 3998-4002.

[23] Ginger, D.; Greenham, N. Charge injection and transport in films of CdSe nanocrystals. *J. Appl. Phys.*, **2000**, *87*, 1361-1368.

[24] Huynh, W.; Dittmer, J.; Alivisatos, A. Hybrid Nanorod-Polymer Solar Cells. *Science,* **2002**, *295*, 2425-2427.

[25] Huynh, W.; Dittmer, J.; Teclemariam, N.; Milliron, D.; Alivisatos, A.; Barnham, K. *Charge transport in hybrid nanorod-polymer composite photovoltaic cells.*, **2003**.

[26] Shang, N.; Papakonstantinou, P.; Wang, P.; Silva, S. Platinum Integrated Graphene for Methanol Fuel Cells. *J. Phys. Chem. C*, **2010**, *114*, 15837-15841.

[27] Qiu, J.; Wang, G.; Liang, R.; Xia, X.; Yu, H. Controllable Deposition of Platinum Nanoparticles on Graphene As an Electrocatalyst for Direct Methanol Fuel Cells. *J. Phys. Chem. C,* **2011**, *115*, 15639-15645.

[28] Lin, Y.; Zhang, K.; Chen, W.; Liu, Y.; Geng, Z.; Zeng, J.; Pan, N.; Yan, L.; Wang, X.; Hou, J. Dramatically Enhanced Photoresponse of Reduced Graphene Oxide with Linker-Free Anchored CdSe Nanoparticles. *ACS Nano,* **2010**, *4*, 3033-3038.

[29] Wang, Y.; Liu, J.; Liu, L.; Sun, D. *High-quality reduced graphene oxide-nanocrystalline platinum hybrid materials prepared by simultaneous co-reduction of graphene oxide and chloroplatinic acid.*, **2011**.

[30] Gong, F.; Wang, H.; Wang, Z. Self-assembled monolayer of graphene/Pt as counter electrode for efficient dye-sensitized solar cell. Phys. Chem. Chem. Phys. 2011, 13, 17676(1-7).

[31] Stoller, M.; Park, S.; Zhu, Y.; An, J.; Ruoff, R. Graphene-Based Ultracapacitors. *Nano Lett.,* **2008**, *8*, 3498-3502.

[32] Lu, D.; Zhang, Y.; Lin, S.; Wang, L.; Wang, C. Sensitive detection of esculetin based on a CdSe nanoparticles-decorated poly(diallyldimethylammonium chloride)-functionalized graphene nanocomposite film. Analyst 2011, 136, 4447(1-7).

[33] Zedan, A.; Sappal, S.; Moussa, S.; El-Shall, M. Ligand-Controlled Microwave Synthesis of Cubic and Hexagonal CdSe Nanocrystals Supported on Graphene. Photoluminescence Quenching by Graphene. *J. Phys. Chem. C,* **2010**, *114*, 19920-19927.

[34] Wang, Y.; Yao, H.; Wang, X.; Yu, S. One-pot facile decoration of CdSe quantum dots on graphene nanosheets: novel graphene-CdSe nanocomposites with tunable fluorescent properties. *J. Mater. Chem.,* **2011**, *21*, 562-566.

[35] Shen, J.; Shi, M.; Li, N.; Yan, B.; Ma, H.; Hu, Y.; Ye, M. Facile synthesis and application of Ag-chemically converted graphene nanocomposite. *Nano Res.,* **2010**, *3*, 339-349.

[36] Mann, J.; Watson, D. Adsorption of CdSe Nanoparticles to Thiolated TiO2Surfaces: Influence of Intralayer Disulfide Formation on CdSe Surface Coverage. *Langmuir,* **2007**, *23*, 10924-10928.

[37] Colvin, V.; Goldstein, A.; Alivisatos, A. Semiconductor nanocrystals covalently bound to metal surfaces with self-assembled monolayers. *J. Am. Chem. Soc.,* **1992**, *114*, 5221-5230.

[38] Eck, M.; Pham, C.; Züfle, S.; Neukom, M.; Sessler, M.; Scheunemann, D.; Erdem, E.; Weber, S.; Borchert, H.; Ruhstaller, B. Improved efficiency of bulk heterojunction hybrid solar cells by utilizing CdSe quantum dot–graphene nanocomposites. *Phys. Chem. Chem. Phys.,* **2014**, *16*, 12251-12260.

[39] Ando, K. Seeking Room-Temperature Ferromagnetic Semiconductors. *Science,* **2006**, *312*, 1883-1885.

[40] Son, D.; Kwon, B.; Park, D.; Seo, W.; Yi, Y.; Angadi, B.; Lee, C.; Choi, W. Emissive ZnO–graphene quantum dots for white-light-emitting diodes. *Nat. Nanotechnol.,* **2012**, *7*, 465-471.

[41] Özgür, Ü.; Alivov, Y.; Liu, C.; Teke, A.; Reshchikov, M.; Doğan, S.; Avrutin, V.; Cho, S.; Morkoç, H. *A comprehensive review of ZnO materials and devices.,* **2005**.

[42] Sun, X.; Huang, J.; Wang, J.; Xu, Z. A ZnO Nanorod Inorganic/Organic Heterostructure Light-Emitting Diode Emitting at 342 nm. *Nano Lett.,* **2008**, *8*, 1219-1223.

[43] Cole, J.; Wang, X.; Knuesel, R.; Jacobs, H. Integration of ZnO Microcrystals with Tailored Dimensions Forming Light Emitting Diodes and UV Photovoltaic Cells. *Nano Lett.,* **2008**, *8*, 1477-1481.

[44] Wang, Z.; Zhan, X.; Wang, Y.; Muhammad, S.; Huang, Y.; He, J. A flexible UV nanosensor based on reduced graphene oxide decorated ZnO nanostructures. *Nanoscale,* **2012**, *4*, 2678-2684.

[45] Song, W.; Kwon, S.; Myung, S.; Jung, M.; Kim, S.; Min, B.; Kang, M.; Kim, S.; Lim, J.; An, K. *High-mobility ambipolar ZnO-graphene hybrid thin film transistors.,* **2014**.

[46] Novoselov, K. Fal'ko, V.; Colombo, L.; Gellert, P.; Schwab, M.; Kim, K. A roadmap for graphene. *Nature,* **2012**, *490*, 192-200.

[47] Bae, S.; Kim, H.; Lee, Y.; Xu, X.; Park, J.; Zheng, Y.; Balakrishnan, J.; Lei, T.; Ri Kim, H.; Song, Y. Roll-to-roll production of 30-inch graphene films for transparent electrodes. *Nat. Nanotechnol.,* **2010**, *5*, 574-578.

[48] Wang, H.; Hu, Y. Graphene as a counter electrode material for dye-sensitized solar cells. *Energy Environ. Sci.,* **2012**, *5*, 8182-8188.

[49] Sahoo, N.; Pan, Y.; Li, L.; Chan, S. Graphene Based Materials for Energy Conversion. *Adv. Mater.,* **2012**, *24*, 4203-4210.

[50] Lee, E.; Kim, J.; Kwon, B.; Jang, E.; An, S. Vacancy filling effect of graphene on photoluminescence behavior of ZnO/graphene nanocomposite. *Phys. Status Solidi Rapid Res. Lett.*, **2014**, *8*, 836-840.

[51] Kavitha, M.; Gopinath, P.; John, H. Reduced graphene oxide–ZnO self-assembled films: tailoring the visible light photoconductivity by the intrinsic defect states in ZnO. *Phys. Chem. Chem. Phys.*, **2015**, *17*, 14647-14655.

[52] Pham, C.; Repp, S.; Thomann, R.; Krueger, M.; Weber, S.; Erdem, E. Charge transfer and surface defect healing within ZnO nanoparticle decorated graphene hybrid materials. *Nanoscale,* **2016**, *8*, 9682-9687.

[53] Kaftelen, H.; Ocakoglu, K.; Thomann, R.; Tu, S.; Weber, S.; Erdem, E. *EPR and photoluminescence spectroscopy studies on the defect structure of ZnO nanocrystals.,* **2012**.

[54] Rao, S.; Stesmans, A.; Wang, Y.; Chen, Y. Direct ESR evidence for magnetic behavior of graphite oxide. *Physica E,* **2012**, *44*, 1036-1039.

[55] Song, N.; Fan, H.; Tian, H. Reduced graphene oxide/ZnO nanohybrids: Metallic Zn powder induced one-step synthesis for enhanced photocurrent and photocatalytic response. *Appl. Surf. Sci.,* **2015**, *353*, 580-587.

[56] Kumar, P.; Bernardi, M.; Grossman, J. The Impact of Functionalization on the Stability, Work Function, and Photoluminescence of Reduced Graphene Oxide. *ACS Nano,* **2013**, *7*, 1638-1645.

CHAPTER 6

Electron Spin Resonance Spectroscopy of Single-Walled Carbon-Nanotube Thin-Films and their Transistors

Kazuhiro Marumoto*

Division of Materials Science, University of Tsukuba, Tsukuba, Ibaraki 305-8573, Japan, Tsukuba Research Center for Energy Materials Science (TREMS), University of Tsukuba, Tsukuba, Ibaraki 305-8571, Japan

Abstract: Carbon nanotubes (CNTs) have been extensively studied due to their excellent properties such as ballistic transport. Electrically induced charge carriers in CNTs and the relation between the spin states and the ballistic transport, however, have not yet been microscopically investigated owing to experimental difficulties. Here we review electron spin resonance (ESR) spectroscopy of semiconductor single-walled CNT (SW-CNT) thin films and their transistors. We have investigated the spin states and the electrically induced charge carriers in the SW-CNTs by utilizing a transistor structure under device operation. The electrically induced ESR method is useful for the microscopic investigation into CNTs because it is capable of directly observing the spins in CNTs. We have observed a clear reverse correlation between the ESR intensity and the transistor current under high charge-density conditions. This result directly demonstrates electrically induced ambipolar spin vanishment in CNTs, providing a first clear evidence of antiparallel spin fillings of the electrically induced charges' spins and the vacancies' spins in CNTs. The ambipolar spin vanishment is considered to improve the transport properties of CNTs because it seems to greatly reduce carrier scatterings. Similar spin vanishment has been observed in single-layer graphene transistors. Thus, this result suggests that the electrically induced ambipolar spin vanishment is a universal phenomenon for carbon materials.

Keywords: Ambipolar spin vanishment, Anisotropy, Antiparallel spin fillings, Carbon nanotubes (CNTs), Charge carriers, Electron spin resonance (ESR), Gate voltage, Number of spins, Single-walled CNTs (SW-CNTs), SW-CNT thin films, SW-CNT transistor, Spin-orbital interaction, Spin states, Spin susceptibility, Temperature dependence, Transport properties, Thin films, Tomonaga-Luttinge--liquid (TLL) states.

* **Corresponding author Kazuhiro Marumoto:** Division of Materials Science, University of Tsukuba, Tsukuba, Ibaraki 305-8573, Japan; Tel: +81-29-853-5117; Fax: +81-29-853-4490; E-mail: marumoto@ims.tsukuba.ac.jp

Dariya Savchenko & Abdel Hadi Kassiba (Eds.)

INTRODUCTION

Carbon nanotubes (CNTs) have been extensively studied because of their various excellent properties such as ballistic transport; the studies of field-effect transistors (FETs) have shown high charge carrier mobility owing to the ballistic transport [1]. The different chirality makes single-walled CNTs (SW-CNTs) to be either metallic (w/o band gap) or semiconducting (w/ band gap) [2, 3]. The correlation between SW-CNTs' structure and interesting electronic properties with diversity is able to fabricate devices with superior performance and functionality compared to silicon-based devices [1, 4 - 9]. For understanding the nature of the ballistic transport completely, a study of the spin states in CNTs is indispensable because the carrier scattering due to spin interactions prevents the ballistic transport to be realized. Notably, a study of the spin states from a microscopic viewpoint under device operation is crucially effective because such study is considered to elucidate the microscopic properties of electrically induced charge carriers in devices and has a direct relation to the device performance. Density functional theory (DFT) calculations with spin polarization have indicated spin formation in SW-CNTs at around their atomic vacancies [10]. The spins due to atomic vacancies seem to be intrinsically formed in CNTs [10]. The relation between the device performance and the spin formation, however, has not yet been studied. The study is considered to provide an insight into the mechanism of the realization of the ballistic transport.

One of the most effective techniques to investigate the spin states in CNT devices is electron spin resonance (ESR) spectroscopy using transistor structures [11 - 16]. The method is a most clean technique because it is capable of directly observing electrically induced charge carriers without chemically changing CNTs. By using the ESR technique, various microscopic properties such as the spin states and the local structures of materials and devices have been clarified [11 - 16]. The ESR study of the spin states in CNT devices under device operation has been expected to be performed for a decade. However, no ESR study of CNT devices has been performed because the experimental difficulties, such as low charge-accumulation ability due to conventional solid insulators and dielectric loss due to metallic thick thin-film CNTs, have not yet been overcome.

In this chapter, we review an ESR study of SW-CNT devices [17]. We utilize an ion-gel insulator in the device structures to overcome the low charge-accumulation ability [13, 14, 18 - 25]. Semiconducting thin-film SW-CNTs are used to prevent the dielectric loss from metallic CNTs [26 - 28]. We perform simultaneous measurements of the ESR and the transistor characteristics, which firstly show a clear reverse correlation between the ESR intensity and the current under high charge density. We directly discover electrically induced ambipolar spin vanishment in the SW-CNTs, which provides a clear evidence of antiparallel spin fillings of electrically induced charges' spins and vacancies' spins in the SW-CNTs. The spin vanishment is considered to improve the transport properties of SW-CNTs because it is capable of reducing carriers' spin scatterings. Our results suggest that the spin vanishment is a universal phenomenon for carbon materials because similar ambipolar spin vanishment has been observed for single-layer graphene transistors [29].

EXPERIMENTAL

SW-CNT Thin Film and Transistor Fabrication

Semiconducting SW-CNTs were used in this study, which were purified from SW-CNTs fabricated by an Arc discharge method (Meijyo Nanocarbon Co., Arc SO). The SW-CNTs were dispersed into 1 or 2% deoxycholate sodium salt (DOC) solutions, and semiconducting SW-CNTs with high quality were selected by a density gradient ultracentrifugation (DGU) method [26 - 28]. The SW-CNT solutions were washed with methanol, hot water, and toluene. Then, the 300-nm-thickness thin films were fabricated on the substrate. The SW-CNT diameters of 1.37–1.4 nm were measured by an atomic force microscope. The further fabrication process is described in previous works [26 - 28].

For the SW-CNT transistor fabrication, a nonmagnetic quartz substrate (Iiyama Precision Glass Co., Ltd.) was used, whose dimension is 30 mm × 3 mm × 1 mm. The SW-CNT thin films with a dimension of 15 mm × 2 mm × 300 nm were fabricated on the substrate. The Ni/Au (3 nm/47 nm) electrodes of the source, drain, and gate were fabricated with a vapor-deposition method on the substrate with the SW-CNTs. The source and drain electrodes have a 0.5 mm channel length and a ~15 mm channel width. An ion-gel insulator was used, which is a solid polymer electrolyte composed of ionic liquid 1-ethyl-3-methylimidazolium bis(trifluoromethylsulfonyl)imide ([EMIM][TFSI]) and a triblock copolymer poly(styrene-*b*-methylmethacrylate-*b*-styrene) (PS-PMMA-PS). The ion gel shows high ionic conductivity and large EDL capacitance [21, 22]. The large EDL capacitances (~10-100 μF cm^{-2}) lead to significant charge accumulation with high on/off current ratios and low voltage [21, 22]. The further details for ion gels are described in previous works [18 - 25]. The ion-gel film was fabricated on the substrate with SW-CNTs using a drop-casting method, which completed the transistor structure. The fabricated device was inserted into an ESR sample tube, which was sealed after wirings under vacuum conditions or under He exchange gas at 100 torr.

ESR and Transfer Characteristic Measurements

The ESR measurements were conducted with a JEOL RESONANCE JES-FA200 X-band spectrometer under vacuum conditions or under He exchange gas at 100 torr. The ESR parameters g factor and linewidth were calibrated with a standard Mn^{2+} marker sample. The spin-susceptibility, χ, was obtained by doubly integrating the ESR signal and by calibrating it with the Mn^{2+} marker sample. The number of spins, N_{spin}, is obtained by doubly integrating the ESR signal and by comparing the N_{spin} with that of the Mn^{2+} markcr sample assuming the Curie law. The N_{spin} of the Mn^{2+} marker sample was calculated with a solution (220 μL) of 4-hydroxy-2,2,6,6-tetramethylpiperidin-1-oxyl (TEMPOL) as a standard. The experimentally obtained g factor was calibrated with a software program of the ESR system of the JEOL RESONANCE considering high second-order correction to the effective resonance field. Its correctness was confirmed using 2,2-diphenyl-1-picrylhydrazyl (DPPH) as another standard sample. Simultaneous measurements of the characteristics ESR and device performance were conducted with a Keithley 2612A source meter.

RESULTS AND DISCUSSION

Gate Voltage Dependence of ESR of SW-CNT Transistors

In this subsection, the ambipolar spin vanishment in SW-CNTs is discussed, which was obtained with a gate voltage (V_G) application. The device structure of the SW-CNT transistor used in this study is shown in Fig. (**1a**). The ion-gel insulator has attracted much attention due to their low voltage operation by the formation of an electric double layer (EDL) at the interfaces between semiconductor and insulator [13, 14, 18 - 25].

The ESR spectra of the SW-CNT transistor at V_G values of 0, 0.6, and 3.2 V with a drain voltage V_D = 0.2 V are shown in Fig. (**2a**). The ESR spectra exhibit a Lorentzian ESR lineshape and have ESR parameters g factor (2.0029 ± 0.0002) and peak-to-peak ESR linewidth ΔH_{pp} (0.6 ± 0.1 mT). These lineshape and parameters show no V_G dependence. As discussed later, the origin of the ESR signal is attributed to atomic vacancies in the SW-CNTs [10]. It is noted that the ESR intensity decreases with increasing the V_G from 0 to 3.2 V.

Electrically Induced Ambipolar Spin Vanishment in SW-CNTs

In this subsection, the correlation between the ESR intensity and the transistor characteristics is discussed as a function of V_G. The number of spins, N_{spin}, is used to discuss the ESR intensity of the SW-CNT transistor. The V_G dependence of the N_{spin} and the drain current (I_D) of the SW-CNT transistor in positive and negative

V_G is shown in Fig. (**2b,c**), which were simultaneously measured with the same device.

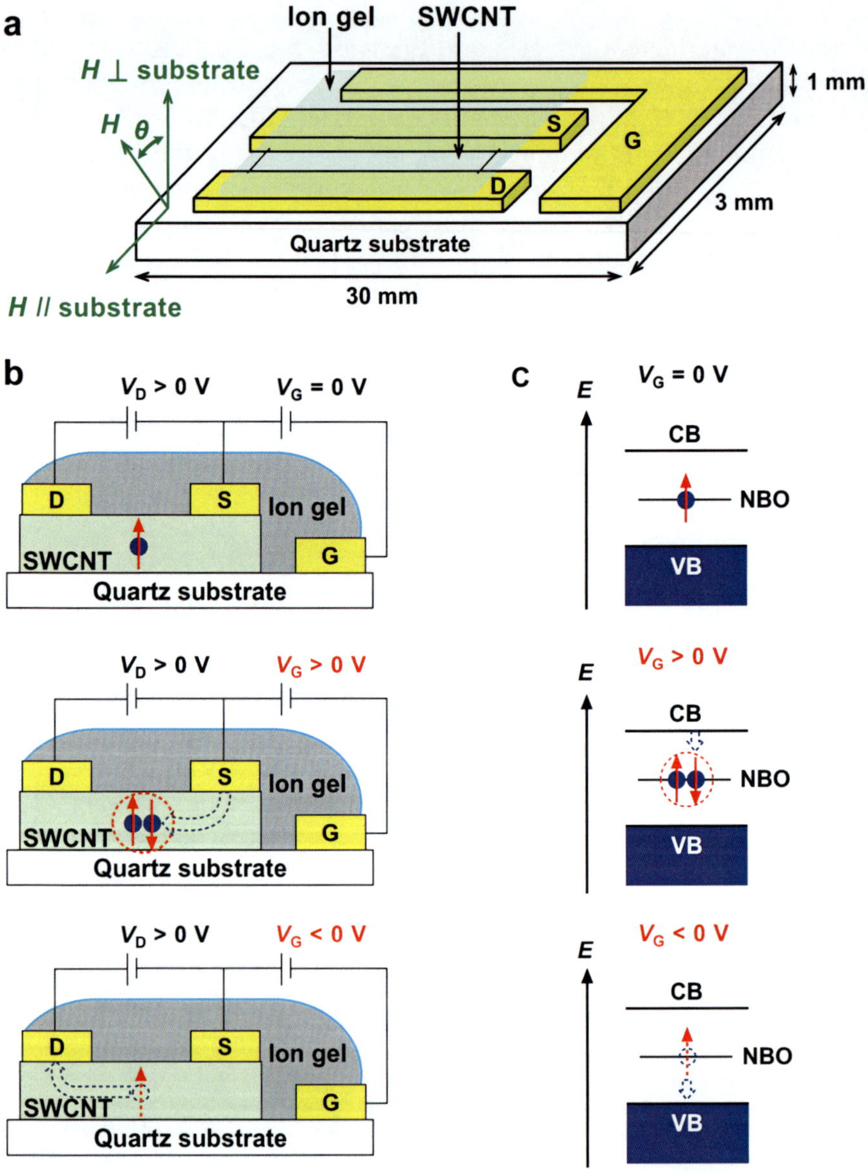

Fig. (1). (a) Schematic of the SW-CNT transistor structure. (b) Schematics of the cross section of the transistor structure with the spin states at $V_G = 0$ V (top), for $V_G > 0$ V (center), and for $V_G < 0$ V (bottom). (c) The energy diagram of SW-CNTs of the conduction band (CB), the valence band (VB), and the non-bonding orbital (NBO) with the spin states at $V_G = 0$ V (top), for $V_G > 0$ V (center), and for $V_G < 0$ V (bottom).

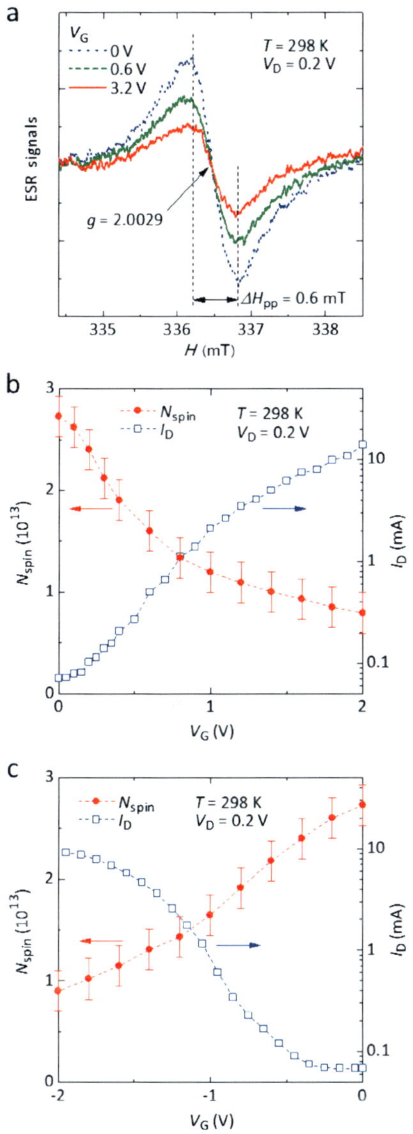

Fig. (2). (a) ESR spectra of the SW-CNT transistor when V_G is 0 V (blue dotted line), 0.6 V (green dashed line), and 3.2 V (red solid line). Here, V_D is 0.2 V, and the external magnetic field H is perpendicular to the substrate ($H_\perp$) at 298 K. (b,c) V_G dependence of the number of spins, N_{spin}, and the I_D of the SW-CNT transistor for (b) $V_G \geq 0$ V and (c) $V_G \leq 0$ V. Here, $V_D = 0.2$ V at 298 K at $H_\perp$, and the I_D data are plotted for a semilogarithm scale.

The I_D increases when $|V_G|$ increases in both positive and negative V_G, which demonstrates ambipolar transistor operation. This behavior is consistent with that of previous studies [18, 19, 23, 25]. In contrast to the I_D increase, it is noted that the N_{spin} clearly decreases when $|V_G|$ increases in both positive and negative V_G. Therefore, a clear reverse correlation between the N_{spin} and the I_D is observed for both electron and hole accumulation in the SW-CNT transistor.

Therefore, electrically induced ambipolar spin vanishment in the SW-CNTs is directly demonstrated. No electrically induced spin vanishment has been reported except for our studies [17, 29]. This result shows the formation of non-magnetic states for vacancies' spins. This is an unexpected result because organic transistors always show that N_{spin} increases when $|V_G|$ increases due to charge accumulation in devices [11 - 14].

Next, the mechanism of the electrically induced ambipolar spin vanishment is discussed. For positive V_G, the spin vanishment can be explained by the antiparallel spin fillings of vacancies' spins and electrically induced electrons' spins by V_G in SW-CNTs, as schematically shown in the center figure of Fig. (**1b,c**). The present result firstly demonstrates a clear evidence of the antiparallel spin fillings of these spins. For negative V_G, the spin vanishment is explained by the discharge of vacancies' spins owing to hole doping by V_G, as schematically shown in the bottom figure of Fig. (**1b,c**).

Notably, the ESR spectrum did not change when the V_D was changed. Supposed that electrically induced charges are the origin of the ESR signal, the ESR intensity should be changed when the V_D is changed because of the V_D dependence of the number of electrically induced charges in a transistor channel under gradual-channel approximation [30, 31]. Thus, no V_D dependence of the ESR signal means that electrically induced charges are not the origin of the observed ESR signal.

Similar spin vanishment has been observed for single-layer graphene transistors [29]. Fig. (**3a,b**) show the dependence of the N_{spin} of the graphene transistor and the SW-CNT transistor on $V_G - V_{CNP}$, respectively, where V_{CNP} is defined as the voltage of the charge-neutral point. The N_{spin} of single-layer graphene with an ideal two-dimensionality has the maximum at the $V_G = V_{CNP}$ and decreases for $V_G < V_{CNP}$ and $V_G > V_{CNP}$, respectively, as similarly observed in the SW-CNTs with a one-dimensionality. Therefore, our results suggest that the electrically induced ambipolar spin vanishment is a universal phenomenon for carbon materials with honeycomb structures regardless of the dimensionality of the electron systems.

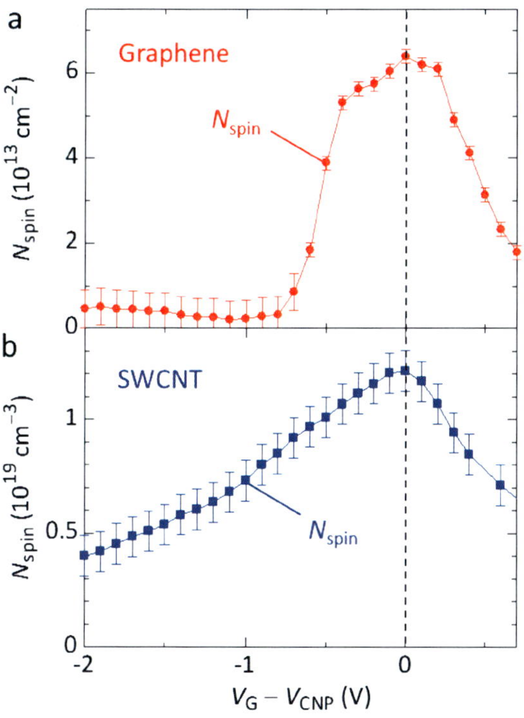

Fig. (3). Dependence of the number of spins, N_{spin}, due to atomic vacancies in the transistors of (a) graphene and (b) SW-CNTs on $V_G - V_{CNP}$.

No ESR Observation in Tomonaga-Luttinger Liquid

In this subsection, we discuss the reason for the non-observation of an ESR signal due to electrically induced charges by considering the formation of Tomonaga-Luttinger-liquid (TLL) states [32, 33]. From the TLL theory with spin-symmetry breaking and electron-electron interactions in one-dimensional (1D) electron system, the ESR linewidth in the TLL states has been shown to become extremely broad, even at low temperatures such as 4 K with the value of the order of 100 mT, and the ESR linewidth is found to be further broadened at higher temperatures [32, 33].

The low-temperature ESR measurements below 4.2 K were performed with V_G application. However, any additional ESR signal due to electrically induced charges was not observed because the ESR linewidth is considered to be extremely broadened as mentioned above. Thus, the formation of the TLL states in the SW-CNT transistor is demonstrated from a microscopic viewpoint by the present ESR study, even for bundled SW-CNTs having relaxed 1D character. It is commented that the origin of the ESR signals of CNTs reported in previous studies may be attributed to localized paramagnetic states (atomic vacancies, tube ends, etc.) and/or extrinsic impurities (catalysts etc.) rather than delocalized charge carriers in the CNTs [32, 34 - 47].

Microscopic Investigation into Atomic Vacancies in SW-CNTs

In this subsection, the microscopic investigation into atomic vacancies, such as the spin concentration, the spin-lattice relaxation time, the motion of vacancies, and the anisotropic spin-orbit interaction, is further discussed. The ESR results of 300-nm-thickness SW-CNT thin films are presented because of the higher signal-to-noise (SN) ratio of the ESR signal than that of the SW-CNT transistor owing to the absence of the dielectric loss from wirings and electrodes. The ESR spectra of the SW-CNT thin film at 15, 60, and 290 K are shown by the solid lines in Fig. (**4a**). The g factor was obtained as 2.0029 ± 0.0002 for each temperature. The ΔH_{pp} value was obtained as 0.36 ± 0.03 mT at 15 K, 0.33 ± 0.03 mT at 60 K, and 0.50 ± 0.05 mT at 290 K, respectively. The least-squares fits with a Lorentzian lineshape have been performed for the observed ESR spectra, which are shown by the open symbols in Fig. (**4a**). For all measured temperatures, the experimental results are well explained by a single Lorentzian lineshape. The previous ESR studies of CNTs have observed a complicated Dysonian lineshape owing to the skin effects of microwave absorption from bulk or thick-film samples [35, 36, 40, 43, 44]. In contrast, the use of thin films in this work is found to be useful for avoiding such skin effects and for performing a precise ESR analysis with a simple Lorentzian lineshape, as described below.

Next, the temperature dependence of the ESR signal of the SW-CNT thin film is discussed by presenting the spin susceptibility, χ, and the ESR linewidth. The temperature dependence of χ and the inverse spin susceptibility, $1/\chi$, are shown in Fig. (**4b**) and its inset, respectively. The least-squares fits using the Curie law for χ and $1/\chi$ are shown by the solid lines in Fig. (**4b**) and its inset, respectively. The experimental results are well explained by the fitting curves. Thus, the χ is found to obey the Curie law. From this result, the origin of the observed ESR signal is further confirmed as magnetically isolated spin species, not as Pauli paramagnetism due to charge carriers. This result is consistent with the isolated spin formation at atomic vacancies in the SW-CNT thin film [10, 46].

The spin concentration of the SW-CNT thin film was evaluated using the N_{spin} and the volume of the film. The spin concentration was obtained as 1.1×10^{18} cm^{-3}. The average distance between spins is estimated to be 9.7 nm from this concentration.

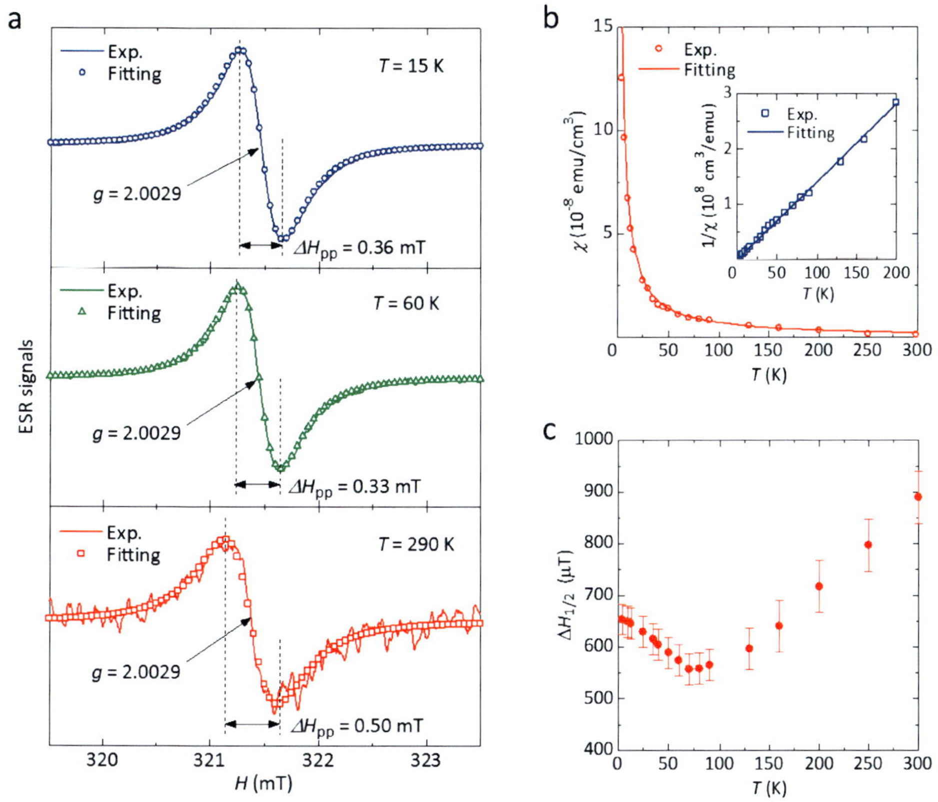

Fig. (4). (a) ESR spectra of the SW-CNT thin film at 15 K (upper), 60 K (center), and 290 K (bottom) at $H_\perp$. The least-squares fits with a Lorentzian lineshape to the observed ESR spectra are shown by circles (upper), triangles (center), and squares (bottom), respectively. (b) Temperature dependence of the spin susceptibility, χ, (main panel) and the inverse spin susceptibility, $1/\chi$, (inset) of the ESR signal of the SW-CNT thin film at $H_\perp$. The least-squares fits using the Curie law to the data are shown by the solid lines. (c) Temperature dependence of the full width at half-maximum, $\Delta H_{1/2}$, of the ESR signal of the SW-CNT thin film at $H_\perp$.

Fig. (**4c**) shows the temperature dependence of the ESR linewidth of the SW-CNT thin film. In this plot, the full width at half maximum of the ESR linewidth, $\Delta H_{1/2}$, is used. The $\Delta H_{1/2}$ increased when the temperature increased above ~70 K (high-temperature region). The increase in the $\Delta H_{1/2}$ with increasing temperature is explained by the decrease in the spin-lattice relaxation time, T_1, of the vacancies' spins because the T_1 value at high temperatures is decreased due to the phonon modes in the SW-CNTs and an ESR linewidth with a Lorentzian lineshape is described by $1/T_1$, that is, $\Delta H_{1/2} \propto 1/T_1$ (Ref [47]). The observed Lorentzian ESR lineshape at high temperatures is reasonably explained by the spin-lattice-relaxation's effects on the ESR linewidth. In contrast, the $\Delta H_{1/2}$ decreased when the temperature increased below ~70 K (low-temperature region). The decrease in the $\Delta H_{1/2}$ with increasing temperature is explained by the motional-narrowing effect on the ESR linewidth owing to the motion of the vacancies' spins, in other words, the motion of vacancies in the SW-CNT thin film [10]. The observed Lorentzian ESR lineshape at low temperatures is reasonably explained by the motional narrowing of the ESR linewidth.

Anisotropic Spin-orbital Interaction in SW-CNTs

In this subsection, the anisotropic ESR spectrum for the direction of the external magnetic field *H* relative to the substrate is discussed to investigate the anisotropic spin-orbital interaction in the SW-CNT thin film. To improve the SN ratio of the ESR spectrum, the measurements were performed at 50 K. The anisotropic *g* factor and $\Delta H_{1/2}$ of the ESR spectrum are shown in Fig. (**5a,b**), respectively. Here θ is the angle between *H* and the normal of the substrate, as shown in Fig. (**1a**). For the *g* factor and the $\Delta H_{1/2}$, clear anisotropy is firstly observed in the ESR studies of the CNTs. The least-squares fits with the following fitting functions for the observed *g* factor and $\Delta H_{1/2}$ are shown by solid lines in Fig. (**5a,b**), respectively:

$$g = \sqrt{g_{\perp}^2 \cos^2\theta + g_{\parallel}^2 \sin^2\theta}, \quad \textbf{(1)}$$

$$\Delta H_{1/2} = \sqrt{\Delta H_{\perp}^2 \cos^2\theta + \Delta H_{\parallel}^2 \sin^2\theta}. \quad \textbf{(2)}$$

Here, $g_{\perp} = 2.002913$ and $\Delta H_{\perp} = 590$ μT at *H* perpendicular to the substrate and $g_{\parallel} = 2.00279$ and $\Delta H_{\parallel} = 455$ μT at *H* parallel to the substrate. The experimental results are well explained by the fitting functions. Thus, this result demonstrates the uniaxial anisotropic *g* factor and $\Delta H_{1/2}$ of the SW-CNT thin film.

Here we define λ, $\boldsymbol{L}$, and $\boldsymbol{S}$ as the spin-orbit coupling constant, the orbital angular momentum, and the spin angular momentum for the vacancies' spins. Then, the observed anisotropic properties reveal an anisotropic spin-orbit interaction $W_{LS} = \lambda(\boldsymbol{L}\cdot\boldsymbol{S})$ of vacancies' spins with respect to SW-CNTs' surface because the *g* factor is determined by the second-order perturbation of W_{LS} and the ESR linewidth is affected by the spin-lattice relaxation due to the W_{LS} [48].

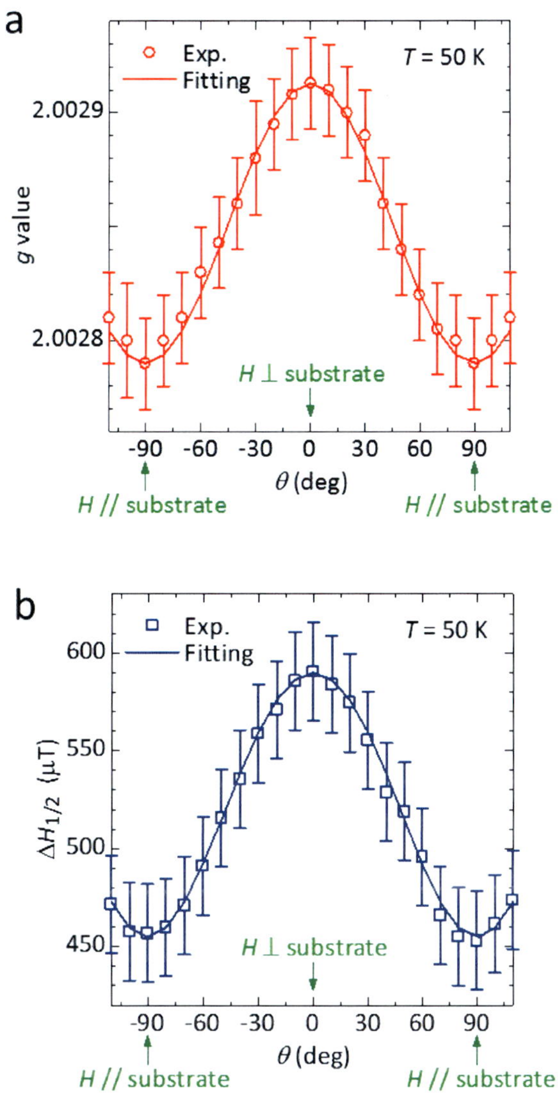

Fig. (5). Dependence of (a) the g factor and (b) the $\Delta H_{1/2}$ of the ESR spectrum of the SW-CNT thin film on the angle between H and the normal of the substrate at 50 K.

It is commented that the observed anisotropy is qualitatively consistent with that of graphite [49]. Thus, the origin of the observed ESR signal is further confirmed as not due to extrinsic impurities such as catalysts, tube ends of the SW-CNTs, *etc.*

CONCLUSIONS

An ESR study of the SW-CNT transistor was performed to investigate the relation between the spin states and the device performance under device operation [17]. The electrically induced ambipolar spin vanishment in SW-CNTs is firstly demonstrated from a microscopic viewpoint. Thus, this demonstration shows the formation of the non-magnetic states of atomic vacancies in the SW-CNT transistor. This non-magnetization is caused by the antiparallel spin fillings of vacancies' spins and electrically induced electrons' spins in positive V_G and by the discharge of the vacancies' spins in negative V_G.

The ambipolar spin vanishment is considered to improve the transport properties of CNTs because it is capable of reducing carrier scatterings from spin interactions. Since similar electrically induced ambipolar spin vanishment has been observed for single-layer graphene transistors [29], this spin vanishment seems a universal phenomenon for carbon materials. The formation of the TLL states in the SW-CNTs explains the non-observation of an ESR signal from electrically induced charges. The microscopic investigation into atomic vacancies, such as the spin concentration, the spin-lattice relaxation time, the motion of vacancies, and the anisotropy of the spin-orbit interaction in the SW-CNT thin film, is further presented. Notably, an anisotropic ESR signal is observed for the SW-CNT thin film for the first time. The microscopic investigation and the ambipolar spin vanishment is useful for understanding the material properties of SW-CNTs and for fabricating high-performance devices based on SW-CNTs.

CONSENT FOR PUBLICATION

Not applicable.

CONFLICT OF INTEREST

The author (editor) declares no conflict of interest, financial or otherwise.

ACKNOWLEDGEMENT

We thank D. Matsumoto, K. Yanagi, T. Takenobu, S. Okada, N. Fujita, Y. Sakurai, K. Kawahara, H. Ago, M. Tsuji, D. Son, Y. Nobusa, and Y. Yomogida for their collaboration. This work was partially supported by JSPS KAKENHI Grant Numbers 24560004 and 15K13329, by JST, PRESTO, and by JST, ALCA.

REFERENCES

[1] Javey, A.; Guo, J.; Wang, Q.; Lundstrom, M.; Dai, H. Ballistic carbon nanotube field-effect transistors. *Nature,* **2003**, *424*(6949), 654-657. [http://dx.doi.org/10.1038/nature01797] [PMID: 12904787]

[2] Dresselhaus, M. Nanotechnology: New tricks with nanotubes. *Nature,* **1998**, *391*, 19-20. [http://dx.doi.org/10.1038/34036]

[3] Ouyang, M.; Huang, J.L.; Lieber, C.M. Fundamental electronic properties and applications of single-walled carbon nanotubes. *Acc. Chem. Res.,* **2002**, *35*(12), 1018-1025. [http://dx.doi.org/10.1021/ar0101685] [PMID: 12484789]

[4] McEuen, P. Nanotechnology: Carbon-based electronics. *Nature,* **1998**, *393*, 15-17. [http://dx.doi.org/10.1038/29874]

[5] Avouris, P. Molecular electronics with carbon nanotubes. *Acc. Chem. Res.,* **2002**, *35*(12), 1026-1034. [http://dx.doi.org/10.1021/ar010152e] [PMID: 12484790]

[6] Graham, A.; Duesberg, G.; Hoenlein, W.; Kreupl, F.; Liebau, M.; Martin, R.; Rajasekharan, B.; Pamler, W.; Seidel, R.; Steinhoegl, W. How do carbon nanotubes fit into the semiconductor roadmap? *Appl. Phys., A Mater. Sci. Process.,* **2005**, *80*, 1141-1151. [http://dx.doi.org/10.1007/s00339-004-3151-7]

[7] Franklin, A.D.; Chen, Z. Length scaling of carbon nanotube transistors. *Nat. Nanotechnol.,* **2010**, *5*(12), 858-862. [http://dx.doi.org/10.1038/nnano.2010.220] [PMID: 21102468]

[8] Avouris, P.; Martel, R. Progress in Carbon Nanotube Electronics and Photonics. *MRS Bull.,* **2010**, *35*, 306-313. [http://dx.doi.org/10.1557/mrs2010.553]

[9] Vijayaraghavan, A.; Hennrich, F.; Stürzl, N.; Engel, M.; Ganzhorn, M.; Oron-Carl, M.; Marquardt, C.W.; Dehm, S.; Lebedkin, S.; Kappes, M.M.; Krupke, R. Toward single-chirality carbon nanotube device arrays. *ACS Nano,* **2010**, *4*(5), 2748-2754. [http://dx.doi.org/10.1021/nn100337t] [PMID: 20408580]

[10] Ma, Y.; Lehtinen, P.; Foster, A.; Nieminen, R. Magnetic properties of vacancies in graphene and single-walled carbon nanotubes. *New J. Phys.,* **2004**, *6*, 68. [http://dx.doi.org/10.1088/1367-2630/6/1/068]

[11] Marumoto, K.; Kuroda, S.; Takenobu, T.; Iwasa, Y. Spatial Extent of Wave Functions of Gate-Induced Hole Carriers in Pentacene Field-Effect Devices as Investigated by Electron Spin Resonance. *Phys. Rev. Lett.,* **2006**, *97*, 256603(1-4). [http://dx.doi.org/10.1103/PhysRevLett.97.256603]

[12] Marumoto, K.; Arai, N.; Goto, H.; Kijima, M.; Murakami, K.; Tominari, Y.; Takeya, J.; Shimoi, Y.; Tanaka, H.; Kuroda, S. Microscopic mechanisms behind the high mobility in rubrene single-crystal transistors as revealed by field-induced electron spin resonance. *Phys. Rev. B.,* **2011**, *83*, 075302. [http://dx.doi.org/10.1103/PhysRevB.83.075302]

[13] Tsuji, M.; Takahashi, Y.; Sakurai, Y.; Yomogida, Y.; Takenobu, T.; Iwasa, Y.; Marumoto, K. Two-dimensional magnetic interactions and magnetism of high-density charges in a polymer transistor. *Appl. Phys. Lett.,* **2013**, *102*, 133301(1-5). [http://dx.doi.org/10.1063/1.4800550]

[14] Takahashi, Y.; Tsuji, M.; Yomogida, Y.; Takenobu, T.; Iwasa, Y.; Marumoto, K. Electron Spin Resonance Study of Organic Interfaces in Ion Gel-Gated Rubrene Single-Crystal Transistors. *Appl. Phys. Express,* **2013**, *6*, 041603. [http://dx.doi.org/10.7567/APEX.6.041603]

[15] Marumoto, K.; Fujimori, T.; Ito, M.; Mori, T. Charge Formation in Pentacene Layers During Solar-Cell Fabrication: Direct Observation by Electron Spin Resonance. *Adv. Energy Mater.,* **2012**, *2*, 591-597. [http://dx.doi.org/10.1002/aenm.201100774]

[16] Nagamori, T.; Marumoto, K. Direct observation of hole accumulation in polymer solar cells during device operation using light-induced electron spin resonance. *Adv. Mater.,* **2013**, *25*(16), 2362-2367.

[http://dx.doi.org/10.1002/adma.201204015] [PMID: 23447406]

[17] Matsumoto, D.; Yanagi, K.; Takenobu, T.; Okada, S.; Marumoto, K. Electrically induced ambipolar spin vanishments in carbon nanotubes. *Sci. Rep.,* **2015**, *5*, 11859. [http://dx.doi.org/10.1038/srep11859]

[18] Siddons, G.; Merchin, D.; Back, J.; Jeong, J.; Shim, M. Highly Efficient Gating and Doping of Carbon Nanotubes with Polymer Electrolytes. *Nano Lett.,* **2004**, *4*, 927-931. [http://dx.doi.org/10.1021/nl049612y]

[19] Shimotani, H.; Kanbara, T.; Iwasa, Y.; Tsukagoshi, K.; Aoyagi, Y.; Kataura, H. Gate capacitance in electrochemical transistor of single-walled carbon nanotube. *Appl. Phys. Lett.,* **2006**, *88*, 073104(1-3). [http://dx.doi.org/10.1063/1.2173626]

[20] Panzer, M.; Frisbie, C. Exploiting Ionic Coupling in Electronic Devices: Electrolyte-Gated Organic Field-Effect Transistors. *Adv. Mater.,* **2008**, *20*, 3177-3180. [http://dx.doi.org/10.1002/adma.200800617]

[21] Cho, J.H.; Lee, J.; Xia, Y.; Kim, B.; He, Y.; Renn, M.J.; Lodge, T.P.; Frisbie, C.D. Printable ion-gel gate dielectrics for low-voltage polymer thin-film transistors on plastic. *Nat. Mater.,* **2008**, *7*(11), 900-906. [http://dx.doi.org/10.1038/nmat2291] [PMID: 18931674]

[22] Lee, J.; Kaake, L.; Cho, J.; Zhu, X.; Lodge, T.; Frisbie, C. Ion Gel-Gated Polymer Thin-Film Transistors: Operating Mechanism and Characterization of Gate Dielectric Capacitance, Switching Speed, and Stability. *J. Phys. Chem. C,* **2009**, *113*, 8972-8981. [http://dx.doi.org/10.1021/jp901426e]

[23] Zaumseil, J.; Ho, X.; Guest, J.R.; Wiederrecht, G.P.; Rogers, J.A. Electroluminescence from electrolyte-gated carbon nanotube field-effect transistors. *ACS Nano,* **2009**, *3*(8), 2225-2234. [http://dx.doi.org/10.1021/nn9005736] [PMID: 19634895]

[24] Yomogida, Y.; Pu, J.; Shimotani, H.; Ono, S.; Hotta, S.; Iwasa, Y.; Takenobu, T. Ambipolar Organic Single-Crystal Transistors Based on Ion Gels. *Adv. Mater.,* **2012**, *24*, 4392-4397. [http://dx.doi.org/10.1002/adma.201290195]

[25] Ha, M.; Seo, J.W.; Prabhumirashi, P.L.; Zhang, W.; Geier, M.L.; Renn, M.J.; Kim, C.H.; Hersam, M.C.; Frisbie, C.D. Aerosol jet printed, low voltage, electrolyte gated carbon nanotube ring oscillators with sub-5 μs stage delays. *Nano Lett.,* **2013**, *13*(3), 954-960. [http://dx.doi.org/10.1021/nl3038773] [PMID: 23394463]

[26] Yanagi, K.; Miyata, Y.; Kataura, H. Optical and Conductive Characteristics of Metallic Single-Wall Carbon Nanotubes with Three Basic Colors; Cyan, Magenta, and Yellow. *Appl. Phys. Express*, **2008**, *1*, 034003. [http://dx.doi.org/10.1143/APEX.1.034003]

[27] Sato, Y.; Yanagi, K.; Miyata, Y.; Suenaga, K.; Kataura, H.; Iijima, S. Chiral-angle distribution for separated single-walled carbon nanotubes. *Nano Lett.,* **2008**, *8*(10), 3151-3154. [http://dx.doi.org/10.1021/nl801364g] [PMID: 18729412]

[28] Yanagi, K.; Udoguchi, H.; Sagitani, S.; Oshima, Y.; Takenobu, T.; Kataura, H.; Ishida, T.; Matsuda, K.; Maniwa, Y. Transport mechanisms in metallic and semiconducting single-wall carbon nanotube networks. *ACS Nano,* **2010**, *4*(7), 4027-4032. [http://dx.doi.org/10.1021/nn101177n] [PMID: 20593841]

[29] Fujita, N.; Matsumoto, D.; Sakurai, Y.; Kawahara, K.; Ago, H.; Takenobu, T.; Marumoto, K. Direct observation of electrically induced Pauli paramagnetism in single-layer graphene using ESR spectroscopy. *Sci. Rep.,* **2016**, *6*, 34966. [http://dx.doi.org/10.1038/srep34966]

[30] Tsuji, M.; Arai, N.; Marumoto, K.; Takeya, J.; Shimoi, Y.; Tanaka, H.; Kuroda, S.; Takenobu, T.; Iwasa, Y. Electron Spin Resonance Study of Interface Trap States and Charge Carrier Concentration in

Rubrene Single-Crystal Field-Effect Transistors. *Appl. Phys. Express,* **2011**, *4*, 085702. [http://dx.doi.org/10.1143/APEX.4.085702]

[31] Tanaka, H.; Hirate, M.; Watanabe, S.; Kaneko, K.; Marumoto, K.; Takenobu, T.; Iwasa, Y.; Kuroda, S. Electron spin resonance observation of charge carrier concentration in organic field-effect transistors during device operation. *Phys. Rev. B.,* **2013**, *87*, 045309(1-7). [http://dx.doi.org/10.1103/PhysRevB.87.045309]

[32] Havlicek, M.; Chernov, A.; Jantsch, W.; Wilamowski, Z.; Yanagi, K.; Kataura, H.; Rümmeli, M.; Malissa, H.; Kuzmany, H. Magnetic phase transition for defect induced electron spins from fully metal-semiconductor separated SWCNTs. *Phys. Stat. Solidi B,* **2012**, *249*, 2562-2567. [http://dx.doi.org/10.1002/pssb.201200426]

[33] Dóra, B.; Gulácsi, M.; Koltai, J.; Zólyomi, V.; Kürti, J.; Simon, F. Electron Spin Resonance Signal of Luttinger Liquids and Single-Wall Carbon Nanotubes. *Phys. Rev. Lett.,* **2008**, *101*, 106408(1-7). [http://dx.doi.org/10.1103/PhysRevLett.101.106408]

[34] Kosaka, M.; Ebbesen, T.; Hiura, H.; Tanigaki, K. Annealing effect on carbon nanotubes. An ESR study. *Chem. Phys. Lett.,* **1995**, *233*, 47-51. [http://dx.doi.org/10.1016/0009-2614(94)01416-S]

[35] Petit, P.; Jouguelet, E.; Fischer, J.; Rinzler, A.; Smalley, R. Electron spin resonance and microwave resistivity of single-wall carbon nanotubes. *Phys. Rev. B,* **1997**, *56*, 9275-9278. [http://dx.doi.org/10.1103/PhysRevB.56.9275]

[36] Claye, A.; Nemes, N.; Jánossy, A.; Fischer, J. Structure and electronic properties of potassium-doped single-wall carbon nanotubes. *Phys. Rev. B,* **2000**, *62*, R4845-R4848. [http://dx.doi.org/10.1103/PhysRevB.62.R4845]

[37] Salvetat, J.; Fehér, T.; L'Huillier, C.; Beuneu, F.; Forró, L. Anomalous electron spin resonance behavior of single-walled carbon nanotubes. *Phys. Rev. B.,* **2005**, *72*, 075440(1-6). [http://dx.doi.org/10.1103/PhysRevB.72.075440]

[38] Náfrádi, B.; Nemes, N.; Fehér, T.; Forró, L.; Kim, Y.; Fischer, J.; Luzzi, D.; Simon, F.; Kuzmany, H. Electron spin resonance of single-walled carbon nanotubes and related structures. *Phys. Stat. Solidi B,* **2006**, *243*, 3106-3110. [http://dx.doi.org/10.1002/pssb.200669218]

[39] Musso, S.; Porro, S.; Rovere, M.; Tagliaferro, A.; Laurenti, E.; Mann, M.; Teo, K.; Milne, W. Low temperature electron spin resonance investigation on SWNTs after hydrogen treatment. *Diamond Related Materials,* **2006**, *15*, 1085-1089. [http://dx.doi.org/10.1016/j.diamond.2005.11.018]

[40] Likodimos, V.; Glenis, S.; Guskos, N.; Lin, C. Antiferromagnetic behavior in single-wall carbon nanotubes. *Phys. Rev. B.,* **2007**, *76*, 075420(1-5). [http://dx.doi.org/10.1103/PhysRevB.76.075420]

[41] Corzilius, B.; Dinse, K.; Hata, K.; Haluška, M.; Skákalová, V.; Roth, S. SWNT probed by multi-frequency EPR and nonresonant microwave absorption. *Phys. Stat. Solidi B,* **2008**, *245*, 2251-2254. [http://dx.doi.org/10.1002/pssb.200879574]

[42] Kombarakkaran, J.; Pietraß, T. Electron spin resonance studies of hydrogen adsorption on single-walled carbon nanotubes. *Chem. Phys. Lett.,* **2008**, *452*, 152-155. [http://dx.doi.org/10.1016/j.cplett.2007.12.043]

[43] Ferrer-Anglada, N.; Monge, A.; Roth, S. Electron spin resonance on single-walled carbon nanotubes obtained from different sources. *Phys. Stat. Solidi B,* **2010**, *247*, 2823-2826. [http://dx.doi.org/10.1002/pssb.201000231]

[44] Havlicek, M.; Jantsch, W.; Rümmeli, M.; Schönfelder, R.; Yanagi, K.; Miyata, Y.; Kataura, H.; Simon, F.; Peterlik, H.; Kuzmany, H. Electron spin resonance from semiconductor-metal separated SWCNTs. *Phys. Stat. Solidi B,* **2010**, *247*, 2851-2854.

[http://dx.doi.org/10.1002/pssb.201000317]

[45] Szirmai, P.; Fábián, G.; Dóra, B.; Koltai, J.; Zólyomi, V.; Kürti, J.; Nemes, N.; Forró, L.; Simon, F. Density of states deduced from ESR measurements on low-dimensional nanostructures; benchmarks to identify the ESR signals of graphene and SWCNTs. *Phys. Stat. Solidi B,* **2011**, *248*, 2688-2691. [http://dx.doi.org/10.1002/pssb.201100191]

[46] Rice, W.D.; Weber, R.T.; Leonard, A.D.; Tour, J.M.; Nikolaev, P.; Arepalli, S.; Berka, V.; Tsai, A.L.; Kono, J. Enhancement of the electron spin resonance of single-walled carbon nanotubes by oxygen removal. *ACS Nano,* **2012**, *6*(3), 2165-2173. [http://dx.doi.org/10.1021/nn204094s] [PMID: 22324937]

[47] Havlicek, M.; Jantsch, W.; Wilamowski, Z.; Yanagi, K.; Kataura, H.; Rümmeli, M.; Tyryshkin, A.; Lyon, S. Indirect exchange interaction in fully metal-semiconductor separated single-walled carbon nanotubes revealed by electron spin resonance. *Phys. Rev. B.,* **2012**, *86*, 045402(1-5). [http://dx.doi.org/10.1103/PhysRevB.86.045402]

[48] Pake, G. *Paramagnetic resonance,* 1st ed; W.A. Benjamin: New York, **1962**, pp. 126-128.

[49] Wagoner, G. Spin Resonance of Charge Carriers in Graphite. *Phys. Rev.,* **1960**, *118*, 647-653. [http://dx.doi.org/10.1103/PhysRev.118.647]

CHAPTER 7

Characterizing the Nature of Surface Radicals in Carbon-Based Materials, Using Gas-Flow EPR Spectroscopy

Ortal Marciano and **Sharon Ruthstein***

Bar-Ilan University, Faculty of Exact Sciences, Department of Chemistry, 5290002, Ramat Gan, Israel

Abstract: Carbon-based materials are highly diverse in terms of their mechanical, electronic, and thermodynamic properties, and as such can be used in a vast array of applications in numerous industries. These materials contain various types of carbon and oxygen radicals. Oxidation processes that influence the composition of these radical populations can have substantial effects on the materials' electronic properties. Therefore, it is important to gain a systematic understanding of oxidation processes in carbonaceous materials at various temperatures and pressures. Electron paramagnetic resonance (EPR) spectroscopy is routinely used to characterize defects in carbon-based materials and the nature of the radicals they contain. Specifically, spin concentrations and the composition and distribution of radicals can be correlated to the electronic properties of carbon-based materials. Accordingly, over the last few years, our group has been using EPR spectroscopy to develop methodologies to explore the oxidation properties of carbon-based materials. Our overarching goal is to produce a toolkit that can correlate between the physical properties of specific carbon-based materials and these materials' sensitivity to oxidation processes. In this chapter, we will describe our findings regarding the oxidation processes of coal and graphene oxide materials. Our data are derived from in-situ EPR experiments in which carbon-based materials were exposed to various atmospheric environments. Our findings have clear practical implications with regard to identifying appropriate storage conditions for carbon-based materials.

Keywords: Bituminous coals, Carbon-centered radicals, Coal, Graphene oxide (GO), EPR, EPR linewidth, Free radicals, In-situ gas-flow EPR, Lorentzian lineshape, Lignites, Low temperature oxidation, NMR, Oxidation process, Oxidation time, Oxygen-centered radicals, Radical concentration, Reduced graphene oxide (rGO), Reduction temperature, Sub-bituminous coals, Surface radicals.

* **Corresponding author Sharon Ruthstein:** Bar-Ilan University, Faculty of Exact Sciences, Department of Chemistry, 5290002, Ramat Gan, Israel; Tel: +972-3-738-4329; Fax: +972-3-738-4053; E-mail: Sharon.ruthstein@biu.ac.il

Dariya Savchenko & Abdel Hadi Kassiba (Eds.)

INTRODUCTION

Carbon-based materials are used in a wide array of applications, including energy conversion and storage, electronics, electrocatalysis, sensors, and medicine. These materials are highly diverse in terms of their mechanical strength, electrical and thermal conductivity, and surface area, such that carbon-based materials with specific properties can be selected to fulfill different sets of requirements [1 - 4] Their molecular structures are considered to be polymer-like, comprising an extensive variety of structural elements and compounds, *e.g.*, aromatic rings, aliphatic carbon atoms, and heterocyclic features. In particular, carbon-based materials have been shown to contain various kinds of carbon and oxygen radicals. These radicals affect the electronic and thermodynamic properties of the carbonaceous materials. Oxidation processes that occur at the surface of a carbon-based material can affect its radical population and thus have a direct effect on the material's electronic properties. In addition, the propensity of carbon-based materials to interact with oxygen and water can result in spontaneous heating and combustion. These phenomena have motivated decades of research on the interactions between molecular oxygen and carbon-based materials [5 - 8]. Oxidation processes of carbon-based materials involve physical adsorption and chemisorption of atmospheric oxygen, which forms surface oxides (including hydroperoxides). These surface oxides can partially decompose to yield low-molecular-weight inorganic gases like carbon oxides (CO, CO_2), water, hydrogen (H_2) and also some low molecular weight organic gases (C_{1-5}) [9, 10]. In the coal macromolecule, for example, the mechanism of the oxidation process is suggested to comprise the following steps:

A. Adsorption

Physical adsorption

(1). $\text{Carbon}_{(solid)} + O_{2(g)} \rightarrow O_{2\,(\text{adsorbed in carbon})}$

Chemical adsorption

(2). $O_{2\,(\text{adsorbed in carbon})} \rightarrow O_{2\,(\text{chemisorbed})}$

B. Formation of surface oxides

(3). $O_{2\,(\text{chemisorbed})} \rightarrow \text{Surface Oxides}$

C. Decomposition of surface Oxides

Primary Process

(4). $\text{Surface Oxides} \rightarrow CO_{2(g)} + H_2O_{(g)}$

Secondary Processes

(5). a- $\text{Surface oxides} \rightarrow CO_{(g)}$
b- $\text{Surface oxides} \rightarrow C_nH_{m(g)}, H_{2(g)}$

Electron paramagnetic resonance (EPR) spectroscopy is one of the most useful tools available for studying the formation of radicals and the changes in their properties during oxidation processes. EPR studies regarding the effects of oxidation on the structure and stability of carbon radicals date back to the early 1980s. However, the primary focus of these early reports was to characterize the structural changes of the investigated materials under extremely harsh conditions (pH or very high temperature) [11 - 13]. In order to obtain a more comprehensive understanding of how oxidation — and particularly oxidation under more typical storage conditions — affects carbon-based materials, their radical content, and the characteristics of these radicals, it is necessary to study oxidation processes at lower temperatures and at atmospheric pressure. Moreover, under extremely harsh conditions, both physical adsorption of oxygen and chemisorption occur at the surface of the carbon-based material, making it difficult to characterize each process individually; a study of oxidation processes under more moderate conditions makes it possible to distinguish the two adsorption mechanisms.

Over the last few years, our group has been utilizing EPR spectroscopy to study the oxidation properties of various carbon-based materials (coals, graphene oxide and reduced graphene oxide). We have characterized the radicals that form at different temperatures, including room temperature. Herein, we will discuss our findings and will compare them to prior findings on similar carbon-based-material systems.

EXPERIMENTAL METHODS

Experimental Setup: *In Situ* Gas Flow EPR Experiments on Carbon-based Materials

We constructed a system in which a consistent environment of gas flows throughout a tube containing a carbon sample (see Fig. **1A** for an illustration). In this setup, the sample can be exposed to nitrogen, carbon dioxide, or helium, or to a vacuum environment. A standard measurement process is composed of the following three steps:

(1) A baseline measurement is established under an air atmosphere for several scans.

(2) The selected gas flows through the system, and the sample is continuously measured until equilibrium is achieved.

(3) The flow is stopped and the sample is exposed to an air atmosphere.

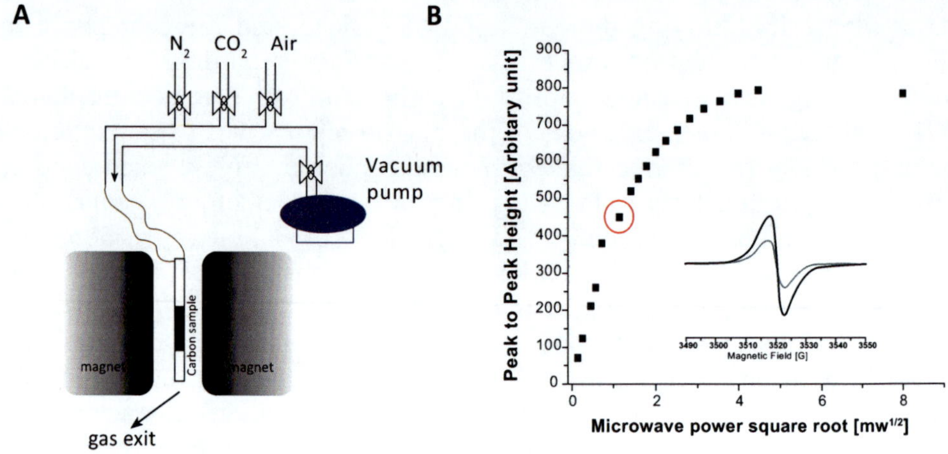

Fig. (1). **A**. The *in-situ* gas flow EPR setup. **B**. The relationship between the square root of microwave power ($mW^{1/2}$) and the peak-to-peak heights of IN (Indonesia) coal. The inset represents the spectra acquired with microwave power of 0.8 mW (gray line), and 16.0 mW (dark line). The circles correspond to a power of 2.0 mW. Reproduced from Ref [14]. with permission from the PCCP Owner Societies.

To determine the saturation power for the carbon-based samples, we carried out a microwave power saturation experiment under anaerobic conditions. For all samples studied here, saturation was obtained at a power larger than 10.0 mW (Fig. **1B**). We used a power of 2.0 mW in all the EPR measurements. EPR spectra were recorded using an E500 Bruker Elexsys spectrometer operating at 9.0–9.5 GHz. The spectra were recorded with modulation amplitude of 0.1 mT, and a time constant of 60 ms. The samples were measured in 3.0 mm quartz tubes. The *g*-values were measured using bisdiphenylene-b-phenylallyl (BDPA) and diphenyl-picrylhydrazyl (DPPH) standards. The spin concentrations were evaluated using known calibrated solutions of DPPH.

General Characteristics of EPR Spectra of Radicals in Carbon-based Materials

The EPR spectra of radicals in the carbonaceous materials considered herein are characterized by a single broad Lorentzian resonance. In such spectra, only the *g*-value, the line-width, and the spin concentration can be obtained. The *g*-value of an EPR spectrum can be used to determine whether a radical is carbon-centered or oxygen-centered [13, 15, 16]. The basic equation for the electron Zeeman interaction:

$$E = h\nu = g\beta_e B_0 \quad (1)$$

defines the g-value [17], where h is the Planck constant, v is the constant mw frequency applied in the experiment, B_0 is the resonance magnetic field, and β_e is the Bohr magneton. For a free electron, this g-value is 2.00232.

Variations in the g-value from 2.00232 are related to magnetic interactions involving the orbital angular momentum of the unpaired electron and its chemical environment. Carbon-centered radicals have the g-values that are close to the free electron g-value 2.0023. Carbon-centered radicals with an adjacent oxygen atom have higher g-values, in the rangc of 2.003-2.004, while oxygen-centered radicals have the g-values above 2.004. In carbon-centered radicals with a nearby oxygen heteroatom, the characteristic range of g-values is 2.0034-2.0039, *i.e.*, slightly higher values than those corresponding to purely carbon-centered radicals. The line-width is governed by the spin-lattice relaxation process. The interaction between adjacent radicals or between a radical and paramagnetic molecular oxygen results in a decrease in the spin-lattice relaxation time, producing an increase in line-width [18 - 20].

EPR ON COAL SAMPLES

Coal Samples

We carried out experiments on three ranks of coals: bituminous, sub-bituminous and lignite. The bituminous coals were from South Africa (denoted SA) and from the US (denoted BA, referring to 'Bailey', the mine where the coal was obtained); the sub-bituminous coal was from Indonesia (denoted IN); and the lignite coals were from Germany (from Hambach and from Lusatia, denoted HA and LUS, respectively). The bituminous coals used in our study serve as the major fossil fuel in pulverized-coal-fired power plants in Israel (more than 60% of the coal consumed in Israel is bituminous). The sub-bituminous Indonesian coal is also fired in Israeli utilities. The lignites are used in German utilities. Selected properties of the various coals are presented in Table **1**.

Table 1. Properties of coals.

#	Sample	Elemental analysis [%wt]			Macropore surface area [m²/gr]	Micropore surface area [m²/gr]
		C	H	O		
1	SA[a]	74.19	4.10	5.59	3.61	101
2	BA[a]	78.07	5.18	5.84	0.96	39.5
3	IN[a]	73.25	4.63	9.05	2.56	117
4	HA[a]	66.12	4.32	23.65	1.41	110
5	LUS[a]	63.24	4.50	27.24	1.53	134

[a] SA=South Africa; BA= USA; IN=Indonesia; HA= Hambach, Germany; LUS=Lusatia, Germany.

We focused on oxidation processes occurring at relatively low temperatures (below 100 °C), on timescales ranging between minutes and weeks. An understanding of such processes can provide insight regarding the effects of storage on the concentration and characteristics of radicals present in coals.

Oxidation Processes Occurring at 95 °C on a Timescale of Weeks

EPR Experiments

For these EPR experiments, the coal used was weathered in glass vials (40 ml), used as batch reactors. Each reactor was charged with 2 g of coal (particle size $74\mu m \leq X \leq 250\mu m$) in an air atmosphere and heated at 95 °C for various periods up to 6 months in a *nüve* oven model FT 300. The oxygen concentration in the oven was held at ~21% throughout the different oxidation periods to simulate the weathering process. Samples were removed from the oven at 2-week intervals and subsequently analyzed.

g-values

Fig. (**2**) shows the *g*-values of each type of a coal sample as a function of oxidation time. It is interesting to note that there is a distinct difference between the g-values of the lower-rank coals (HA, LUS) and those of the higher-rank coals (BA, IN, SA). Specifically, the lower-rank coals exhibit *g*-values between 2.0037 and 2.0041; as noted above, these values are characteristic of carbon-centered radicals with a nearby oxygen atom. The higher-rank coals exhibit *g*-values between 2.0031 and 2.0033, a range characterizing carbon-centered radicals.

Furthermore, the measured *g*-values for all coals are effectively constant over time, at around 2.0032 ± 0.00015 for the higher-rank coals, and 2.0039 ± 0.0002 for the lower-rank coals. The variation of the *g*-values of the lower-rank coal is a bit larger than that of the higher-rank coal. The types of radicals stabilized in the individual coals were not affected by the oxidation treatment, though the larger fluctuation in the g-value of the lower-rank coals around the mean (2.0039) suggests that, during the weathering process, the carbon-centered radicals with nearby oxygen were exposed to a more dynamic environment compared with the carbon-centered radicals with no nearby oxygen.

Line-width

Although the *g*-values of the various coals remained relatively constant, we observed changes over time in the line-widths of the signals (see Fig. **3**). As noted above, the line-width is inversely proportional to the spin-lattice relaxation time. A change in the line-width is indicative of interactions of the radicals either with the surrounding matrix (trace paramagnetic impurities) or with oxygen molecules [19 - 21].

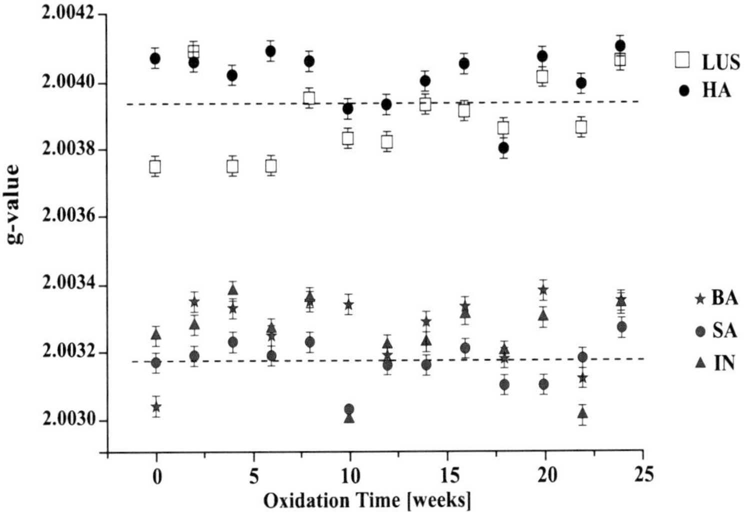

Fig. (2). *g*-values of the various samples as a function of oxidation time at 95 °C. Reproduced from Ref [14]. with permission from the PCCP Owner Societies.

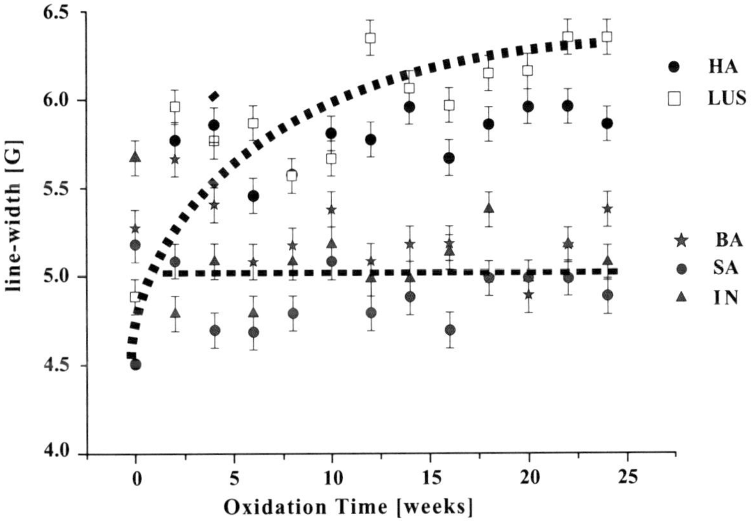

Fig. (3). Line-width values of the various samples as a function of oxidation time. (Microwave power = 2.0 mW, modulation amplitude = 1.0 G). Reproduced from Ref [14]. with permission from the PCCP Owner Societies.

Fig. (**3**) shows a clear difference between lower- and higher-rank coals. For HA and LUS—which, as discussed above, contain carbon-centered radicals with a nearby oxygen—the line-width increases over time, especially in the first 6 weeks, from 4.5 G to 6.0 G ± 0.2 G. For SA, BA, and IN—which contain carbon-centered radicals—the line-width is constant, around 5.0 ± 0.25 G. The differences within each class (*i.e.*, the differences between HA and LUS, or the differences between the SA, BA, and IN) are probably attributable to differences in the percentage composition of carbon, and in the characteristics of the carbon radicals.

In the absence of paramagnetic metal ions or sulfur atoms, coal samples have been shown to exhibit the following Lorentzian lines [13, 21 - 23]:

Lorentzian 1: with line-width of 3.0-7.0 G and *g*-values of 2.0035-2.0039 (1)
Lorentzian 2: with line-width of 1.6-7.0 G and *g*-values of 2.0028-2.0034 (2)
Lorentzian 3: with line-width of 0.8-1.5 G and *g*-values of 2.0027-2.0034 (3)

Our results suggest that the lower-rank coals (HA and LUS) exhibit Lorentzian type one, whereas the higher-rank coals are characterized by Lorentzian type two. Indeed, previous research has suggested that the type of Lorentzian for a given coal strongly depends on the rank of the coal [13, 23].

Radical Concentrations

A change in the line-width of a given coal sample may result in a change in radical concentration. The free radical densities (concentrations) of the different coals are presented in Fig. (**4**). As can be observed in Fig. (**4**), the concentration of stable free radicals in the higher-rank coals increases steadily over time, from 1×10^{20} spins/cm^3 to $\approx 5 \times 10^{20}$ spins/cm^3, and up to $\approx 7 \times 10^{20}$ spins/cm^3 for the SA sample. This increase indicates that, in these coals, radical concentration is dependent on the low-temperature oxidation process. For the lower-rank coals, the radical concentration is almost unaffected and remains constant over time within the experimental error, though it is possible to make out an increasing trend in the concentration of free radicals in the first 12 weeks (Fig. **4** inset), followed by a slight reduction in the radical density.

Comparison with Previous Results

Retcovsky *et al.* [21] investigated the structural and chemical properties of various American coals. In their experiments, all coals were dried overnight at 100 °C. They found that for coals with a carbon content of up to 94%, the spin concentration increased with carbon content. For higher percentages of carbon content, the spin concentration decreased dramatically.

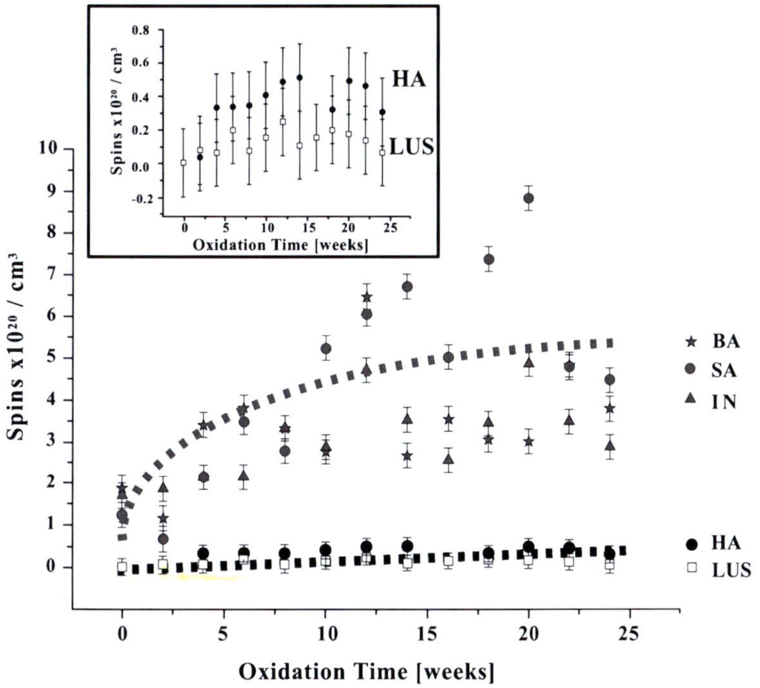

Fig. (4). Free radical densities (spins/cm^3) of the various samples as a function of oxidation time. The inset is a magnification for HA and LUS plots. Reproduced from Ref [14]. with permission from the PCCP Owner Societies.

Similarly, for coals with a carbon content of up to 80%, the line-width increased with carbon content, and for coals with >80% carbon the line-width decreased with carbon content. In our case, the higher-rank coals contained 70%-80% carbon (Table **1**), and their spin concentrations were 5-7 times higher than those of lower-rank coals, with the carbon content between 60% and 70%. The line-width at time zero is a bit larger for higher-rank coals than for lower-rank coals, as noted also by Retcovsky *et al.* [21] However, for the lower-rank coal, the line-width increases with the oxidation time, whereas for the higher-rank coal, the line-width remains constant with oxidation time.

Summary of EPR Spectroscopy Results

Our EPR spectroscopy experiments indicate that, at 95 °C, the effects of oxidation processes are observed primarily in the spin concentrations of higher-rank coals. In contrast, in the lower-rank coals, the spin concentrations remained constant, though the line-widths increased as a function of the oxidation time. The latter observations suggest that, in low-rank coals, oxidation processes at 95 °C do not generate additional radicals; rather, they increase the extent to which radicals that are already present in the material interact with paramagnetic oxygen molecules, resulting in an increase in line-width. In higher-rank coals, however, oxidation processes at 95 °C generate new radicals, which do not interact with the molecular oxygen.

Nuclear Magnetic Resonance (NMR) Studies

In order to further analyze structural differences between the lower- and higher-rank coals, we carried out solid-state NMR measurements, which are described in Fig. (**5**) and Table **2**.

Table 2. NMR peak analysis.

#	Sample	A1[a] [%]		A2[a] [%]		A3[a] [%]	
		Fresh	22 weeks	Fresh	22 weeks	Fresh	22 weeks
1	SA	63.0	70			37.0	30.0
2	BA	60.0	64.6			40.0	35.4
3	IN	57.2	59.7			42.8	40.3
4	HA	39.1	41.7	12.4	9.45	48.4	48.8
5	LUS	41.4	43.4	25.0	23.6	33.5	33.0

[(a)]A1 – area of peaks between 100 and 160 ppm, A2- area of peaks between 60 and 100 ppm, A3 – area of peaks between 0 and 60 ppm (see Fig. **5**). 22 weeks – samples after 22 weeks of oxidation at 95 °C. The error is ±0.5%.

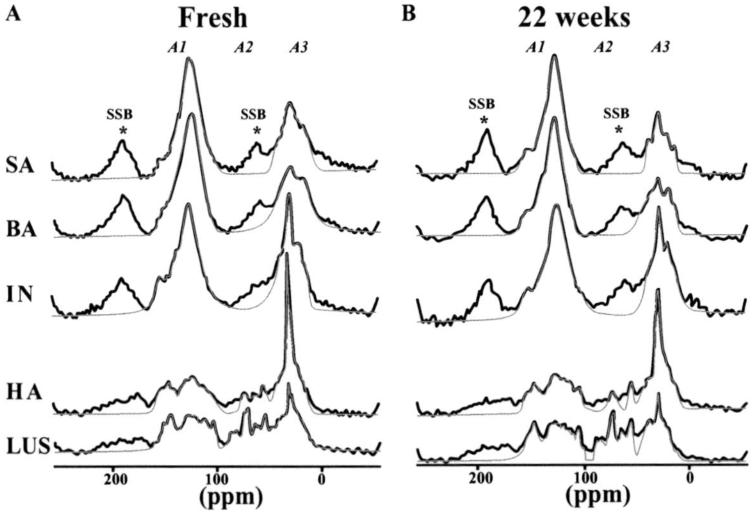

Fig. (5). ^{13}C CPMAS NMR spectra of the various coal samples. (A) Fresh samples. (B) After 22 weeks of oxidation at 95 °C. Side bands (SSB) are denoted by asterisks. Reproduced from Ref [15]. with permission from the PCCP Owner Societies.

Fig. (**5**) shows the ^{13}C NMR spectra of the various coal samples for fresh samples, and after an oxidation period of 22 weeks at 95 °C. Signals between 100 ppm and 160 ppm represent aromatic carbons (A1 area), signals between 60 and 100 ppm represent aliphatic carbons with an adjacent oxygen atom (A2 area), whereas signals at 0-60 ppm represent aliphatic carbons (A3 area). The data were fitted using the DmFit program (gray lines) [24], and the fitting parameters are listed in Table **2**.

The NMR results suggest that oxidation processes at 95 °C appreciably increased the aromatic nature and decreased the aliphatic content of the higher-rank coals (Table **2** shows a clear increase in A1 for the SA, BA, and IN coals and a decrease in A3). These dimensions are not appreciably affected in the lower-rank coals (*i.e.*, the lignites, HA and LUS), which contain higher aliphatic content. Moreover, the spectra of HA and LUS are characterized by resolved phenolic peaks at 147 ppm, whereas the spectra of the higher-rank coals are not. For HA, the percentage of aliphatic groups is not affected by the oxidation process; there is, however, a more significant decrease in the A2 signal for the HA coal than for the LUS coal. Taken together with our EPR results, these observations suggest, during the 95 °C oxidation process, the coal samples that undergo larger structural changes during are also those that generate new carbon radicals over time.

The NMR data further indicate that after oxidation treatment at 95 °C, the A3 region undergoes a chemical shift from 30 ppm to 28 ppm. This shift was observed for all coals, suggesting that after oxidation there are more C-C bonds than C-H bonds.

These results suggest that, in higher-rank coals that underwent weathering treatment, the increase in the concentration of stable radicals (observed in our EPR experiments) was caused by the cleavage of aliphatic carbon due to oxidation. The % of the increase in radical density correlates with the decrease in the NMR aliphatic carbon signal as well as the relativc increase in the percentage of C-C aliphatic bonds and aromaticity (A1 area, Table **2**). The observation that for HA and LUS the total radical concentration did not change after treatment is consistent with the smaller change in the aromaticity for those samples.

Summary and Conclusions

Taken together, the NMR and EPR data suggest that the following mechanism occurs during the oxidation process at 95 °C, and that the existence of this mechanism depends on the coal rank:

$$\text{C-H} \rightarrow \text{C}\cdot + 1/2\text{H}_2$$
$$2\,\text{C}\cdot \rightarrow \text{C-C}$$

Tian *et al.* [25] observed a decrease of up to 15% in the spin concentrations of burned coal samples stored for two months at room temperature. They suggested that some radicals are shorter-lived than others. Their observations, together with our own, indicate that in carbon-based materials stored for long periods of time (timescale of weeks), even at relatively low temperatures, not only physical adsorption can occur, but also chemical adsorption, and the production or elimination of radical species.

IN-SITU OXIDATION MEASUREMENTS AT ROOM TEMPERATURE (RT) ON A TIMESCALE OF MINUTES

The oxidation process can also occur at RT in an atmospheric environment over short periods of time. In most cases, under these conditions, only physical absorption of oxygen occurs; nevertheless, such processes can have significant practical implications regarding the storage conditions of the coals.

EPR Experiments

We performed *in-situ* gas flow EPR experiments at RT and atmospheric pressure. Interestingly, during the gas flow (N_2/CO_2/He or vacuum), a second species at appeared at $g \sim 2.0028$ (corresponding to Lorentzian type 3). Fig. (**6A**) presents two scans for the HA sample: at $t = 0$ under aerobic conditions and at $t = 1900$ s after exposure to CO_2; the second species is clearly present in the later scan. Notably, the total spin concentration for each sample remained constant within our experimental error (±10%). The time of formation of this second radical species varied across samples, in the range of 100–5000 s. Under the exposure to CO_2, this second species appeared quickly in HA, BA, and SA coals, whereas in IN and LUS, the EPR spectra of the second species were not detected within the 4,000 sec of the CO_2 gas flow. The formation rates for these second-species radicals were analyzed and are presented in Fig. (**6B**). Interestingly, the time required for species 2 to reach equilibrium (which is indicative of removing physically bound O_2) increases with radical density.

As some of the coal samples have similar surface areas but differ in their kinetic behavior (*i.e.* HA vs. LUS), it is clear that the rate of formation of species 2 is not purely a function of surface area. Rather, it is also dependent on the extent to which the material's surface comprises micropores *versus* macropores. This is because, first, the diffusion of molecular oxygen is significantly affected by the pore size. Confinement in a micropore system, such that the rate of exchange (bulk vs. pore) is hindered, can drive O_2 molecules to closer interaction with the surface; in contrast, in a macropore system, the higher partial pressure of O_2 is required. Second, when the micropore surface area is smaller (*e.g.*, the micropore surface area of BA is ~2.6 times smaller than that of the SA and HA coals; Table

1), the radical concentrations measured are more densely packed. The differences in the profiles of the release and binding of molecular oxygen by the different coals (which is indicated by presence/absence of a species 2 signal) could indicate different levels of binding/adsorption energies of molecular oxygen.

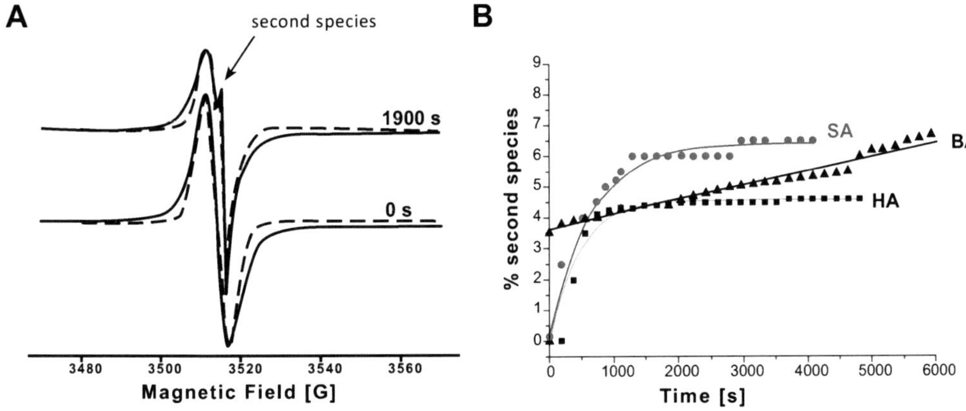

Fig. (6). (A) EPR spectra (solid lines) and simulations (dashed lines) of the HA sample under aerobic conditions, 298 K, (t = 0 s), and after exposure to CO_2 for 1900 s. The formation of the second species is marked in the figure. (B) The formation of the second species during the interaction with N_2 gas, for HA, BA, and SA coals. Reproduced from Refs [15, 26]. with permission from the PCCP Owner Societies.

Fig. (**6B**) clearly shows that there are two stages of formation for the radical species 2 in the BA coal: Stage 1 is characterized by the appearance of an immediate high concentration of species 2 (3.5%), whereas in stage 2, there is a gradual and slower rate of formation until equilibrium is achieved. No such pattern is observed in the SA and HA coals. It is possible that, due to the much higher concentration of radicals in BA coal, the effects of different levels of physically adsorbed O_2 (surface, subsurface) are observed. Areas, where the radical density is lower, are more prone to oxygen displacement by the gas phase, whereas in areas with a higher radical density, the exchange is less thermodynamically favorable.

Comparison with Prior Results

Konchits *et al.* [27] used EPR spectroscopy to explore the properties of natural coals, which were not treated at high pressure or high temperature. The EPR signal from coal in the presence of air was much broader. However, when the oxygen was pumped out, the EPR spectrum was characterized by two signals. The first one was broad, and the second was very narrow, and was consistent with the second species observed in our system (g ~ 2.0028). Interestingly, as gas continued to be pumped out of the system for several hours, the EPR spectrum narrowed, and was ultimately characterized by a single narrow signal. This suggests that the EPR spectrum of the carbon-centered radicals strongly depends on the porosity of the material, and the interaction with molecular oxygen. This conclusion is also supported by the study of Retcovsky *et al.* [21] who investigated about 20 different coals of various ranks, and observed the second species only in some of the coals, with a carbon content of 70%-80%.

EPR ON GRAPHENE OXIDE-BASED MATERIALS

In recent years, graphene, a two-dimensional graphite layer, has been identified as a promising form of carbon due to its mechanical strength, high electrical and thermal conductivity, and very high surface area [28 - 31]. Interest in and use of graphene has increased in recent years, as methods of producing the material have been simplified, thereby reducing costs and enabling manufacturing to be upscaled [32, 33].

Graphene oxide (GO) is a two-dimensional material derived from graphene with covalent C-O bonds. The bulk form of GO, conventionally named graphite oxide. However, GO unlike graphene, it does not possess sp^2 hybridization or high electrical conductivity. Hence, GO is considered to be a good insulator [34, 35]. Reduction of GO can restore the aromatic structure of graphene as well as the sp^2 hybridization, thereby restoring conductivity, in addition to other properties of graphene [34 - 36]. However, the reduction process also leads to defects such as vacancies that occur as a result of the departure of oxidized carbon groups such as CO_2 and CO [34, 35, 37].

EPR Spectroscopy of Graphene Oxide Reduced at Different Temperatures

We carried out EPR experiments on GO samples that were reduced under hydrogen at various temperatures between 500 °C and 1000 °C. Several interesting features were observed in the EPR spectra (Fig. 7) of the various samples of reduced GO (rGO).

First, the line-width of a given rGO sample is greatly dependent on the reduction level of that sample: The EPR spectra of rGO samples reduced at lower temperatures (*e.g.*, rGO reduced at 500 °C; rGO@500°C) are much broader than those of rGO samples reduced at a higher temperature (*e.g.* rGO@1000°C). Second, the various rGO samples were characterized by different g-values; specifically, lower *g*-values were observed for samples reduced at higher temperatures (*e.g.*, $g = 2.06$ for rGO@500 °C, vs. $g = 2.004$ for rGO@1000 °C). The latter observation suggests that GO samples reduced at higher temperatures have lower concentrations of oxygen molecules on their surfaces. Finally, the radical concentration was much higher for rGO reduced at 1000 °C than for rGO reduced at 500 °C. Table **3** shows the properties of the various rGO samples.

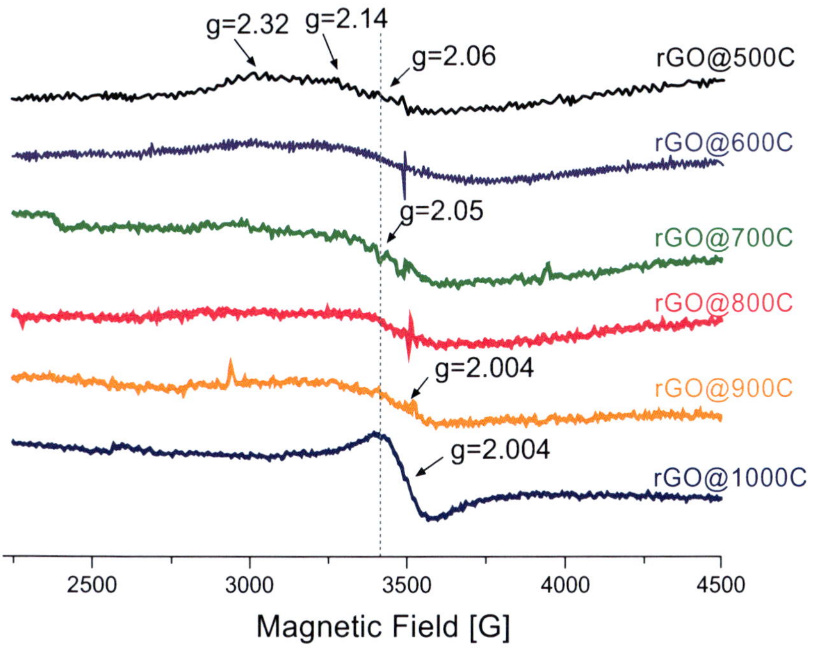

Fig. (7). CW-EPR spectra of rGO reduced at various temperatures between 500 °C and 1000 °C. Reproduced from Ref [38]. with permission from the ACS.

Table 3. Properties of rGO samples.

Sample	Elemental analysis [%wt]		Number of spins (n/cm^3)
	O	C	
rGO@500°C	12.85	74.63	$1.9 \pm 0.2 \times 10^{17}$
rGO@600°C	14.34	74.93	$2.3 \pm 0.3 \times 10^{17}$
rGO@700°C	5.509	73.96	$3.6 \pm 0.4 \times 10^{17}$
rGO@800°C	5.625	81.12	$3.8 \pm 0.2 \times 10^{17}$
rGO@900°C	2.543	79.83	$4.5 \pm 0.2 \times 10^{17}$
rGO@1000°C	2.184	87.58	$10.0 \pm 0.1 \times 10^{18}$

The comparatively narrow line-width for rGO@1000°C, together with the fact that there are more radicals in this sample than in rGO@500°C, indicates that, as the reduction temperature increases, the free radicals become more spread apart, and their distribution becomes more homogeneous.

The fact that there is also a difference in the *g*-values suggests that, as reduction temperature increases, a shift takes place such that the quantity of oxygen-centered radicals decreases, and that of carbon-centered radicals increases, and the latter exceeds the former. The anisotropic EPR spectra observed for rGO reduced at low temperatures indicate that these oxygen-centered radicals are localized with a low tendency to migrate. Conductivity measurements showed that rGO@1000°C samples are seven times more conductive than rGO@500°C samples [38]; this suggests a correlation between the concentration of radicals, their tendency to migrate, and conductivity.

Gudkov *et al.* [39] observed that under strong oxidation, the concentration of paramagnetic centers in GO decreases significantly (4-5 fold). Such a process causes the formation of a covalent C-O bond in the periphery of the graphite structures, resulting in an appearance of a new paramagnetic center on one of the three neighboring carbon atoms. When more carbon atoms become covalently bound to oxygen atoms, converting the hybridization of the latter from sp^2 to sp^3, the radicals formed are more stable, and thus less prone to migrate and recombine. In this case, the conductivity of the material should decrease as well.

In-situ Oxidation Measurements at RT for Graphene Oxide Materials

In order to explore the stability of the various rGO samples to an oxidizing environment, we performed *in situ* gas flow EPR experiments. Fig. (**8A**) shows the line-widths corresponding to rGO samples reduced at different temperatures as a function of exposure time to an N_2 atmosphere. None of the samples show substantial changes in line-width over time, though rGO samples that were reduced at lower temperatures exhibit larger fluctuations in the line-width.

The fact that no significant changes are observed for rGO samples when exposed to nitrogen, after the adsorbed oxygen is removed, suggests that the surface of the rGO samples is quite packed and dense, and that there are relatively few adsorbed oxygen molecules on the surface. Interestingly, in the rGO@500°C sample, a new narrow signal ($g = 2.0028$, similar to species 2 in the coal samples) arises over the course of nitrogen exposure (see Fig. **8B**).

The intensity of this signal increases, until a steady state is achieved around 600 s (Fig. **8C**). This observation suggests that rGO samples reduced at lower temperatures are less packed and can undergo physical adsorption of oxygen molecules.

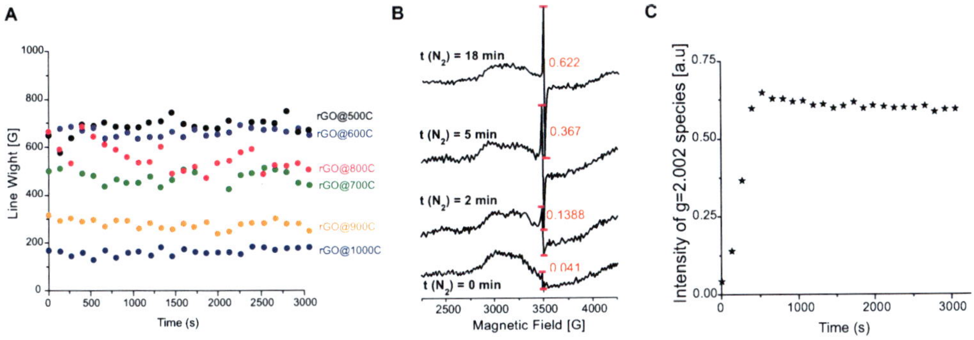

Fig. (8). (A) Line-width for the various rGO samples as a function of exposure time to N_2 gas. (B) CW-EPR spectra of rGO reduced at 500 °C, for various periods of exposure to N_2 gas. (C) Signal intensity over time for the narrow component signal with $g = 2.002$, during exposure to N_2 atmosphere. Reproduced from Ref [38]. with permission from the ACS.

OUTLOOK

For several years, our group has been studying the effects of oxidation on the composition and characteristics of radicals in various carbon-based materials, including coals and GO materials. It has been suggested that three types of unpaired electrons can be detected in carbon materials by EPR measurement: conduction electrons, electrons localized on structural defects, and delocalized isolated electrons [40]. All these radicals are averaged in the Lorentzian line; however, characteristic features of EPR signals can provide indications of the nature of the paramagnetic species that are present in a given material, and the nearby atoms. Herein, we aimed to correlate between the composition of the paramagnetic centers and their capacity to interact with molecular oxygen. We examined oxidation processes that took place at RT over short periods of time (a timescale of minutes), in addition to oxidation processes that took place at higher temperatures for longer periods of time (days/weeks). In the RT oxidation processes, only physical adsorption of oxygen molecules occurred. In materials exposed to higher temperatures for longer periods of time, oxidation affected the composition of the radical population and the quantities of the various radical species.

We observed that for both coals and GO materials, materials with higher carbon content are characterized by higher concentrations of carbon-centered radicals. In coals, we attributed this property to the higher aromaticity of higher-rank materials (70%-80% carbon content) as compared to lower-rank materials (60%-70% carbon content) [41]. We also showed that, when weathered at 95 °C for several weeks, higher-rank coals are more sensitive than lower-rank coals to oxidation processes and can form new sets of radicals upon oxygen chemisorption. Specifically, such oxidation processes led to cleavage of aliphatic radicals. In our experiments with GO materials (reduced at temperatures ranging from 500°C to 1000°C), we observed that, as the reduction temperature increased, the concentration of oxygen-centered radicals decreased, and that of carbon-centered radicals increased. The latter experiments further suggested that the presence of carbon-centered radicals, which are less localized to their environment have a stronger tendency to migrate compared with oxygen-centered radicals, increased the conductivity of the samples. Therefore, higher-rank materials that are characterized by higher quantities of carbon-centered radicals are characterized by higher conductivity.

Compared with oxidation processes that took place at higher temperatures, oxidation processes that took place at RT were less dependent on the ranks of the carbon-based materials. At RT, the porosity of the sample and the density of the radicals determined whether the carbon-based materials would interact with molecular oxygen. Fig. (**9**) presents a summary of our results thus far.

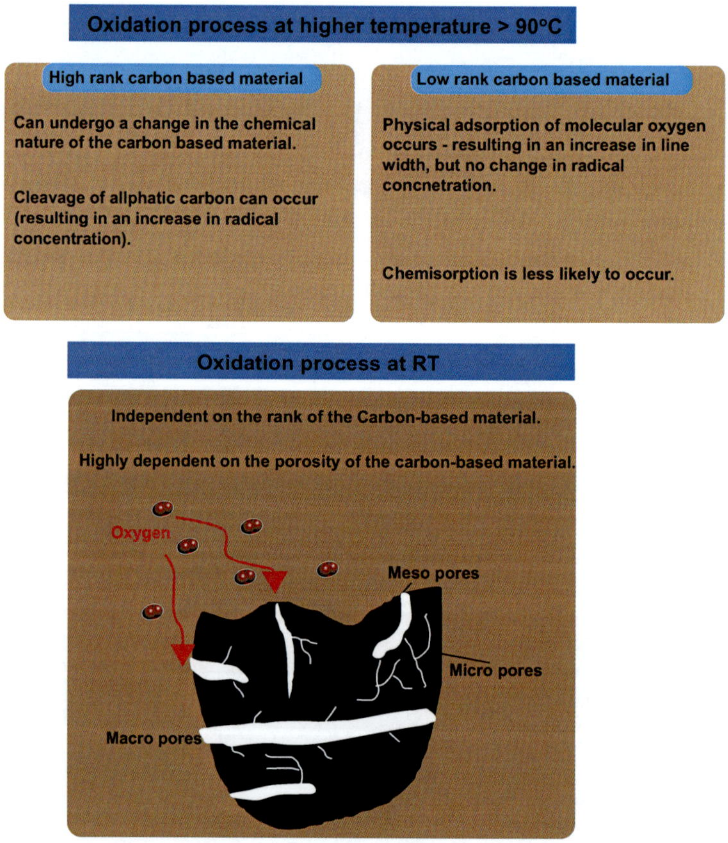

Fig. (9). A summary of the conclusions of this study.

Over the last 50 years, much progress has been made towards understanding oxidation processes and their effects on the radical content of carbon-based materials. Yet, systematic studies are still required in order to establish the effects of oxidation processes that take place in different carbon-based materials, over broad temperature ranges, and at different timescales. We note that owing to the rapid relaxation time of radicals during oxidation processes, spectroscopic methods are limited in their capacity to provide a detailed molecular-level understanding of the various stages of the oxidation processes. Computational studies have the potential to close these gaps.

CONSENT FOR PUBLICATION

Not applicable.

CONFLICT OF INTEREST

The author (editor) declares no conflict of interest, financial or otherwise.

ACKNOWLEDGEMENT

Declare none.

REFERENCES

[1] Song, L.W.; Watkins, G.D. EPR identification of the single-acceptor state of interstitial carbon in silicon. *Phys. Rev. B Condens. Matter,* **1990**, *42*(9), 5759-5764. [http://dx.doi.org/10.1103/PhysRevB.42.5759] [PMID: 9996162]

[2] Wang, W.; Ma, Y.; Li, S.; Teng, S. Effect of temperature on the EPR properties of oil sahle pyrolysates. *Energy Fuels,* **2016**, *30*, 830-834.

[3] Concheso, A.; Santamaria, R.; Menendez, R.; Jimenez-Mateos, J.M.; Alcantara, R.; Lavela, P.; Tirado, J.L. Effect of oxidative stabilization on the electrochemical performance of carbon mesophases as electrode materials for lithium batteries. *J. Solid State Electrochem.,* **2005**, *9*, 627-633. [http://dx.doi.org/10.1007/s10008-004-0621-y]

[4] Zhu, Y.; Murali, S.; Cai, W.; Li, X.; Suk, J.W.; Potts, J.R.; Ruoff, R.S. Graphene and graphene oxide: synthesis, properties, and applications. *Adv. Mater.,* **2010**, *22*(35), 3906-3924. [http://dx.doi.org/10.1002/adma.201001068] [PMID: 20706983]

[5] Deng, J.; Xiao, Y.; Li, Q.Z.; Lu, J.; Wen, H. Experimental studies of spontaneous combustion and anerobic cooling of coal. *Fuel,* **2015**, *157*, 261-269. [http://dx.doi.org/10.1016/j.fuel.2015.04.063]

[6] 6. Green, U.; Aizenshtat, Z.; Metzger, L.; Cohen, H. Field and laboratory simulation study of hot spots in stockpiled bituminous coal. *Energy Fuels,* **2012**, *26*, 7230-7235. [http://dx.doi.org/10.1021/ef301474t]

[7] Liang, Y.; Tian, F.; Luo, H.; Tang, H. Characteristics of coal re-oxidation based on microstructural and spectral observation. *Int. J. Min. Sci. Technol.,* **2015**, *25*, 749-754. [http://dx.doi.org/10.1016/j.ijmst.2015.07.008]

[8] Shao, Z.; Wang, D.; Wang, Y.; Zhong, X. Theory and application of magnetic and self-potential methods in the detection of the HEshituoluogai coal fire, China. *J. Appl. Geophys.,* **2014**, *104*, 64-74.

[http://dx.doi.org/10.1016/j.jappgeo.2014.02.014]

[9] Grossman, S.; Davidi, S.; Cohen, H. Evolution Of Molecular Hydrogen During The Atmospheric Oxidation Of Coal. *Fuel,* **1991**, *70*, 897-898. [http://dx.doi.org/10.1016/0016-2361(91)90202-L]

[10] Davidi, S.; Grossman, S.; Cohen, H. Organic Volatiles Emissions Accompanying The Low-Temperature Atmospheric Storage Of Bituminous Coals. *Fuel,* **1995**, *74*, 1357-1362. [http://dx.doi.org/10.1016/0016-2361(95)00088-M]

[11] Jezierski, A.; Czechowski, F.; Jerzykiewicz, M.; Chen, Y.; Drozd, J. Electron paramagnetic resonance (EPR) studies on stable and transient radicals in humic acids from compost, soil, peat and brown coal. *Spectrochim. Acta A Mol. Biomol. Spectrosc.*, **2000**, *56A*(2), 379-385. [http://dx.doi.org/10.1016/S1386-1425(99)00249-8] [PMID: 10727156]

[12] Ottaviani, M.F.; Mazzeo, R.; Turro, N.J.; Lei, X. EPR study of the adsorption of dioxin vapours onto microporous carbons and mesoporous silica. *Microporous Mesoporous Mater.,* **2011**, *139*, 179-188. [http://dx.doi.org/10.1016/j.micromeso.2010.10.040]

[13] Pilawa, B.; Wieckowski, A.B.; Pietrzak, R.; Wachowska, H. Multi-component EPR spectra of coals with different carbon content. *Acta Phys. Pol. A,* **2005**, *108*, 403-407. [http://dx.doi.org/10.12693/APhysPolA.108.403]

[14] Green, U.; Aizenshtat, Z.; Ruthstein, S.; Cohen, H. Stable radicals formation in coals undergoing weathering: effect of coal rank. *Phys. Chem. Chem. Phys.,* **2012**, *14*(37), 13046-13052. [http://dx.doi.org/10.1039/c2cp41696d] [PMID: 22886081]

[15] Green, U.; Keinan-Adamsky, K.; Attia, S.; Aizenshtat, Z.; Goobes, G.; Ruthstein, S.; Cohen, H. Elucidating the role of stable carbon radicals in the low temperature oxidation of coals by coupled EPR-NMR spectroscopy - a method to characterize surfaces of porous carbon materials. *Phys. Chem. Chem. Phys.,* **2014**, *16*(20), 9364-9370. [http://dx.doi.org/10.1039/c4cp00791c] [PMID: 24718808]

[16] Pol, S.V.; Pol, V.G.; Gedanken, A. Encapsulating ZnS and ZnSe nanocrystals in the carbon shell: a RAPET approach. *J. Phys. Chem. C,* **2007**, *111*, 13309-13314. [http://dx.doi.org/10.1021/jp072877r]

[17] Kivelson, D. Theory of ESR linewidths of free radicals. *J. Chem. Phys.,* **1960**, *33*, 1094-1106. [http://dx.doi.org/10.1063/1.1731340]

[18] Kweon, D-H.; Kim, C.S.; Shin, Y-K. Regulation of neuronal SNARE assembly by the membrane. *Nat. Struct. Biol.,* **2003**, *10*(6), 440-447. [http://dx.doi.org/10.1038/nsb928] [PMID: 12740606]

[19] Merianos, H.J.; Cadieux, N.; Lin, C.H.; Kadner, R.J.; Cafiso, D.S. Substrate-induced exposure of an energy-coupling motif of a membrane transporter. *Nat. Struct. Biol.,* **2000**, *7*(3), 205-209. [http://dx.doi.org/10.1038/73309] [PMID: 10700278]

[20] Xu, Y.; Zhang, F.; Su, Z.; McNew, J.A.; Shin, Y-K. Hemifusion in SNARE-mediated membrane fusion. *Nat. Struct. Mol. Biol.,* **2005**, *12*(5), 417-422. [http://dx.doi.org/10.1038/nsmb921] [PMID: 15821745]

[21] Retcofsky, H.L.; Stark, J.M.; Friedel, R.A. Electron Spin Resonance in american coals. *Anal. Chem.,* **1968**, *40*, 1699-1704. [http://dx.doi.org/10.1021/ac60267a011]

[22] Pilawa, B.; Wieckowski, A.B.; Pietrzak, R.; Wachowska, H. Microwave saturation of EPR spectra of oxidised coal. *Cent. Eur. J. Chem.,* **2007**, *5*, 330-340.

[23] Pilawa, B.; Wieckowski, A.B. Groups of paramagnetic centers in coal samples with different carbon contents. *Res. Chem. Intermed.,* **2007**, *33*, 825-839. [http://dx.doi.org/10.1163/156856707782169336]

[24] Massiot, D.; Fayon, F.; Capron, M.; King, I.; Le Calve, S.; Alonso, B.; Durand, J.O.; Bujoli, B.; Gan, Z.; Hoatson, G. Modelling one and two-dimensional solid-state NMR spectra. *Magn. Reson. Chem.*, **2002**, *40*, 70-76.
[http://dx.doi.org/10.1002/mrc.984]

[25] Tian, L.; Koshland, C.P.; Yano, J.; Yachandra, V.K.; Yu, I.T.S.; Lee, S.C.; Lucas, D. Carbon-centered free radicals in particulate matter emissions from wood and coal combustion. *Energy Fuels,* **2009**, *23*(5), 2523-2526.
[http://dx.doi.org/10.1021/ef8010096] [PMID: 19551161]

[26] Green, U.; Aizenshtat, Z.; Ruthstein, S.; Cohen, H. Reducing the spin-spin interaction of stable carbon radicals. *Phys. Chem. Chem. Phys.*, **2013**, *15*(17), 6182-6184.
[http://dx.doi.org/10.1039/c3cp50533b] [PMID: 23518921]

[27] Konchits, A.A.K.; Shanina, B.D.; Valakh, M.Y.; Yanchuk, I.B.; Yukhymuchuk, V.O.; Alexeev, A.D.; Vasilenko, T.A.; Molchanov, A.N.; Kirillov, A.K. Local structure, paramagnetic properties, and porosity of natural coals: spectroscopic studies. *J. Appl. Phys.*, **2012**, *112*, 043504-043509.
[http://dx.doi.org/10.1063/1.4745015]

[28] Ning, G.; Fan, Z.; Wang, G.; Gao, J.; Qian, W.; Wei, F. Gram-scale synthesis of nanomesh graphene with high surface area and its application in supercapacitor electrodes. *Chem. Commun. (Camb.),* **2011**, *47*(21), 5976-5978.
[http://dx.doi.org/10.1039/c1cc11159k] [PMID: 21475753]

[29] Lee, C.; Wei, X.; Kysar, J.W.; Hone, J. Measurement of the elastic properties and intrinsic strength of monolayer graphene. *Science,* **2008**, *321*(5887), 385-388.
[http://dx.doi.org/10.1126/science.1157996] [PMID: 18635798]

[30] Novoselov, K.S.; Geim, A.K.; Morozov, S.V.; Jiang, D.; Zhang, Y.; Dubonos, S.V.; Grigorieva, I.V.; Firsov, A.A. Electric field effect in atomically thin carbon films. *Science,* **2004**, *306*(5696), 666-669.
[http://dx.doi.org/10.1126/science.1102896] [PMID: 15499015]

[31] Balandin, A.A.; Ghosh, S.; Bao, W.; Calizo, I.; Teweldebrhan, D.; Miao, F.; Lau, C.N. Superior thermal conductivity of single-layer graphene. *Nano Lett.,* **2008**, *8*(3), 902-907.
[http://dx.doi.org/10.1021/nl0731872] [PMID: 18284217]

[32] Wang, G.; Wang, B.; Park, J.; Wang, Y.; Sun, B.; Yao, J. Highly efficient and large-scale synthesis of graphene by electrolytic exfoliation. *Carbon,* **2009**, *47*, 3242-3246.
[http://dx.doi.org/10.1016/j.carbon.2009.07.040]

[33] Kim, K.S.; Zhao, Y.; Jang, H.; Lee, S.Y.; Kim, J.M.; Kim, K.S.; Ahn, J-H.; Kim, P.; Choi, J-Y.; Hong, B.H. Large-scale pattern growth of graphene films for stretchable transparent electrodes. *Nature,* **2009**, *457*(7230), 706-710.
[http://dx.doi.org/10.1038/nature07719] [PMID: 19145232]

[34] Marcano, D.C.; Kosynkin, D.V.; Berlin, J.M.; Sinitskii, A.; Sun, Z.; Slesarev, A.; Alemany, L.B.; Lu, W.; Tour, J.M. Improved synthesis of graphene oxide. *ACS Nano,* **2010**, *4*(8), 4806-4814.
[http://dx.doi.org/10.1021/nn1006368] [PMID: 20731455]

[35] Dreyer, D.R.; Park, S.; Bielawski, C.W.; Ruoff, R.S. The chemistry of graphene oxide. *Chem. Soc. Rev.,* **2010**, *39*(1), 228-240.
[http://dx.doi.org/10.1039/B917103G] [PMID: 20023850]

[36] Tung, V.C.; Allen, M.J.; Yang, Y.; Kaner, R.B. High-throughput solution processing of large-scale graphene. *Nat. Nanotechnol.,* **2009**, *4*(1), 25-29.
[http://dx.doi.org/10.1038/nnano.2008.329] [PMID: 19119278]

[37] Schniepp, H.C.; Li, J-L.; McAllister, M.J.; Sai, H.; Herrera-Alonso, M.; Adamson, D.H.; Prud'homme, R.K.; Car, R.; Saville, D.A.; Aksay, I.A. Functionalized single graphene sheets derived from splitting graphite oxide. *J. Phys. Chem. B,* **2006**, *110*(17), 8535-8539.
[http://dx.doi.org/10.1021/jp060936f] [PMID: 16640401]

[38] Marciano, O.; Gonen, S.; Levy, N.; Teblum, E.; Yemini, R.; Nessim, G.D.; Ruthstein, S.; Elbaz, L. Modulation of oxygen content in graphene surfaces using temperature programmed reductive annealing: Electron Paramagnetic Resonance (EPR) and electrochemical study. *Langmuir,* **2016**, *32*(44), 11672-11680.
[http://dx.doi.org/10.1021/acs.langmuir.6b02987] [PMID: 27728764]

[39] Gudkov, M.V.; Melnikov, V.P. EPR study of graphite oxide thermal reduction: the evolution of paramagnetism and conductivity. *Nanosystems: Phys. Chem. Math.*, **2016**, *7*, 244-252.

[40] Tadyszak, K.; Augustyniak-Jablokow, M.A.; Wieckowski, A.B.; Najder-Kozdrowska, L.; Strzelczyk, R.; Andrzejewski, B. Origin of electron parmagnetic resonance signal in anthracite. *Carbon,* **2015**, *94*, 53-39.
[http://dx.doi.org/10.1016/j.carbon.2015.06.057]

[41] Pilawa, B.; Wachowska, H.; Kozlowski, H.; Wieckowski, A.B.; Najder-Kozdrowska, L. Groups of parmagnetic centers in reduced and methylated low rank coal. *Acta Phys. Pol. A,* **2015**, *128*, 264-267.
[http://dx.doi.org/10.12693/APhysPolA.128.264]

CHAPTER 8

Application of the Two-Temperature EPR Measurement Method to Carbonaceous Solids

Andrzej B. Więckowski[1,2,*] and **Grzegorz P. Słowik**[1]

[1] *Institute of Physics, Faculty of Physics and Astronomy, University of Zielona Góra, Szafrana 4a, 65-516 Zielona Góra, Poland*

[2] *Institute of Molecular Physics of the Polish Academy of Sciences, Smoluchowskiego 17, 60-179 Poznań, Poland*

Abstract: The two-temperature measurement method has found use in electron paramagnetic resonance (EPR) spectroscopy for determining the fraction of two kinds of paramagnetic species of different nature. The method can particularly be applied when the temperature dependence of EPR line intensity shows deviations from the Curie law. By applying this method it was possible to determine the fractions of paramagnetic centres responsible for the Curie- and the Pauli-like paramagnetism in carbon-based materials. The method has also found application when the origin of the EPR line was originated due to paramagnetic centres in spin doublet states ($S = 1/2$) and in excited spin triplet states ($S = 1$).

Keywords: Carbonaceous solids, Carbon-based materials, Carbon black, Coal, Conduction electron spin resonance (CESR), Curie law, Curie paramagnetism, Decomposition of the EPR spectra, EPR, EPR line intensity, EPR linewidth, Macerals, Excited triplet states, Multi-walled carbon nanotubes (MWNT), Paramagnetic centres, Pauli paramagnetism, Pyrolytic carbon, Spin doublet states, Spin triplet states, Two-temperature EPR measurement method.

INTRODUCTION

One of the most important issues during the examination of carbon-based materials is the characterization of their physicochemical properties and the relationship between the latter and the structure. In the electron paramagnetic resonance (EPR) spectra of carbonaceous solids, we very often deal with only single lines. They reveal the dependence of the line intensity I on temperature.

* **Corresponding author Andrzej B. Więckowski:** Institute of Physics, Faculty of Physics and Astronomy, University of Zielona Góra, Zielona Góra, Poland; Tel: +48 (68) 320-2439; Fax: +48 (68) 326-5449; E-mail: Andrzej.Wieckowski@vp.pl

Dariya Savchenko & Abdel Hadi Kassiba (Eds.)

This temperature dependence is in some cases caused by the presence of two types of mechanisms of the interaction between paramagnetic centres. This property has been used as an opportunity to distinguish between the two systems of electron spins and to determine their concentration.

Groups of paramagnetic centres fulfilling the Curie law, which is given in the form $I_C(T) = C/T$, and centres not-fulfilling the Curie law $I_{NC}(T)$ are present in different carbonaceous materials, *e.g.* coal macerals, exinite and vitrinite, in multi-walled carbon nanotubes (MWNT), ancient silk textiles, pyrolytic graphite, pulverized coals, *etc.*

The aim of this chapter is to give an overview of studies on carbon-containing materials performed by the two-temperature EPR measurements. This method is used for determining the relative contributions of spin systems present in paramagnetic carbon-based samples. Measurements of the EPR line intensities I at two different temperatures T_1 and T_2 make it possible to determine the contribution X of paramagnetic centres not-fulfilling the Curie law present in the samples studied at a given temperature.

THE TWO-TEMPERATURE MEASUREMENT METHOD

The derivation of the formula for the two-temperature measurement method, aimed at determining the relative contribution X of paramagnetic centres not-fulfilling the Curie law, is given below.

The total EPR line intensity $I(T)$ (dependent on temperature T) is the sum of partial line intensities of paramagnetic centres fulfilling the Curie law (given in the form $I_C(T) = C/T$) and centres not-fulfilling the Curie law $I_{NC}(T)$:

$$I(T) = I_{NC}(T) + I_C(T) = I_{NC}(T) + C/T \tag{1}$$

After simplifying the symbols $I(T_1) = I_1$ and $I(T_2) = I_2$, the following relationship is obtained for the ratio I_2/I_1:

$$I_2/I_1 = \frac{I_{NC}(T_2) + C/T_2}{I_{NC}(T_1) + C/T_1} = \frac{I_{NC}(T_2)}{I_{NC}(T_1) + C/T_1} + \frac{C/T_2}{I_{NC}(T_1) + C/T_1} \tag{2}$$

When assuming $T_2 = T_1$ we obtain:

$$I_2 / I_1 = I_1 / I_1 = \frac{I_{NC}(T_1) + C / T_1}{I_{NC}(T_1) + C / T_1} = \frac{I_{NC}(T_1)}{I_{NC}(T_1) + C / T_1} + \frac{C / T_1}{I_{NC}(T_1) + C / T_1} = \\ = X + \frac{C / T_1}{I_{NC}(T_1) + C / T_1} = 1, \tag{3}$$

where

$$X = \frac{I_{NC}(T_1)}{I_{NC}(T_1) + C / T_1} \tag{4}$$

and

$$1 - X = \frac{C / T_1}{I_{NC}(T_1) + C / T_1} \tag{5}$$

The eq. (4) and (5) can also be written as:

$$X / I_{NC}(T_1) = \frac{1}{I_{NC}(T_1) + C / T_1} \tag{4a}$$

$$\frac{1 - X}{C / T_1} = \frac{1}{I_{NC}(T_1) + C / T_1} \tag{5a}$$

By inserting eq. (4a) and (5a) into the eq. (2) we obtain:

$$I_2 / I_1 = \frac{I_{NC}(T_2)}{I_{NC}(T_1) + C / T_1} + \frac{C / T_2}{I_{NC}(T_1) + C / T_1} = \frac{I_{NC}(T_2)}{I_{NC}(T_1)} X + \frac{C / T_2}{C / T_1}(1 - X) \tag{6}$$

By denoting the parameter

$$D_{NC} = I_{NC}(T_2) / I_{NC}(T_1) \tag{7}$$

and introducing it into the eq. (6) we get

$$I_2 / I_1 = D_{NC} X + (T_1 / T_2)(1 - X) \tag{8}$$

Finally, after the transformation of eq. (8) the formula for calculating the fraction X of paramagnetic centres not-fulfilling the Curie law is obtained:

$$X = \frac{I_2 / I_1 - T_1 / T_2}{D_{NC} - T_1 / T_2} \tag{9}$$

CARBONACEOUS SOLIDS WITH PAULI PARAMAGNETISM

Carbon Black

In order to obtain information on possible deviations from the Curie law behavior by the determination of the ratio of line intensities in the EPR spectra of heat-treated carbons, Antonowicz [1 - 3] has conducted two-temperature measurements (at room temperature and at liquid-nitrogen temperature) for P33 carbon black, cokes and carbonized sugar. His studies gave an insight into the behavior of the spin centres. When the Curie law is obeyed, the ratio of the line intensity at liquid nitrogen temperature (77 K) to the line intensity at room temperature (293 K) should be nearly equal to about 4 (293/77 = 3.8). It was found that for heat-treated samples of P33 carbon black, the values of this line intensity ratio were in the range of 0.4-2.1. In the case of P33 carbon black, he came to a conclusion that unpaired spins undergo a variation from localized to conduction band electron type. The spin centres in carbonized sugar and various cokes obeyed nearly exactly the Curie law [1 - 3]. However, after the adsorption of oxygen, the ratio of the line intensity at liquid nitrogen temperature to that at room temperature was found to be 2.3 [3].

The results of the two-temperature measurements obtained by Antonowicz [1 - 3] were verified and confirmed by Yokozawa [4] for P33 and Thermax carbon blacks. Since Antonowicz [1 - 3] has measured the temperature dependence of the magnetic resonance absorption line intensity only at two temperatures, namely at liquid nitrogen and room temperature, Yokozawa [4] decided to obtain more complete data and clarify the nature of changes occurring with the degree of graphitization of the carbon black samples by studying them in a continuous way – from Curie law behavior to metallic conduction. He measured carbon samples in the range of 130-300 K and found that for samples heat-treated at lower temperatures, the resonance line intensity I increases continuously with a decrease in temperature in accordance with the Curie law, as it was expected for localized spin centres. However, in the case of samples heat-treated at higher temperatures, the resonance line intensity increased slowly with an increase in temperature, which is the behavior typical of the paramagnetism on conduction carriers in graphite. In the samples, an increase in the temperature of heat treatment resulted in the transition from localized to delocalized character.

Mrozowski [5] has studied the temperature dependence of the EPR signal in carbons and graphites. The abovc author used an incompletely graphitized P33 carbon black that showed the presence of both conduction and localized spin centres because the temperature dependence was intermediate between metallic and Curie type. He has also performed variable-temperature EPR studies [6] of a series of variously heat-treated samples of neutron-irradiated, incompletely graphitized, P33 carbon black – heat treatment temperature (HTT): 2600 °C (EPR measurement temperature range: 80-300 K). The temperature (T) dependence of the EPR line intensity (I) was measured for the aforementioned samples. It has been assumed that both the conduction carriers and the localized spin centres contribute independently to the EPR line intensity. The individual contributions can be obtained by resolving the line intensity I *versus* temperature T curve into two parts, a pure Curie part for localized electrons with $I_L = C/T$ and a part for conduction electrons, in the case of which the temperature dependence in the form $I_c = A + BT$ is assumed. Measurements carried out for graphite have shown that the intensity at 80 K decreased to 88% relative to the value at the 300 K. For each of the carbon black samples examined by EPR in Ref [6], the values of the fraction X of the total intensity at room temperature T_{RT} due to conduction carriers can be calculated from the mix formula:

$$R = 0.88 \cdot X + 3.65 \cdot (1 - X), \tag{10}$$

where $R = I_{LN}/I_{RT}$ denotes the ratio of the line intensities I_{LN} and I_{RT}, at liquid nitrogen temperature T_{LN} = 80 K and at room temperature T_{RT} = 293 K, respectively. Mrozowski [6] has stated that it is sufficient and fully justified to measure the EPR line intensities at two temperatures to determine the two independent contributions from conduction carriers and localized spins. It has been found that at room temperature (293 K) 81.5% of the EPR line intensity of the original P33 carbon black is due to conduction carriers and only 18.5% due to localized spins.

In his review paper [7] Mrozowski reported that localized spin defects and/or conduction carriers were suggested as possible absorption centres in carbons. In his earlier studies [5, 6] on P33 carbon black (HTT: 2600°C), performed in the temperature range of 78-300 K, the clarification was achieved. The EPR line intensity can be split into two components: I_L that strictly fulfills the Curie law $I_L = C/T$, and I_c – the Pauli temperature-independent part (or slightly dependent on temperature as follows from theoretical band calculations, *e.g.* for graphite). In the equation for the total line intensity:

$$I = I_c + I_L = X \cdot I + (1-X) \cdot I \tag{11}$$

the quantity X is the fraction of the EPR line intensity due to the conduction carriers at a given temperature (the meaning of the symbols in eq. (11) is the same as in eq. (1); the indices in eq. (11) were taken from the paper by Mrozowski [7]). Paramagnetism, which is due to localized spin centres, follows the Curie law, whereas the free carrier spin centres give rise to the Pauli paramagnetism. To find the fraction X_{RT} at room temperature, Mrozowski [7] has presented the generalized form of eq. (10):

$$R = I_{80} / I_{RT} = \mu \cdot X_{RT} + 3.65 \cdot (1 - X_{RT}), \tag{10a}$$

where the coefficient μ is equal to 0.88 for graphite, and 1 for a pure Pauli paramagnetism case (the ideal case of total absence of localized spin centres) and for instance $T_{RT} = 293$ K and $T_{LN} = 80$ K.

By using the Mrozowski's [7] empirical additivity law for paramagnetism, Delhaes and Carmona [8] have calculated (for two temperatures of EPR measurements: $T_{LHe} = 4.2$ K and $T_{RT} = 300$ K) the respective proportions of Curie and Pauli paramagnetisms in P33 carbon black. To interpret the thermal variations of the g-factor and the linewidth the authors examined different macroscopic models of two-spin systems in strong interaction which have explained the additivity of the g-factor and linewidth for localized centres and free carriers. According to Mrozowski [7], the additivity of g-factor and linewidth ΔB_{pp} can be written similarly to eq. (11) in the form:

$$g = X \cdot g_c + (1-X) \cdot g_L \tag{12}$$

and

$$\Delta B_{pp} = X \cdot \Delta B_{pp,c} + (1-X) \cdot \Delta B_{pp,L}, \tag{13}$$

respectively. The aforementioned additivity of the EPR properties due to the existence of two-spin systems helps to understand physical properties, *e.g.* such as the negative magnetoresistance.

Pyrolytic carbon

In the study on negative magnetoresistivity in pyrolytic carbon, Rozpłoch [9] carried out EPR measurements at temperatures $T_1 = 100$ K and $T_2 = 295$ K. He has determined the total number of spins N present in the samples studied (for $T_2 =$ 295 K) heated at different temperatures in the temperature range of 1973-2873 K and the number of spins N_L in localized states. Rozpłoch [9] employed, after the work by Mrozowski [7], the formula for the determination of the fractional part of unpaired spins of charge carriers from stretched states

$$I(T_1)/I(T_2) = w \cdot X + (T_2/T_1) \cdot (1-X), \tag{14}$$

where $I(T_1)$ and $I(T_2)$ are areas under the EPR absorption line measured at temperatures T_1 and T_2 (with numerical factor w for graphite equal to 0.88), respectively. In the case of samples heated at 1973 K and 2873 K, the fractions X were equal to 0.762 and 0.996, respectively.

Multi-walled Carbon Nanotubes (MWCNTs)

The two-temperature EPR measurements were also carried out by Szroeder, Rozpłoch and Marciniak [10] who studied the macroscopic bundles of multi-walled carbon nanotubes (MWCNTs). The aim of these experiments was to determine the relative contributions of the conduction π-electron spins and the localized spin centres. Depending on the chirality and diameter, the individual nanotubes are predicted to reveal either metallic or semiconducting properties. For metallic nanotubes, the defects of bonds between cylindrical sheets can be trapping sites for conduction electrons. The localized spin centres near defect states show Curie-like temperature dependence, by contrast the EPR temperature dependence of conduction electron spins is close to that predicted by the Pauli paramagnetism. The comparison of EPR signal intensity using the two-temperature measurement method permits to determine the relative contributions of both types of spin centres.

The shapes of the experimental EPR line in bulk carbon nanotubes were strongly asymmetric due to the skin effect. An analysis by using a modified Dyson's theory led to the conclusion that the resonance lines reflect conduction electron spin resonance (CESR) phenomena within MWCNTs bundles as a whole.

The intensity of the absorptive part of the Dysonian resonance line is contributed to both the conduction electrons and localized spin centres at the tube tips and defects such as deformed azulene structures (heptagon-pentagon rings) on the graphene cylinders. Two-temperature measurements allowed the subdivision of the total intensity of the absorptive part of the resonance line I into two components that display the Pauli- ($I_{\mathrm{Pauli}} = A + B \cdot T$) and Curie-like ($I_{\mathrm{Curie}} = C/T$) temperature dependence, respectively, where $I = I_{\mathrm{Pauli}} + I_{\mathrm{Curie}}$. If $X = I_{\mathrm{Pauli}}/I$ denotes the contribution of delocalized electron states then,

$$I(T_{LN})/I(T_{RT}) = \frac{A+BT_{LN}}{A+BT_{RT}} X + (T_{RT}/T_{LN})(1-X) \quad \textbf{(15)}$$

Neglecting that π-electrons contribute to the creation of holes in the valence band, it was estimated using the equation (15) that ~80% of spin states are delocalized and Pauli-like temperature behavior occurs. Much lower X in the bundles of MWCNTs in comparison to that in the annealed pyrolytic carbon can be interpreted as a consequence of the tubular geometry of graphene layers being conducive to the creation of trapping sites for conduction π-electrons. This result suggests that more defects are present in MWCNTs than in pyrolytic carbon or graphite. Topological imperfections and open edges of graphite domains on the tips cause an increase in the trapped electron density. That leads to a decrease in charge carrier mobility.

COAL MACERALS WITH THERMALLY EXCITED TRIPLET STATES

In all the previously mentioned papers, the two-temperature EPR measurement method was used for determining the contribution of electron spin centres responsible for the Pauli and the Curie law paramagnetism.

Recently, Słowik and Więckowski [11] applied the two-temperature EPR measurement method to coal macerals (exinite and vitrinite) for distinguishing between the paramagnetic centres in the doublet ($S = 1/2$) and thermally excited triplet ($S = 1$) states, which fulfill and do not fulfill the Curie law, respectively. The concept of macerals is the most often applied one when describing the petrographic structure of coal. The macerals – exinite, vitrinite, and inertinite, are the basic petrographic components of coal. Their properties are dependent on the coal rank with disordered internal structure. Exinite, vitrinite, and inertinite are microscopically distinguishable on the basis of the reflectivity of light and their colour.

In our study [11] of the coal macerals – exinite and vitrinite, performed by the two-temperature EPR measurement method, we have aimed at determining the relative contribution X of the paramagnetic centres not-fulfilling the Curie law and comparing them with the X values for thermally decomposed coal macerals. The existence of paramagnetic centres in the ground doublet state $S = 1/2$, for which the temperature dependence of intensity is described by the Curie law ($I = C/T$) and paramagnetic centres in excited triplet states ($S = 1$) was evidenced in coal and coal macerals.

The EPR spectra of the three macerals – exinite, vitrinite, and inertinite were subjected to the decomposition into four component lines of Gaussian (G) and/or Lorentzian (L1, L2, L3) lineshape with different linewidths ΔB_{pp} fulfilling the inequalities $\Delta B_{pp}(G) > \Delta B_{pp}(L1) > \Delta B_{pp}(L2) > \Delta B_{pp}(L3)$ (Fig. **1**). The EPR spectra of exinite and vitrinite consist of three lines, one line with Gaussian lineshape (G) and two lines with Lorentzian lineshape (L1, L3). The EPR spectrum of inertinite consists of two lines with Lorentzian lineshape (L2, L3). It is only the Lorentzian line L1 from these four component lines (G, L1, L2, L3) that does not fulfill the Curie law.

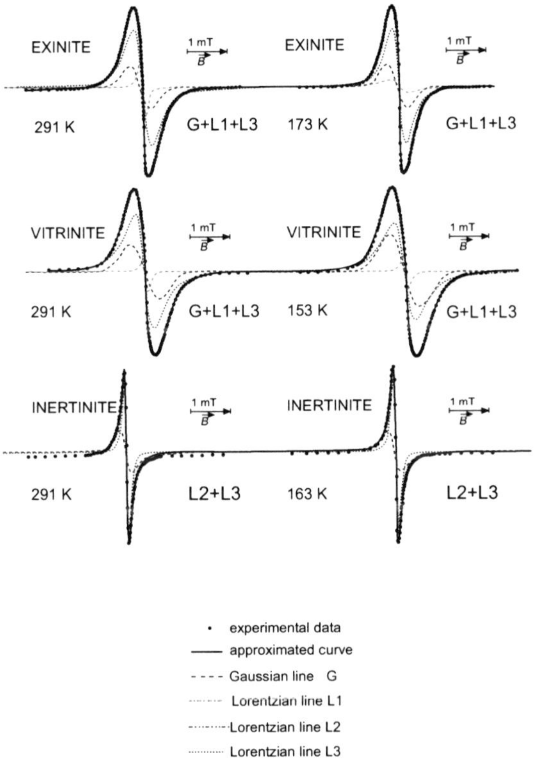

Fig. (1). Decomposition of the EPR spectra of exinite, vitrinite and inertinite.

Because inertinite does not contain paramagnetic centres not-fulfilling the Curie law, it was not studied further. In the total group of paramagnetic centres generating component lines in the form of Lorentzian L1 shape in EPR spectra of coal macerals – exinite and vitrinite, there is a part of centres in thermally excited triplet state (S = 1). Their contribution to the total line intensity I_{NC} is described with the formula derived by Bleaney and Bowers [12]:

$$I_{NC} = \frac{B}{T\left[\exp(J/kT)+3\right]}, \quad (16)$$

where I_{NC} – line intensity generated by spin centres in the triplet state (S = 1) [arb.u.], T – temperature [K], J – excitation singlet-triplet energy [J], k – Boltzmann constant [J/K].

The two-temperature EPR measurements were carried out at T_1 = 293 K and T_2 = 173 K for exinite and at T_1 = 293 K and T_2 = 153 K for vitrinite, respectively. The calculated values of parameter D_{NC} from eq. (7):

$$D_{NC} = I_{NC}(T_2)/I_{NC}(T_1) = (T_1/T_2)\frac{\exp(J/kT_1)+3}{\exp(J/kT_2)+3} \quad (17)$$

for the studied macerals are 0.689 for exinite and 0.125 for vitrinite, and they denote the line intensity ratio D_{NC} of paramagnetic centres which do not fulfill the Curie law, measured at temperatures T_1 and T_2. In our case the eq. (8) assumes the form:

$$I_2/I_1 = D_{NC}X + (T_1/T_2)(1-X) = $$
$$= (T_1/T_2)\frac{\exp(J/kT_1)+3}{\exp(J/kT_2)+3}X + (T_1/T_2)(1-X) \quad (18)$$

The calculated values of the relative contribution X from eq. (9):

$$X = \frac{I_2/I_1 - T_1/T_2}{D_{NC} - T_1/T_2} = \frac{I_2/I_1 - T_1/T_2}{(T_1/T_2)\left\{\left[\exp(J/kT_1)+3\right]/\left[\exp(J/kT_2)+3\right]\right\} - T_1/T_2} = \quad (19)$$
$$= \frac{(I_2/I_1)(T_2/T_1)-1}{\left\{[\exp(J/kT_1)+3]/[\exp(J/kT_2)+3]\right\}-1}$$

of paramagnetic centres which do not fulfill the Curie law in exinite and vitrinite at room temperature $T_1 = 293$ K are $X = 0.36$ for exinite and $X = 0.17$ for vitrinite. The existence of paramagnetic centres in thermally excited triplet state ($S = 1$) in the whole group of paramagnetic centres present in exinite and vitrinite is one of several models explaining the observed temperature dependence of EPR line intensity.

Pilawa [13] has applied the method of two-temperature measurement to study the system of paramagnetic centres responsible for the broad resonance line component having the Lorentzian lineshape (denoted L1) occurring in the two coal macerals – exinite and vitrinite. She obtained numerical values of the relative contribution X for paramagnetic centres which do not fulfill the Curie law in exinite and vitrinite, thermally decomposed at 573 K (300 °C) and 923 K (650 °C). The values of X determined were 0.32 and 0.28 (for exinite heated at 573 K and 923 K) and 0.41 and 0.50 (for vitrinite heated at 573 K and 923 K), respectively. Pilawa [13] came into the conclusion that delocalized π-electrons of aromatic units are responsible for deviation from the Curie law in exinite and vitrinite.

CONCLUSION

The two-temperature EPR measurement method is a valuable tool for the determination of relative contributions of two spin systems with different types of mechanisms of interaction between paramagnetic centres present in carbon-based materials. The method was exemplified on carbonaceous samples containing paramagnetic centres responsible for the Curie- and the Pauli-like paramagnetism. It has also been applied to determine the concentration of paramagnetic centres in coal macerals being in spin doublet states ($S = 1/2$) and in excited spin triplet states ($S = 1$). The two-temperature EPR measurement method is based on the existing temperature dependence of the EPR line intensity I of a single line or a decomposed component of a multicomponent signal. Measurements of the EPR line intensities I at two different temperatures T_1 and T_2 (*e.g.* at room temperature and at liquid-nitrogen temperature) make it possible to distinguish between the two spin systems and to determine at a given temperature the relative contribution X, defined by eq. (4), of paramagnetic centres not-fulfilling the Curie law present in the samples studied.

CONSENT FOR PUBLICATION

Not applicable.

CONFLICT OF INTEREST

The author (editor) declares no conflict of interest, financial or otherwise.

ACKNOWLEDGEMENT

The authors are grateful to Professor Franciszek Rozpłoch (Faculty of Physics, Astronomy and Applied Informatics, Nicolaus Copernicus University in Toruń) for his valuable information on the two-temperature EPR measurement method.

REFERENCES

[1] Antonowicz, K. Formation of Spin Centers in Carbons by Oxidation. *J. Chem. Phys.*, **1962**, *36*, 2046-2053.
[http://dx.doi.org/10.1063/1.1732826]

[2] Antonowicz, K. Formation by Chemical Attack and Nature of Spin Centers in Carbon *Proceedings of the Fifth Carbon Conference, Volume I,* **1962**, pp. 56-62.
[http://dx.doi.org/10.1016/B978-0-08-009707-7.50011-X]

[3] Antonowicz, K. The Interaction of Strongly Adsorbed Oxygen with Free Spin Centers in Carbon. *Carbon,* **1964**, *1*, 111-115.
[http://dx.doi.org/10.1016/0008-6223(64)90064-8]

[4] Yokozawa, Y. Temperature Dependence of Magnetic Resonance Absorption in Polycrystalline Carbons and Graphite. *J. Chem. Phys.*, **1962**, *37*, 204-205.
[http://dx.doi.org/10.1063/1.1732966]

[5] Mrozowski, S. ESR in Doped and in Irradiated Polycrystalline Graphites. *Bull. Am. Phys. Soc.*, **1963**, *8*, 344.

[6] Mrozowski, S. Electron Spin Resonance in Neutron Irradiated and in Doped Polycrystalline Graphite – Part I. *Carbon,* **1965**, *3*, 305-320.
[http://dx.doi.org/10.1016/0008-6223(65)90065-5]

[7] Mrozowski, S. Specific Heat Anomalies and Spin-Spin Interactions in Carbons: A Review. *J. Low Temp. Phys.*, **1979**, *35*, 231-298.
[http://dx.doi.org/10.1007/BF00115580]

[8] Delhaes, P.; Carmona, F. Etude par resonance paramagnetique electronique entre 4,2 et 300 K d'une famille de noirs de carbone. *Carbon,* **1972**, *10*, 677-690.
[http://dx.doi.org/10.1016/0008-6223(72)90075-9]

[9] Rozpłoch, F. *Negative magnetoresistivity in pyrolytic coal. Dr. hab*; Nicolaus Copernicus University, **1984**, pp. 45-, 56-57, 74. (in Polish)

[10] Szroeder, P.; Rozpłoch, F.; Marciniak, W. Two-Temperature EPR Measurements of Multi-Walled Carbon Nanotubes, Interfacial Effects and Novel Properties of Nanomaterials. *Diffus. Defect Data Solid State Data Pt. B Solid State Phenom.*, **2003**, *94*, 275-278.
[http://dx.doi.org/10.4028/www.scientific.net/SSP.94.275]

[11] Słowik, G.P.; Więckowski, A.B. Determination of the fraction of paramagnetic centers not-fulfilling the Curie law in coal macerals by the two-temperature EPR measurement method. *Nukleonika,* **2015**, *60*, 389-393.
[http://dx.doi.org/10.1515/nuka-2015-0069]

[12] Bleaney, B.; Bowers, K.D. Anomalous Paramagnetism of Copper Acetate. *Proc. R. Soc. Lond. A Math. Phys. Sci.*, **1952**, *214*, 451-465.
[http://dx.doi.org/10.1098/rspa.1952.0181]

[13] Pilawa, B. *Evolution of paramagnetic centres during thermal decomposition of coal macerals. Dr. hab*; Pedagogical University of Zielona Gora, **2000**, pp. 109-110. (in Polish)

CHAPTER 9

Paramagnetic Defects and Impurities in Nanodiamonds as Studied by Multi-frequency CW and Pulse EPR Methods

Victor Soltamov [1,2,*], **George Mamin**[2], **Sergei Orlinskii**[2] and **Pavel Baranov**[1]

[1] *Microwave Spectroscopy of Crystals Laboratory, Ioffe Institute, RAS, St Petersburg, 194021, Russian Federation*

[2] *Department of Quantum electronics and radiospectroscopy, Kazan Federal University, Kazan, 420000, Russian Federation*

Abstract: Spin properties of defects in carbon nanostructures are one of the fundamental directions in the physics of nanomaterials. The problem of doping nanostructures and creating intrinsic and extrinsic defects in such structures as a result of various actions (heat treatment, ionizing radiation, chemical action, *etc.*) is playing the central role in the further implementation of these nanostructures in real devices. Electron paramagnetic resonance (EPR) is known to be one of the most informative methods to study the intrinsic and extrinsic defects at the molecular level. The use of different frequency bands (low frequency X-band or high frequency W-band) and different regimes of the EPR signal detection (continuous or pulsed) allows one to get an access not only to the identification an electronic and microscopic structure of the defects in the crystalline matrix, but also to study coherence properties of the defects' spin. In this chapter, by means of EPR, we provide the direct observation of paramagnetic impurities in the crystalline core of nanodiamonds and we also show that nitrogen impurities in nanodiamonds interact with the diamond lattice in a similar way as in the bulk diamond crystals. We also present the results of observation of high-density NV defect ensembles created directly by high-pressure high-temperature (HPHT) sintering procedure of the detonation nanodiamonds and show that the spin ensemble of the NV defects is characterized by the long spin-lattice and spin-spin relaxation times. The latter is important for bioimaging and quantum sensing applications.

Keywords: Angular dependence, Detonation nanodiamonds, EPR, Electron spin echo (ESE), ESE detected EPR, Euler angles, Fine structure, High-frequency EPR, EPR spectra simulation, HPHT nanodiamonds, Hyperfine interaction,

* **Corresponding author Victor Soltamov:** Microwave Spectroscopy of Crystals Laboratory, Ioffe Institute, RAS, St Petersburg, 194021, Russian Federation and Department of Quantum electronics and radiospectroscopy, Kazan Federal University, Kazan, 420000, Russian Federation; Tel: +7 (812) 292-7320; Fax: +7 (812) 297-1017; E-mail: victrosoltamov@gmail.com

Dariya Savchenko & Abdel Hadi Kassiba (Eds.)

Multifrequency EPR, Natural diamond nanocrystals, Nitrogen centers, Nitrogen pairs, NV centers, Spin Hamiltonian, Size effect, Symmetry, Synthetic nanodiamonds.

INTRODUCTION

Electron paramagnetic resonance (EPR) was discovered by Zavoisky, in 1944 [1]. EPR related methods such as Electron Nuclear Double Resonance, Double Electron-Electron Resonance and Optically Detected Magnetic Resonance, are one of the most informative tools for detecting point defects and impurities in solids. These methods offer a uniquely sensitive probe of chemical identification, local structural order, the atomic-scale environment of the defect, can provide details of electron density distributions. The microscopic structure of many intrinsic and impurity defects in solids was studied and established by EPR.

Spin is a purely quantum-mechanical object and spin phenomena begin to play a crucial role in the development of various instruments and devices based on nanostructures. Methods of magnetic resonance are the basic techniques for studying the spin phenomena in condensed matter and biological systems [2 - 4].

Understanding the structure and constituents of defects in nanostructures is important since their presence can greatly affect the properties of the material. Pulsed high field EPR spectroscopies were shown to be excellent tools for the investigation of nanostructures [5, 6] and the electronic properties of semiconductor quantum dots (QDs) (see [7] and references therein).

This chapter covers the investigations of diamond nanoparticles by multi-frequency CW and pulse EPR methods.

STUDY OF NITROGEN CENTRES IN NANODIAMONDS

Nanodiamonds (NDs) have attracted increasing attention worldwide after a number of breakthroughs in synthesis, purification, isolation and their surface modification techniques achieved in the late 1990s [8]. At present, NDs can be synthesized at a relatively low cost by various synthesis techniques: detonation, laser ablation, high-energy ball milling of high-pressure high-temperature (HPHT) diamond microcrystals, plasma-assisted chemical vapour deposition (CVD), autoclave synthesis from supercritical fluids, chlorination of carbides, ion irradiation of graphite, electron irradiation of carbon onions [9].

Here we will consider the several ND particles, formed by the detonation of strong explosives, the so-called detonation nanodiamond (DND), natural diamond nanocrystals and NDs synthesized by HPHT technique.

Nitrogen is the main impurity in diamonds and the form in which nitrogen is present in diamonds largely determines their properties and serves as the leading factor of the diamond classification. Nitrogen creates various paramagnetic centers in a diamond and exists as individual atoms and nitrogen clusters. Particularly the presence of the paramagnetic nitrogen impurities as well as the size-dependent concentration of nitrogen paramagnetic centres in HPHT nanodiamonds will be shown in the diamond core of each type of NDs.

Let us start from the W-band (≈ 94 GHz) EPR experiments performed to study natural nanodiamond crystals with an average size in the range from 0 to 250 nm [10]. Fig. (**1a**) presents the W-band CW EPR spectrum of the sample. Spectrum exhibit an intense central line corresponding to an electron *g* factor of 2.0027 and weak satellites located symmetrically with respect to the central line (shown enlarged).

The intense central line is commonly attributed to the broken bonds on the surface of diamond micro and nanocrystals [11, 12]. Interpretation of the satellite lines is difficult, because of the overlapping of these satellites and the intense signal from the surface centers. Meanwhile, the satellites nearest to the central line were previously observed in natural diamond microcrystal powders with a size of greater than 1 μm and had been attributed to individual nitrogen atoms N^0 [11]. It is obvious that to perform the qualitative interpretation of the satellites observed in nanosized diamond crystals, one need to suppress the signal arises due to the surface broken bonds. It should be mentioned that all of the lines shown in Fig. (**1a**) can be observed in a wide temperature range up to room temperature so that the respective centers could not be separated by temperature variation.

One of the advantages of the high-frequency pulsed technique used in the study is that it allowed us to separate contributions from the paramagnetic centers presented in the nanodiamond crystalline core and the surface centers spectrally, leading to the enhancing visibility of the satellite signals. So, using long interpulse distance the contribution from surface centers, *i.e.* broken bonds, was almost suppressed due to the short relaxation time of the latter. The result of such approach is presented in Fig. (**1b**). Looking ahead, such EPR spectrum is explained by the presence of two types of nitrogen centers in the nanocrystal diamond core: the single substitutional nitrogen centers in a neutral charge state N^0, so-called P1 centers [13] and nitrogen pairs N_2^+ [14].

A single nitrogen atom in a zero charge state N^0 substituting for a carbon atom is the most studied center in a diamond. It is characterized by a deep donor level with the energy of 1.7 eV relative to the conduction band [15], which is caused by strong lattice relaxation. N_2^+ is the simplest center among the nitrogen clusters and represents a pair of two equivalent nitrogen atoms in adjacent carbon positions with one common unpaired electron (W24 center [13, 16]).

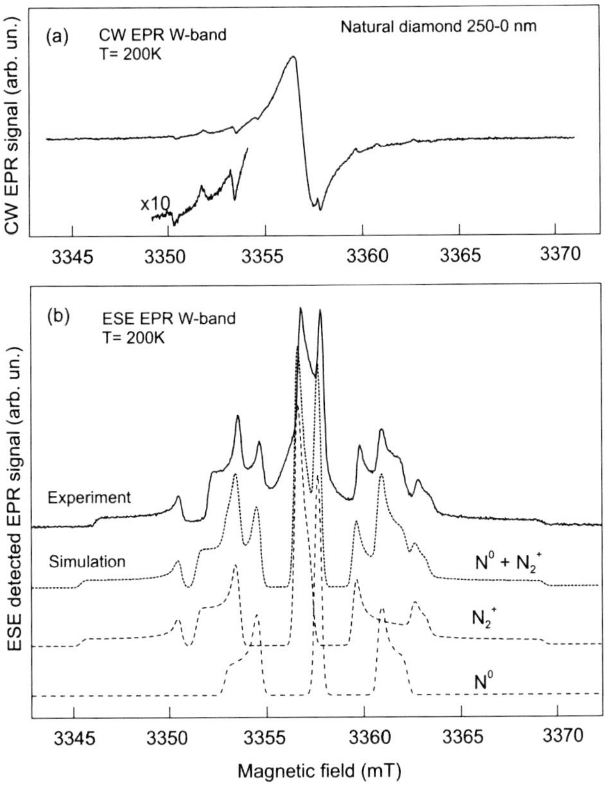

Fig. (1). (a) CW and (b) pulsed ESE W-band EPR spectra of the natural nanodiamond powder. The dotted lines are the simulated EPR spectra of the N^0 and N_2^+ centers and their sum. Partially Reproduced from Ref [10] with the Permission from Pleiades Publishing, Ltd. 2010.

The EPR spectra of the N^0 and N_2^+ centers are described by the following spin Hamiltonians:

$$\hat{H} = \beta_e \vec{B}\vec{g}_1\vec{S} + \vec{S}\vec{A}_1\vec{I}(N^0) \quad \textbf{(1)}$$

$$\hat{H} = \beta_e \vec{B}\vec{g}_2\vec{S} + \vec{S}\vec{A}_2(\vec{I}_1 + \vec{I}_2)(N_2^+), \quad \textbf{(2)}$$

where the first term in each expression corresponds to the Zeeman interaction of the magnetic field B with the spin $S = 1/2$ center having the electron g factor in the form of $\vec{g}_1$ and $\vec{g}_2$ tensors for the N^0 and N_2^+ centers, respectively; the second term is the hyperfine interaction in the form of the $\vec{A}_1$ and $\vec{A}_2$ tensors between the unpaired electron and one or two equivalent ^{14}N nuclei (nuclear spin $I = 1$) for the N^0 and N_2^+ centers, respectively. The simulated EPR spectra of the N^0 and N_2^+ centers, as well as the resultant sum of both simulated spectra, are shown by the dashed lines in Fig. (**1b**).

The parameters of the simulation are as follows: linewidth of 3 G and the parameters $g_{||1} = g_{\perp 1} = 2.0024$, $A_{||1} = 4.07$ mT, and $A_{\perp 1} = 2.9$ mT for the N^0 centers and $g_{||2} = 2.00245$, $g_{\perp 2} = 2.0030$, $A_{||2} = 5.538$ mT, and $A_{\perp 2} = 2.9$ mT for the N_2^+ centers. The good agreement between the simulated and experimental spectra is seen definitely confirming the proposed interpretation of the experimental EPR spectrum and may serve as a reference to control the concentration of paramagnetic nitrogen in diamond nanocrystals. The significant anisotropy in the *g* factor of nitrogen pairs and the absence of the anisotropy for atomic nitrogen centers seen from the experiment should be noted. Thus, using a high frequency pulsed EPR technique, we were able to resolve the EPR signals of the N^0 and N_2^+ centers and have determined a difference in their *g* factors.

We then focused on studying the paramagnetic centers in the core of the nanodiamonds formed by the explosive technique (DNDs) [18], which were produced through the detonation of strong explosives and have a size of 4-5 nm [17].

The ESE detected EPR signal at a frequency of 94 GHz in the DND sample is shown in Fig. (**2**). Since the detonation nanodiamond has a complex structure in the form of a diamond crystalline core and a shell consisting of sp^2 hybridized carbon atoms [19, 20], the EPR signals should present all of the features of the structure of the detonation nanodiamond. As was previously discussed, the separation of the ESE signals from the diamond core and surface shell can be performed by using different time sequences of microwave pulses, as shown in the Fig. (**2**) inset.

Spectrum 1 recorded at a temperature $T = 10$ K with a short interval $\tau = 230$ ns between the first and second pulses exhibits an intense double line with the g factors of 2.0030 and 2.0004, which was attributed to broken bonds on the surface shell of the detonation nanodiamond [12, 18].

The spectrum 1 shown in the enlarged scale (×5) demonstrates four additional lines on both sides of the central line. Enhancement of the line intensity becomes possible by increasing the time interval between the pulses, the fact clearly seen from spectrum 2, which was recorded at a temperature of 10 K with the interval $\tau = 800$ ns between the pulses (Fig. **2**). Owing to the suppression of the intense central lines associated with the shell, the spectrum consisting of five lines marked by the vertical lines, became observable ($g = 2.0037$ for the central line). Moreover, additional lines were seen in the fields of 3350–3360 mT. These lines were apparently due to the interaction with individual nitrogen atoms N^0. The simulation of the EPR spectrum, arising from the N^0 centers is shown by the dashed line and can be done in the same way as it was previously discussed spectrum. The EPR spectrum consisting of five lines (spectrum 2 in Fig. **2**) likely belongs to the multivacancy complex. A defect with the spin $S = 3/2$ consisting of three unpaired electrons attached to broken bonds, *i.e.*, the so-called R8 center [21], is the most appropriate candidate.

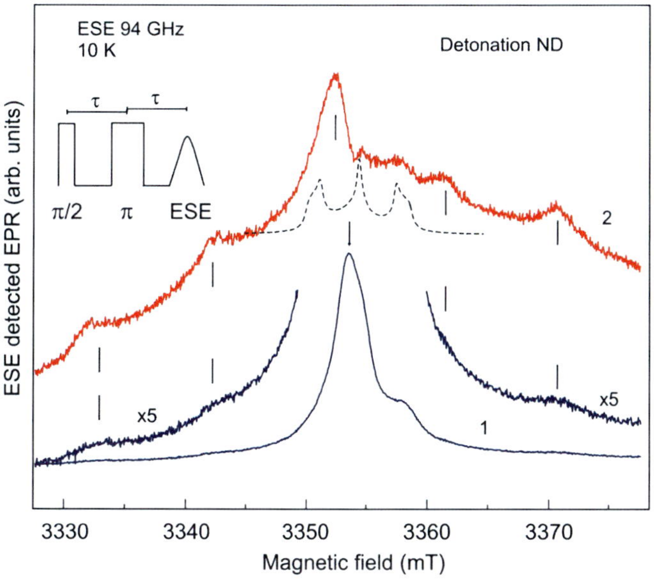

Fig. (2). The ESE detected EPR signal at a frequency of 94 GHz in a detonation nanodiamond at a temperature of 10 K. The interval between the pulses τ = (*1*) 230 and (*2*) 800 ns. The dashed line corresponds to the simulation of individual nitrogen centers N^0. Inset shows the pulse sequence. Partially Reproduced from Ref [18] with the Permission from Pleiades Publishing, Ltd. 2010.

The interesting feature of the nitrogen centers in nanodiamonds is that the concentration of the centers strongly depends on the nanoparticle size [22]. To demonstrate this, a series of the size-calibrated nanodiamonds produced by Tomei Diamond, Japan have studied: seven samples with the average size from 1 μm down to 10 nm, produced by HPHT synthesis from the diamond crystals with the high nitrogen content in the form of single substitutional nitrogen (type Ib diamonds). The sizes (*d*) of the nanodiamonds and their notations are presented in Table **1**. To identify paramagnetic centers of different origin, contributing to the EPR spectra the W-band CW EPR technique was used. To measure the concentration of the nitrogen centers in each sample notated in Table **1** conventional X-band spectroscopy was employed. As a reference MgO:Mn crystal powder with the concentration of paramagnetic species of 5 x 10^{14} spins per sample was used. Modeling of EPR spectra was realized using EasySpin toolbox for Matlab [23].

Table 1. Size of the diamond samples.

Sample notation	**Median size *d*, nm**
MD10	5-15
MD20	15-25
MD50	40-64
MD100	90-129
MD200	180-229
MD500	480-529
MD1000	980-1029

The W-band EPR spectra and their simulations under the assumption of the presence of three paramagnetic centers with S = 1/2 (denoted as N^0, "Surface center 1" and "Surface center 2") are shown in Fig. (**3**).

The signals "Surface center 1" and "Surface center 2" are out of the scope of this work, so they are not considered further.

The N^0 paramagnetic center shows up as a five lines powder EPR spectrum due to the anisotropic hyperfine interaction with ^{14}N (I = 1, 99.6% natural abundance), described by the spin Hamiltonian of the form:

$$\hat{H} = g\beta_e BS + A_{\parallel} S_z I_z + A_{\perp}(S_x I_x + S_y I_y) \quad \textbf{(3)}$$

with g-factor of 2.0024 and hyperfine interaction constants $A_{\perp}$ = 2.92 mT; $A_{\parallel}$ = 4.08 mT. The parameters are in a good agreement with the previously reported on the N^0 in the monocrystalline and powder diamond samples [13, 24].

To measure the number of the N^0 centers, X- and W-band EPR technique was used as follows. First, in X-band, the total concentration of spins in the sufficient amount of MD1000 (ca. 15 mg) was determined and then compared to the concentration in the reference sample. A comparison was done by exploiting the double integration of the CW EPR spectra. In order to keep the microwave tuning parameters unchanged, we used only MD1000: the samples with the smaller sizes shifted the tuning conditions away from reference sample conditions.

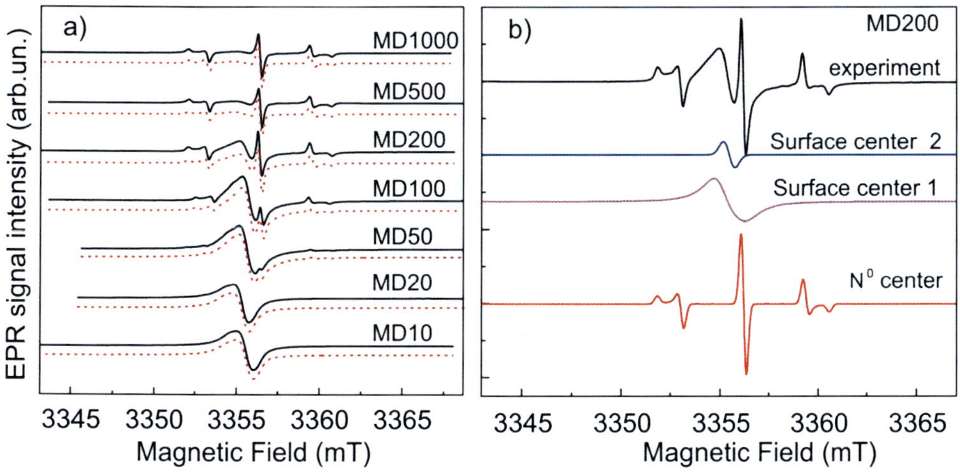

Fig. (3). (a) W-band CW EPR spectra of nanodiamond samples (solid lines) and their corresponding simulations (red dashed lines). (b) Example of the spectral decomposition of the EPR spectrum for MD200 assuming the presence of three contributions. The spectra are detected at room temperature. Reproduced from Ref [22] with the Permission from Kazan University.

Second, the determined total spin concentration was re-calculated to the mass of the MD1000 used for the W-band experiments. The total number of measured spins was equal to 3.2×10^{18} spin per gram. No changes in the tuning-matching conditions with the MD sample sizes were obtained in the W-band experiments that probably could be explained by the small amount of substances within the W-band tubes (less than 1 mg) and, therefore, the influence of the species on the microwave field distribution in the cavity is negligible.

Third, we have extracted the relative concentrations of the observed paramagnetic centers in a series of size-calibrated samples using MD1000 as a reference for the W-band. The results of the numerical measurements are presented in Table **2**.

Table 2. Paramagnetic species concentration.

Sample	**MD1000**	**MD500**	**MD200**	**MD100**	**MD50**	**MD20**	**MD10**
Concentration (spin per gram)	$1.4(1)\times10^{18}$	$1.2(1)\times10^{18}$	$0.8(1)\times10^{18}$	0.2×10^{18}	0.1×10^{18}	-	-

The interesting behaviour of the N^0 concentration follows from Fig. (**3**) and summarized in Table **2**. Namely, the lack of EPR signal from N^0 centers in the samples with the median size lower than 40-50 nm. The observation could be explained by either conversion of the neutrally charged nitrogen N^0 to paramagnetic-silent charge state or due to the nitrogen flushing outside of the diamond lattice region of the particle due to its energy-induced preferences [25]. Indeed, there are some reports for the nitrogen-vacancy (NV) charge conversion with the nanoparticle size [26] or induced by surface chemical treatment [27]. Thus it is shown that the concentration of the N^0 centers becomes negligible in HPHT diamonds characterized by dimensions less than 40-50 nm, probably this is due to their conversion to the non-paramagnetic state.

Up to this point, the important conclusion driven from the observation of both nitrogen in the neutral charge state and vacancy centers is that in its initial form, detonation nanodiamond has defects necessary for the formation of nitrogen-vacancy (NV) defects in diamond. The NV center in its negative charge state (NV^-) is playing one of the central roles in the nowadays development of quantum sensing and bioimaging applications, because it exhibits long ground-state spin coherence and optically induced spin polarization over a broad range of wavelengths in ambient conditions [28 - 30].

The next section is devoted to the study NV^- ensembles created directly by the HPHT sintering procedure of detonation nanodiamond (DND) particles.

NV CENTERS IN DIAMONDS CREATED BY SINTERING PROCEDURE OF DETONATION NANODIAMONDS

In this section, results of the high-density NV ensembles study, created directly by the HPHT sintering procedure, are presented [10, 31]. The sintering procedure can be done in two steps: purification of commercially available DND followed by the HPHT sintering procedure. The DND were purified to reduce the metal inclusions. Before sintering, DND were modified by C_{60} fullerenes with 99.5% purity and then sintered in a toroidal high-pressure chamber at temperature 800^0 C and pressure 6 GPa for 11 s. After the sintering, the powder was individual arrays with a linear size of about 10 μm. The sample image is shown in the right inset of Fig. (**4a**).

It should be noted here, that usually NV centers are produced by irradiation of bulk diamond, micro and nanodiamonds with ions, neutrons or high-energy electrons [32 - 34], aimed to the vacancies creation. Annealing at the temperatures of about 750–900^0 C (which are optimal for the diffusion of the vacancies) leads to the trapping of the vacancies to the nitrogen impurity atoms and creation of NV centers. In our research, the samples were not subjected to any irradiation procedure. Study of the paramagnetic centers spin properties was performed by means of W-band EPR technique.

Fig. (**4**) shows that the W-band ESE detected EPR spectra at a frequency of 94

GHz at room temperature for several orientations of the magnetic field in a single 10 × 10 × 10 μm array. The spectra were recorded at a microwave power of about 100 mW with a $\pi/2$ pulse of 48 ns, an interval between the pulses of 260 ns, and a pulse repetition time of 2 ms. Intense anisotropic EPR signals were observed. Two different groups of the signals can be identified; the group consisting of three lines in the range of 3345–3360 mT with a relatively small anisotropy and the group consisting of EPR lines with high anisotropy, covering wide magnetic field range of 3240–3460 mT. The important conclusion follows from the observation of the EPR signals anisotropy, namely the diamond structure produced through the HPHT synthesis is oriented. The central group of lines, shown in the inset for one of the orientations, was discussed in the previous section and was attributed the N^0 centers. The angular dependencies of the EPR spectra detected in the range of 0°–180° with a step of 10° are shown by the closed circles in Fig. (**4b**). The ESE spectra are also shown for several angles for illustration. To reveal the nature of these lines, we calculated the angular dependences of the EPR spectra with the spin Hamiltonian

$$\hat{H} = g\beta_e BS + D\left[S_z^2 - 1/3S(S+1)\right], \qquad \textbf{(4)}$$

where the first term presents the Zeeman interaction in the magnetic field for a center with the electron spin S and isotropic electron g factor, is the Bohr magneton, D is the parameter characterizing the splitting of the fine structure in the axial crystal field, and z is the defect axis. The observed angular dependencies are characteristic of the system with the electron spin $S = 1$ in the axial field with the C_{3v} symmetry.

The initial point for the calculation of the EPR spectra is the presence of a signal in which two pairs of lines are observed: one pair with the maximum splitting near 200 mT and the other pair with a large integral intensity and a smaller splitting. Assuming that this orientation is close to the orientation of the magnetic field along the <111> axis of the diamond (this spectrum in Figs. (**4a** and **4b**) is marked as $\theta = 0°$), for the calculation we took the following Euler angles for the four orientations of the <111> axis, which exist in the diamond crystal: $\alpha = 0°$, $\beta = 0°$, $\gamma = 0°$; $\alpha = 0°$, $\beta = 110°$, $\gamma = 0°$; $\alpha = 120°$, $\beta = 110°$, $\gamma = 0°$; and $\alpha = 240°$, $\beta = 110°$, $\gamma = 0°$.

The calculation was performed at the azimuth angle $\phi = 45°$. Results are shown by the solid lines in Fig. (**4b**). It should be noted that following the Fig. (**4b**) not all of the experimental points can be described with the above Euler angles and that the number of these points is twice as large. Explanation of the existence of additional lines is the formation of twin regions in the diamond system. Angular dependences of twins can be calculated by adding the four centers with the Euler angles $\alpha = 0°, \beta = 0°, \gamma = 0°$; $\alpha = 0°, \beta = 250°, \gamma = 0°$; $\alpha = 120°, \beta = 250°, \gamma = 0°$; and $\alpha = 240°, \beta = 250°, \gamma = 0°$. Such calculated angular dependencies for twins are shown by dashed lines. The good agreement of the measured and calculated angular dependences is seen thus approve the correctness of the chosen model. Parameters of the Hamiltonian used for simulation are as follows: $g = 2.0028$ and $D = 0.0958$ cm^{-1} and coincide well with the known parameters for the nitrogen-vacancy defects in bulk diamond crystals [35].

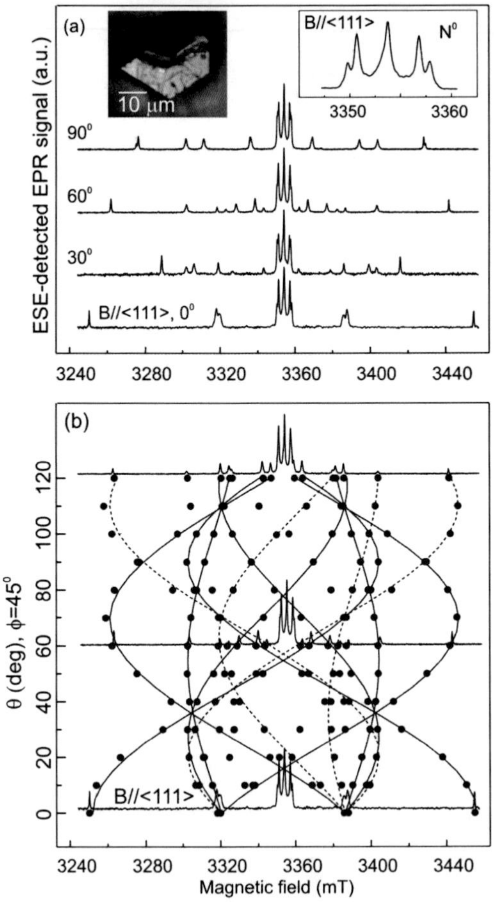

Fig. (4). (a) The W-band EPR signal detected at room temperature in the sample obtained by sintering DNDs at $T = 800$°C and $P = 6$ GPa for several orientations of the magnetic field with respect to the <111> axis. Inset in the right shows the image of the sintered diamond sample used. The left inset shows the central part of the EPR spectrum for $B \parallel$ <111> corresponding to the N^0 centers. (b) The (points) measured and (lines) calculated angular dependences of the EPR spectra on the direction of the magnetic field B with respect to the <111> axis ($\phi = 45°$). Partially reproduced from Ref [18] with the Permission from Pleiades Publishing, Ltd. 2010.

Thus, by means of EPR technique, the presence of the NV^- centres in HPHT sintered DNDs was shown. Moreover, the presence of the vacancies and N^0 centers in the initial DND source allows one to obtain the resultant microsized diamonds with a high concentration of the NV^- centers without any irradiation procedure.

CONCLUSIONS AND OUTLOOK

By means of multi-frequency CW and pulse EPR methods the nitrogen-related paramagnetic centers were studied in natural and sintered nanodiamonds. The presence of the isolated nitrogen centers N^0 and nitrogen pairs N_2^+ have been detected and identified, and their structure has been unambiguously determined. Based on the combination of the conventional X- and W-band EPR measurements size-dependent effects were studied in HPHT nanodiamonds. Namely, it was shown that the reduction of the diamond size down to the 50 nm give rise to the conversion of the neutral charge state of the N^0 centers into the non-paramagnetic state.

Study of the detonation nanodiamonds sintered at high pressure and temperature revealed the presence of the negatively charged nitrogen-vacancy centers and showed the possibility to obtain the high concentration of the latter without the irradiation procedure. Moreover, the observed angular dependence of the NV centers explicitly corresponds to that observed in the single crystalline diamond, thus the novel way of fabrication the micron-sized diamond crystals by means of HPHT synthesis of nanosized diamond particles was demonstrated.

CONSENT FOR PUBLICATION

Not applicable.

CONFLICT OF INTEREST

The author (editor) declares no conflict of interest, financial or otherwise.

ACKNOWLEDGEMENT

G.M. and S.O.acknowledge support by the Russian Science Foundation (Project № 18-13-00149).

REFERENCES

[1] Zavoisky, E. The paramagnetic absorption of a solution in parallel fields. *J. Phys.-USSR.*, **1944**, *8*, 377-380.

[2] Yavkin, B.V.; Mamin, G.V.; Orlinskii, S.B.; Gafurov, M.R.; Salakhov, M.Kh.; Biktagirov, T.B.; Klimashina, E.S.; Putlayev, V.I.; Tretyakov, Y.D.; Silkin, N.I. Pb3+ radiation defects in

$Ca_9Pb(PO_4)_6(OH)_2$ hydroxyapatite nanoparticles studied by high-field (W-band) EPR and ENDOR. *Phys. Chem. Chem. Phys.*, **2012**, *14*(7), 2246-2249. [http://dx.doi.org/10.1039/c2cp23601j] [PMID: 22241285]

[3] Gafurov, M.; Biktagirov, T.; Yavkin, B.; Mamin, G.; Filippov, Y.; Klimashina, E.; Putlayev, V.; Orlinskii, S. Nitrogen-containing species in the structure of the synthesized nano-hydroxyapatite. *JETP Lett.*, **2014**, *99*, 196-203. [http://dx.doi.org/10.1134/S0021364014040079]

[4] Burlaka, A.; Selyuk, M.; Gafurov, M.; Lukin, S.; Potaskalova, V.; Sidorik, E. Changes in mitochondrial functioning with electromagnetic radiation of ultra high frequency as revealed by electron paramagnetic resonance methods. *Int. J. Radiat. Biol.*, **2014**, *90*(5), 357-362. [http://dx.doi.org/10.3109/09553002.2014.899448] [PMID: 24597749]

[5] Orlinskii, S; Schmidt, J.; Baranov, P.; Hofmann, D.; de Mello Donegá, C.; Meijerink, A. *Probing the Wave Function of Shallow Li and Na Donors in ZnO Nanoparticles.*, **2004**. [http://dx.doi.org/10.1103/PhysRevLett.92.047603]

[6] Kutin, Y.S.; Mamin, G.V.; Orlinskii, S.B. Identification of Fe^{3+}-Li^{+} complexes in ZnO by means of high-frequency EPR/ENDOR spectroscopy. *J. Magn. Reson.*, **2013**, *237*, 110-114. [http://dx.doi.org/10.1016/j.jmr.2013.09.014] [PMID: 24184711]

[7] Baranov, P.G.; Orlinskii, S.B.; de Mello Donegá, C.; Schmidt, J. High-Frequency EPR and ENDOR Spectroscopy on Semiconductor Quantum Dots. *Appl. Magn. Reson.*, **2010**, *39*(1-2), 151-183. [http://dx.doi.org/10.1007/s00723-010-0151-y] [PMID: 20936163]

[8] Osawa, E. Disintegration and Purification of Crude Aggregates of Detonation Nanodiamond.*Synthesis, Properties and Applications of Ultrananocrystalline Diamond*; Gruen, D.; Shenderova, O.; Vul', A., Eds.; Springer: Dordrecht, **2005**, pp. 231-240. [http://dx.doi.org/10.1007/1-4020-3322-2_17]

[9] Osawa, E. Recent progress and perspectives in single-digit nanodiamond. *Diam. Relat. Mater.*, **2007**, *16*, 2018-2022. [http://dx.doi.org/10.1016/j.diamond.2007.08.008]

[10] Baranov, P.; Il'in, I.; Soltamova, A.; Vul', A.; Kidalov, S.; Shakhov, F.; Mamin, G.; Orlinskii, S.; Salakhov, M. Electron spin resonance detection and identification of nitrogen centers in nanodiamonds. *JETP Lett.*, **2009**, *89*, 409-413. [http://dx.doi.org/10.1134/S0021364009080074]

[11] Samsonenko, N.; Sobolev, E. Electron paramagnetic resonance caused by a diamond surface. *JETP Lett.*, **1967**, *5*, 250-253.

[12] Shames, A.; Panich, A.; Kempiński, W.; Alexenskii, A.; Baidakova, M.; Dideikin, A.; Osipov, V.; Siklitski, V.; Osawa, E.; Ozawa, M. Defects and impurities in nanodiamonds: EPR, NMR and TEM study. *J. Phys. Chem. Solids*, **2002**, *63*, 1993-2001. [http://dx.doi.org/10.1016/S0022-3697(02)00185-3]

[13] Smith, W.; Sorokin, P.; Gelles, I.; Lasher, G. Electron-Spin Resonance of Nitrogen Donors in Diamond. *Phys. Rev.*, **1959**, *115*, 1546-1552. [http://dx.doi.org/10.1103/PhysRev.115.1546]

[14] Wyk, J.; Woods, G. Electron spin resonance of excited states of the H3 and H4 centres in irradiated type Ia diamonds. *J. Phys. Condens. Matter*, **1995**, *7*, 5901-5911. [http://dx.doi.org/10.1088/0953-8984/7/29/017]

[15] Graeff, C.; Nebel, C.; Stutzmann, M.; Flöter, A.; Zachai, R. Characterization of textured polycrystalline diamond by electron spin resonance spectroscopy. *J. Appl. Phys.*, **1997**, *81*, 234-237. [http://dx.doi.org/10.1063/1.363987]

[16] Tucker, O.D.; Newton, M.E.; Baker, J.M. EPR and ^{14}N electron-nuclear double-resonance measurements on the ionized nearest-neighbor dinitrogen center in diamond. *Phys. Rev. B Condens.*

Matter, **1994**, *50*(21), 15586-15596.
[http://dx.doi.org/10.1103/PhysRevB.50.15586] [PMID: 9975922]

[17] Aleksenski, A.; Baidakova, M.; Osipov, V.; Vul, A. The Fundamental Properties and Characteristics of Nanodiamonds.*Nanodiamonds: Applications in Biology and Nanoscale Medicine*; Ho, D., Ed.; Springer: New York, **2009**, pp. 55-77.

[18] Soltamova, A.; Il'in, I.; Shakhov, F.; Kidalov, S.; Vul', A. Yavkin, B.; Mamin, G.; Orlinskii S.; Baranov, P. Electron Paramagnetic Resonance Detection of the Giant Concentration of Nitrogen Vacancy Defects in Sintered Detonation Nanodiamonds. *JETP Lett.,* **2010**, *92*, 102-106.
[http://dx.doi.org/10.1134/S0021364010140067]

[19] Fang, X.; Mao, J.; Levin, E.M.; Schmidt-Rohr, K. Nonaromatic core-shell structure of nanodiamond from solid-state NMR spectroscopy. *J. Am. Chem. Soc.,* **2009**, *131*(4), 1426-1435.
[http://dx.doi.org/10.1021/ja8054063] [PMID: 19133766]

[20] Baidakova, M.; Vul', A. New prospects and frontiers of nanodiamond clusters. *J. Phys. D Appl. Phys.,* **2007**, *40*, 6300-6311.
[http://dx.doi.org/10.1088/0022-3727/40/20/S14]

[21] Ammerlaan, C. *Paramagnetic Centers in Diamond,* 1st ed; Springer: Berlin, **2001**, p. 6.

[22] Yavkin, B.; Mamin, G.; Gafurov, M.; Orlinskii, S. Size-dependent concentration of N0 paramagnetic centres in HPHT nanodiamonds. *Magn. Reson. Solids,* **2015**, *17*(15101), 1-7.

[23] Stoll, S.; Schweiger, A. EasySpin, a comprehensive software package for spectral simulation and analysis in EPR. *J. Magn. Reson.,* **2006**, *178*(1), 42-55.
[http://dx.doi.org/10.1016/j.jmr.2005.08.013] [PMID: 16188474]

[24] Orlinskii, S.; Bogomolov, R.; Kiyamova, A.; Yavkin, B.; Mamin, G.; Turner, S.; Van Tendeloo, G.; Shiryaev, A.; Vlasov, I.; Shenderova, O. Identification of Substitutional Nitrogen and Surface Paramagnetic Centers in Nanodiamond of Dynamic Synthesis by Electron Paramagnetic Resonance. *Nanosci. Nanotechnol. Lett.,* **2011**, *3*, 63-67.
[http://dx.doi.org/10.1166/nnl.2011.1121]

[25] Gamarnik, M.Y. Energetical preference of diamond nanoparticles. *Phys. Rev. B Condens. Matter,* **1996**, *54*(3), 2150-2156.
[http://dx.doi.org/10.1103/PhysRevB.54.2150] [PMID: 9986065]

[26] Rondin, L.; Dantelle, G.; Slablab, A.; Grosshans, F.; Treussart, F.; Bergonzo, P.; Perruchas, S.; Gacoin, T.; Chaigneau, M.; Chang, H. *Surface-induced charge state conversion of nitrogen-vacancy defects in nanodiamonds.,* **2010**.
[http://dx.doi.org/10.1103/PhysRevB.82.115449]

[27] Hauf, M.; Grotz, B.; Naydenov, B.; Dankerl, M.; Pezzagna, S.; Meijer, J.; Jelezko, F.; Wrachtrup, J.; Stutzmann, M.; Reinhard, F. *Chemical control of the charge state of nitrogen-vacancy centers in diamond.,* **2011**.
[http://dx.doi.org/10.1103/PhysRevB.83.081304]

[28] Kucsko, G.; Maurer, P.C.; Yao, N.Y.; Kubo, M.; Noh, H.J.; Lo, P.K.; Park, H.; Lukin, M.D. Nanometre-scale thermometry in a living cell. *Nature,* **2013**, *500*(7460), 54-58.
[http://dx.doi.org/10.1038/nature12373] [PMID: 23903748]

[29] Doherty, M.; Struzhkin, V.; Simpson, D.; McGuinness, L.; Meng, Y.; Stacey, A.; Karle, T.; Hemley, R.; Manson, N.; Hollenberg, L. Electronic Properties and Metrology Applications of the Diamond NV− Center under Pressure. *Phys. Rev. Lett.***2014**, 112, 047601(1-5).

[30] Arai, K.; Belthangady, C.; Zhang, H.; Bar-Gill, N.; DeVience, S.J.; Cappellaro, P.; Yacoby, A.; Walsworth, R.L. Fourier magnetic imaging with nanoscale resolution and compressed sensing speed-up using electronic spins in diamond. *Nat. Nanotechnol.,* **2015**, *10*(10), 859-864.
[http://dx.doi.org/10.1038/nnano.2015.171] [PMID: 26258549]

[31] Baranov, P.G.; Soltamova, A.A.; Tolmachev, D.O.; Romanov, N.G.; Babunts, R.A.; Shakhov, F.M.;

Kidalov, S.V.; Vul', A.Y.; Mamin, G.V.; Orlinskii, S.B.; Silkin, N.I. Enormously high concentrations of fluorescent nitrogen-vacancy centers fabricated by sintering of detonation nanodiamonds. *Small,* **2011**, *7*(11), 1533-1537.
[http://dx.doi.org/10.1002/smll.201001887] [PMID: 21520495]

[32] Chang, Y.R.; Lee, H.Y.; Chen, K.; Chang, C.C.; Tsai, D.S.; Fu, C.C.; Lim, T.S.; Tzeng, Y.K.; Fang, C.Y.; Han, C.C.; Chang, H.C.; Fann, W. Mass production and dynamic imaging of fluorescent nanodiamonds. *Nat. Nanotechnol.,* **2008**, *3*(5), 284-288.
[http://dx.doi.org/10.1038/nnano.2008.99] [PMID: 18654525]

[33] Boudou, J.; Curmi, P.; Jelezko, F.; Wrachtrup, J.; Aubert, P.; Sennour, M.; Balasubramanian, G.; Reuter, R.; Thorel, A.; Gaffet, E. High yield fabrication of fluorescent nanodiamonds. *Nanotechnology,* **2009**, *20*(235602), 1-11.
[http://dx.doi.org/10.1088/0957-4484/20/23/235602]

[34] Acosta, V.; Bauch, E.; Ledbetter, M.; Santori, C.; Fu, K.; Barclay, P.; Beausoleil, R.; Linget, H.; Roch, J.; Treussart, F. Diamonds with a high density of nitrogen-vacancy centers for magnetometry applications. *Physical Review B.,* **2009**, *80*, 115202.
[http://dx.doi.org/10.1103/PhysRevB.80.115202]

[35] Loubser, J.; Wyk, J. Electron spin resonance in the study of diamond. *Rep. Prog. Phys.,* **1978**, *41*, 1201-1248.
[http://dx.doi.org/10.1088/0034-4885/41/8/002]

CHAPTER 10

EPR and FMR of SiCN Ceramics and SiCN Magnetic Derivatives

Sushil K. Misra[1,*] and **Sergey I. Andronenko**[1,2]

[1] *Department of Physics, Concordia University, Montreal, H3G 1M8, Canada*

[2] *Institute of Physics, Kazan Federal University, Kazan 420008, Russian Federation*

Abstract: Silicon nitro carbide, SiCN, exhibits excellent high-temperature properties. It can withstand temperatures of up to 1800° C, which is superior to those of Si, SiC and Si_3N_4. Magnetic composites, as well as electrically conductive ceramics on the basis of SiCN, can be developed. Therefore, SiCN constitutes a new class of materials for high-temperature electronics. SiCN ceramics, doped with the transition metal ions exhibiting superparamagnetic features are promising in building high-temperature magnetic and pressure sensors. EPR (electron paramagnetic resonance) and FMR (ferromagnetic resonance) techniques can provide important information on the properties of SiCN and its magnetic derivatives, in conjunction with structural, magnetic and electric measurements. In the present work, EPR signals due to sp^2–hybridized carbon-related dangling bonds were recorded over the 4 - 300 K range. SiCN ceramics consist of nanoparticles of SiCN and a free carbon phase. The two EPR signals, which were only resolved at the higher frequencies of W (95 GHz) and G (170 GHz) bands are due to carbon-related dangling bonds present as (i) defects on the free-carbon phase and (ii) within the bulk of SiCN ceramic network. SiCN magnetic ceramics, doped with the Fe ions were synthesized at different pyrolysis temperatures in the range 600° - 1600°C. Several magnetic phases in SiCN/Fe composite are detected by EPR/FMR technique. The main sources of magnetism in these samples are: (i) superparamagnetic nanoparticles of Fe_3Si, (T_C = 800°C), (ii) nanoparticles of Fe_5Si_3 (T_C = 393°C), which appear above 1000°C in single-domain state and (iii) nanoparticles of $Fe_{70}Si_xC_{30-x}$ (620°C).

Keywords: Antiferromagnetic interaction, Curie law, Curie temperature, Dangling bonds, EPR, EPR linewidth, Ferromagnetic nanoparticles, FMR, Fluctuations of the magnetization, High-frequency EPR, Multi-frequency EPR, Nanoparticles, Phase-transition, Silicon nitro carbide (SiCN) ceramics, SiCN/Fe ceramics, SiCN/Mn ceramics, Superparamagnetic nanoparticles, Superparamagnetism, Temperature dependence.

[*] **Corresponding author Sushil K. Misra:** Department of Physics, Concordia University, Montreal, H3G 1M8, Canada; Tel: (514)482-3690; E-mail: sushil.misra@concordia.ca

Dariya Savchenko & Abdel Hadi Kassiba (Eds.)

INTRODUCTION

Contemporary electronics widely uses silicon carbide MEMS (Micro- Electro-Mechanical Systems) technology (SiC-MEMS). SiC is a semiconductor with good mechanical and thermal stability, and a wide band-gap for stable electronic properties at elevated temperatures [1]. However, the processes associated with the fabrication of SiC-MEMS have been known to be time-consuming, costly and technically very challenging. A new class of polymer-derived ceramic, which essentially consists of Silicon Carbon Nitride (SiCN), was proposed recently. A starting material for SiCN synthesis is a liquid polymer, which can be easily shaped using micro-moulds or microphotolitography, and this is one of the advantages of SiCN. The commercial precursor CERASET™ (KiON group AG) (polyureasilazane) for the synthesis of SiCN ceramics is available. The liquid polymer is cross-linked at 400°C in nitrogen gas flow to create a solid transparent polymer, whose free-standing forms are then pyrolyzed, yielding a black-colored ceramic, which can withstand temperatures above 1800°C [2, 3]. This SiCN ceramic exhibits outstanding creep, hardness and oxidation resistance, which are superior to those of Si, SiC and Si_3N_4.

Electrically conductive SiCN polymer-derived ceramics were developed by Liew *et al.* [4], and its conductivity was further modified by boron-doping (SiBCN) or Al-doping (SiAlCN) [5]. It is, thus possible to produce complex isolator-conductive material structures and develop new MEMS-SiCN technique. The addition of polymers, containing different magnetic transition metal ions, to initial silazane precursor leads to the formation of superparamagnetic SiCN ceramics, which can be used as magnetic sensors [6]. SiCN ceramics possess excellent piezoelectric properties [7] and can be used in pressure sensors. The investigation of SiCN ceramics and its conductive and magnetic derivatives is of great interest for developing high-temperature sensor applications [8, 9]. Formation of the free-carbon phase, which influences physical properties, was investigated in polymer-derived SiCN ceramics in detail [10 - 14].

Li *et al.* [15] exploited FTIR, Raman and XRD techniques for the investigation of SiCN ceramics obtained by pyrolysis from CERASET™ precursor. It was observed that the free C phase is formed owing to fast heating upon pyrolysis. Latest molecular dynamics simulations show that aromatic bonded carbon (C) ions are described by graphene network structure in the SiCN ceramics [16]. The existence of free C in SiCN structure and its ordering affect both mechanical and electrical features of polymer-derived SiCN ceramics, remarkably.

The EPR investigations on SiCN and SiBCN ceramics derived from non-commercial polymer precursors were reported [11 - 13, 17]. The origin of the

EPR line observed in *a*-C:H, turbostratic carbon, multiwall carbon nanotubes and irradiated multiwall carbon nanotubes, possibly related to the free C phase, was also studied [18 - 20]. Multifrequency EPR technique was applied at 9.6, 95 and 170 GHz to study SiCN ceramics at different temperatures from 4 K up to 300 K [21]. The intense EPR line was observed in this ceramics, which are ascribed to carbon-related sp^2-dangling bonds [21]. Latest verification of such identification of this EPR signal was obtained [22, 23]. The EPR signal in UV-irradiated SiCN films was associated with nitrogen-related dangling bonds [24]. The EPR signals in SiCN films were ascribed to sp^2-carbon-related dangling bonds, threefold-coordinated Si dangling bonds and the trapped holes at Si atom, as it was found by Savchenko *et al.* [25].

It is possible to tailor the magnetic properties of polymer-derived SiCN/Fe composites from paramagnetic to superparamagnetic, and then to ferromagnetic, with a variation of synthesis conditions, pyrolysis temperature, silazane precursor and Fe-containing precursor. Superparamagnetic features of SiCN/Fe and SiCN/Mn ceramics could be used to construct spintronic devices [26, 27], exploited at extremely high temperatures. Magnetic Fe-doped SiCN ceramics (SiCN/Fe) can be synthesized by adding different Fe-containing compounds to the initial polymer precursor. The SiCN ceramics synthesized at $T > 1000°C$ possess good thermal/mechanical features for high-temperature and high-pressure applications. SiCN ceramics doped with Mn (SiCN/Mn) can also be synthesized by adding different Mn-containing polymers to the initial precursor. Such SiCN ceramic composites are very promising for MEMS applications. Therefore, SiCN ceramics have been extensively investigated in the past few years [2 - 25, 28 - 30]. They are composed of nanoparticles having the mean size of 1.3 nm [11, 12, 28]. The crystallization and formation of such SiCN nanoparticles started above 1300°C as found by small angle X-ray scattering (SAXS) [11]. The existence of carbon nanoparticles was verified by the intensity ratio of D and G bands in Raman spectra for the samples pyrolyzed at 1000°C [13]. The size of carbon nanoparticles is also about 1 nm [11, 13]. The nanostructure of SiCN ceramics, consisting of free C phase and SiCN ceramic network, plays a very significant role in the magnetization of SiCN/Fe materials, as considered below. The free C phase can be organized in the so-called "cage" structure, as it was suggested for SiCO ceramics [29, 30]. The SiCN/Fe nanocrystalline structure is composed of different nanoparticles of various sizes. The magnetic coupling of the Fe magnetic moments inside little nanoparticles than gives rise to the superparamagnetism. These Fe-containing nanocrystallites can exhibit ferromagnetic behavior if their size is large enough. Recently great attention was paid to the synthesis and the investigations of SiCN polymer derived ceramics doped with iron-containing compounds [6, 31 - 49]. Saha *et al.* [31] firstly obtained polymer-derived magnetic SiCN/Fe ceramic composites from CERASET™ solution using Fe_2O_3

powder. SiCN ceramics with the addition of ferrocene were synthesized by Dimitru *et al.* [32, 33]. The superparamagnetic behavior of SiCN/Fe magnetic ceramics was also observed by Li *et al.* [34, 35]. SiCN/Fe ceramics was investigated by FMR/EPR, IR and Raman spectroscopy, XRD and by measurements of magnetic features by Andronenko *et al.* [49]. FMR/EPR spectroscopy was utilized successfully to separate various types of magnetisms in SiCN/Fe ceramics, revealing at least three of them:

i. ferromagnetic nanoparticles of $Fe_5Si_3C_x$,
ii. ferromagnetic clusters distributed in the free C phase and, probably,
iii. new Fe-containing nanocrystallites can appear at pyrolysis temperature above 1250° C and FMR/EPR spectra reflect such transformation.

Several kinds of Fe-containing crystallites, such as α-Fe, Fe_3Si and Fe_5Si_3 were found in SiCN/Fe composites [46 - 48]. Recent investigations of magnetic properties of polymer-derived SiCN/Fe ceramics clearly show their predominantly superparamagnetic behavior with blocking temperature from 20 K [6] to 100 K [34, 35], the absence of coercive fields, remanent magnetization and rather wide distribution of nanoparticle sizes.

Yan *et al.* [37] found that Mn_5Si_3 nanocrystallites determine the magnetic properties of polymer-derived SiCN/Mn composite obtained by the addition of Mn metal powder to liquid CERASET™. This result agrees with the conclusion that $Mn_5Si_3C_x$ nanoparticles are the main source of magnetism in SiCN/Mn composites, synthesized by the use of manganese (II) acetylacetonate [50].

The EPR technique can be applied successfully to study magnetic SiCN ceramics activated with transition metal ions. It is the purpose of this review to summarize the results of EPR investigations of carbon-related dangling bonds in polymer-derived SiCN nanoceramics at several frequencies, as high as 170 GHz. Furthermore, this review will report the results of the EPR/FMR investigation on SiCN/Fe and SiCN/Mn ceramics obtained from liquid polyureasilazane precursor [21, 49, 50], to present detailed data on the origin of the various sources of magnetism contributed in these ceramics.

SYNTHESIS AND STRUCTURE

The fabrication procedure of SiCN ceramics is described elsewhere [2, 3]. Liquid polymer precursor CERASET™ (polyureasilazane) was crosslinked at 400°C for 60 min to fabricate transparent solid polymers. Thereafter, the crosslinked polymer was pyrolyzed at temperatures of 1000°, 1050°, 1100°, 1150° and 1200°C under nitrogen-gas flow to yield solid ceramic samples. The aim of pyrolysis was to exclude the C excess in the form of CH_4 molecules and to precipitate free C, resulting in the incorporation of some amount of C ions into the mixed tetrahedra $Si(C_{1-x}N_x)_4$. The composition of SiCN ceramics pyrolyzed at 1100° C was determined to be $SiC_{0.74}N_{0.67}$; 30 wt % of C was in the free C phase as derived by elementary chemical analysis and relative intensity of NMR signals from the free-C and bound C ions.

SiCN/Fe Ceramics

The samples of magnetic SiCN/Fe ceramics were prepared by utilizing liquid polyureasilazane (CERASET™) precursor [49], to which 5-10 wt % of Fe (III) acetylacetonate was added. To fabricate a transparent dark-red solid polymer, the samples were cross-linked at 400°C for 90 min. Thereafter, the cross-linked polymer was pyrolyzed at T = 600°C-1600°C in N gas flow. The composition of polymer-derived ceramics, as derived from elementary chemical analysis, was determined to be $SiC_{0.68}N_{0.41}$ at pyrolysis temperatures of 1100°C. The content of N decreased to 0.15 at high pyrolysis temperatures, accompanied by 30 wt % of C in the free C phase. The amount of Fe content determined by neutron-activated analysis of SiCN/Fe composite was to be of 0.2 wt %. The X-ray powder patterns for SiCN samples synthesized in this way at various pyrolysis temperatures were discussed by Andronenko *et al.* [49]. These X-ray patterns were similar to those, obtained for pure SiCN ceramics [51, 52] synthesized at 1450°C. The presence of peaks due to Fe_5Si_3 particles having a hexagonal structure was revealed in high-resolution X-ray patterns [53]. A weak peak owing to α-Fe particles was also detected.

SiCN/Mn Ceramics

The samples of magnetic SiCN/Mn ceramics were prepared [50], using liquid polyureasilazane (CERASET™) precursor, with an addition of 1 wt % of Mn (II) acetylacetonate $Mn(C_5H_7O_2)_2$. The thermal treatment was fulfilled in the same way as for SiCN/Fe ceramics [50].

PURE SICN: EPR OF CARBON RELATED DANGLING BONDS

A Bruker X-band (9.6 GHz) EPR spectrometer (ER-200D) equipped with oxford helium gas-flow temperature controller at Concordia University was utilized to measure the EPR spectra in the different SiCN ceramics in the temperature interval from 4 K to 300 K [21]. The EPR spectra measured at T = 295 K in SiCN samples, pyrolyzed at different temperatures, are shown in Fig. (**1**).

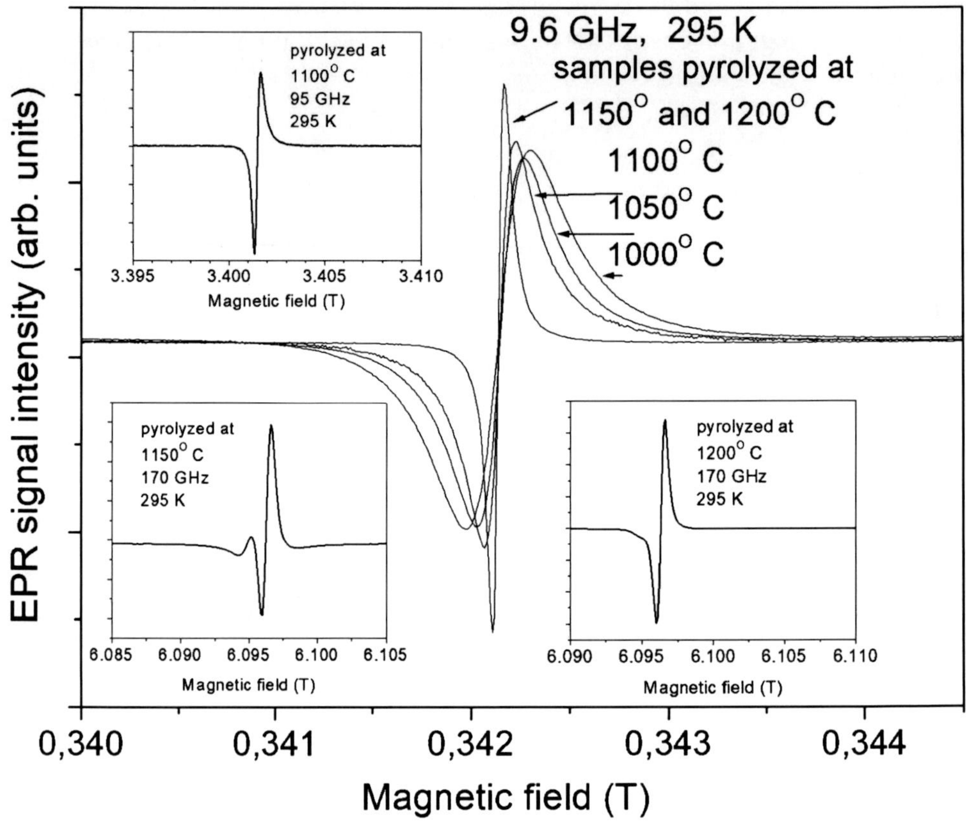

Fig. (1). EPR spectra of C-related dangling bonds in the different SiCN ceramics at 295 K. The inset shows W-band (95 GHz) EPR spectrum at 295 K. Reused from [21] with the Permission from AIP Publishing LLC.

Room-temperature EPR studies of pure SiCN samples were also performed at W-band (95 GHz) and G-band (170 GHz) at Cornell University. Only one intense EPR signal at $g = 2.0027 \pm 0.0001$ was detected at room temperature at X-band for all investigated samples. The *g*-value was measured using 2.2-Diphenyl-1-Picrylhydrazyl (DDPH) as a reference sample. NMR Bruker gaussmeter was used for the magnetic field calibration for all investigated SiCN samples. The obtained *g*-value proved unambiguously that the observed EPR line in the investigated samples is caused by C-related defects [54, 55], rather than that from the Si dangling bonds ($g = 2.005$) [55].

The exchange narrowing of EPR linewidth (0.07 mT) was found for pure SiCN samples synthesized at 1150°C and 1200°C at X-band frequencies and was explained by the increase of the spin concentration of dangling bonds emerged in the free C phase. Moreover, the spin concentration of the carbon dangling bonds is unchanged in the samples as there is no oxidation process and the H atoms are completely removed (above 1150°C), as confirmed by chemical analysis. This narrow EPR signal can be utilized as a *g*-marker at X-band. The dependences of the X-band EPR linewidth and wt % of H content versus pyrolysis temperature are represented in Fig. (**2**).

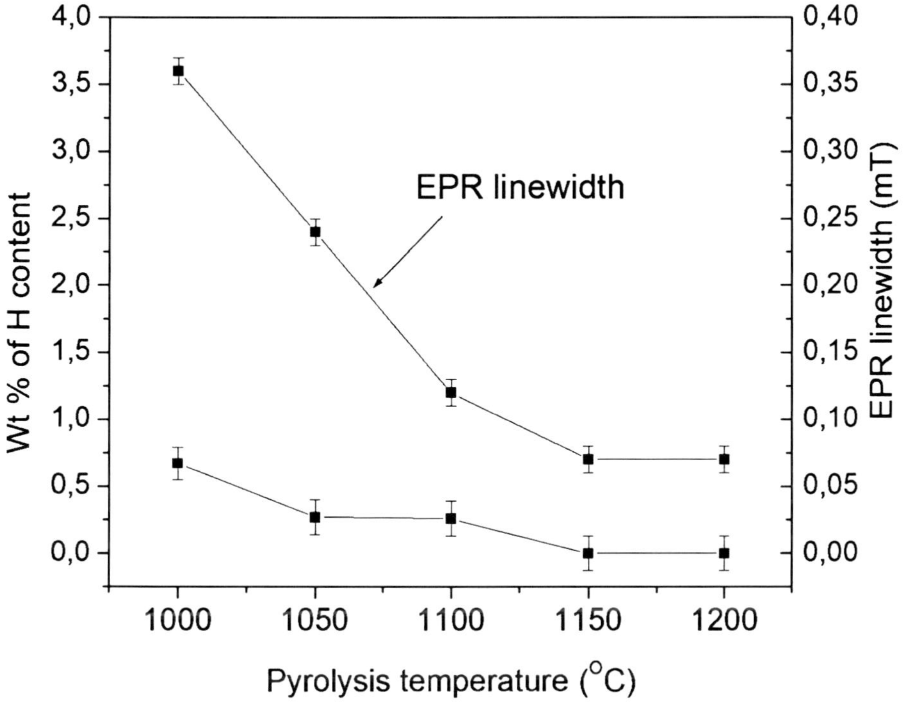

Fig. (2). The plot of the EPR linewidth and hydrogen content depending on the pyrolysis temperature of the samples. Reused from [21] with the Permission from AIP Publishing LLC.

The 95-GHz, W-band and EPR spectrum at 295 K agreed with that recorded at X-band, whereas the G-band (170 GHz) room-temperature spectrum showed the existence of second, less intense signal with larger *g*-factor, which can be associated with the shoulder revealed in X-band spectra at 4 K. The EPR spectra in the temperature interval of 4-295 K for the SiCN ceramics have mostly Lorentzian lineshape. The visible deviation from Lorentzian lineshape at X-band found for the samples pyrolyzed at 1150°C and 1200°C, can be interpreted by the overlapping of two EPR signals, which were distinguished at G-band. This takes place when a minimum of one signal overlaps with the maximum of the other signal. Total EPR lineshape of these two overlapping signals turned out to be similar to Dysonian lineshape [56] at X- and W-bands.

A "shoulder" on the EPR line recorded at 4 K on the sample of SiCN pyrolyzed at 1150° C was observed and assigned to another EPR line, also related to carbon-related dangling bonds with g = 2.0032. They both decreased with a rise of the pyrolysis temperature, assuming that the amount of dangling bonds enhances with removing more H ions. The increase in the spin concentration of the dangling bonds results in the formation of the clusters and finally in exchange narrowing of the EPR line. The spin concentration of the dangling bonds was evaluated by comparison of the EPR signal integral intensity with that of a "strong-pitch" standard sample with known spin concentrations at X-band frequencies. The average number of spins was found to be of 10^{18} per cm^3 in all SiCN samples.

The EPR lineshape was revealed to be almost symmetrical for the samples pyrolyzed at $T < 1100$°C. The linewidth at W-band for the sample synthesized at 1100°C is 0.32 mT at 295 K, which is twice larger than observed at X-band (= 0.14 mT). For the samples synthesized at 1150°C and 1200°C, two EPR signals were resolved at G-band with $g = 2.0027 \pm 0.0001$ and $g = 2.0032 \pm 0.0002$ which have different intensity and linewidth. The enhancement in the linewidth at higher frequencies can be explained by g-strain broadening, which becomes apparent with frequency increase.

The intense EPR signal at g = 2.0027 is caused by the dangling bonds that exist in the surface defects of the free C phase, while the less intense EPR signal with slightly higher g-factor (2.0032) is from the carbon-related dangling bonds that exist in the Si-C-N tetrahedra. The temperature-independent g = 2.0027 is similar to the g-value observed for glassy graphite samples caused by the high density of localized states at the Fermi level [18]. At the same time, it is far from the EPR signals at g = 2.018, 2.02 and 2.012, which were observed in planar graphite phase [18], turbostratic carbon [19] and multiwall carbon nanotube phase [20], correspondingly, and have ordered networks. Therefore, it was assumed that the SiCN ceramics under investigation do not contain ordered C networks. Alternatively, the EPR line with g = 2.0032 can be assigned to sp^3-carbon related dangling bonds, associated with bulk "free" carbon phase.

Temperature Variation of the EPR Linewidth and Estimation of Exchange Interaction

The EPR linewidth has strong temperature dependence as represented in Fig. (**3**) for X-band. The temperature dependence of the EPR linewidth in the liquid-helium temperature range was well described by the exchange interactions in amorphous materials between carbon dangling bonds as deduced by Misra [57]. Accordingly,

$$\Delta B = A' + B'{*}T + C'/(1 + \exp(J/T)). \qquad \textbf{(1)}$$

In this expression, A', B', C' are constants and T is the temperature. The second term describes the linewidth broadening due to exchange coupling between spins modulated by vibrations via the spin-orbit coupling of the spins. This linear contribution (B') is negligible for the EPR linewidth in pure SiCN. The third term here, represents the contribution to the linewidth due to cross-relaxation wherein, an electron spin relaxes by simultaneous transitions between its energy levels and that of an exchange-coupled pair.

The value of antiferromagnetic interaction (J) was rather large, as it was found by fitting to EPR linewidth data, being anywhere from –12.5 to –15.1 K, as listed in Table **1**, which also contains the values of the parameters A', B' and C'.

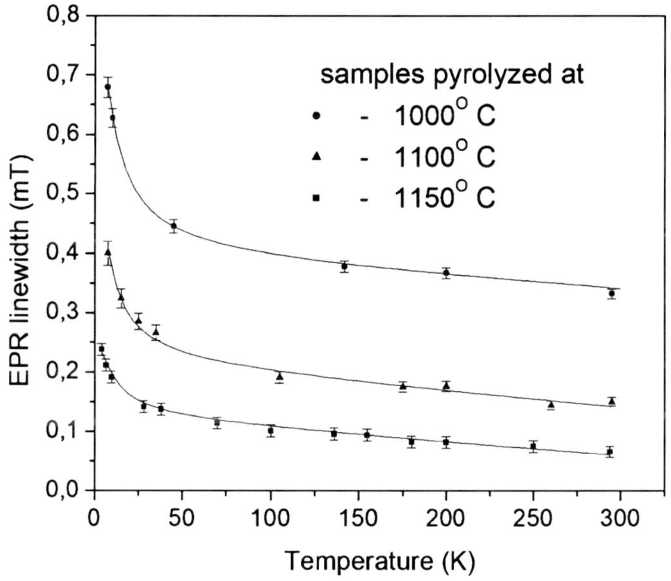

Fig. (3). Temperature variation of the EPR linewidth in the different SiCN ceramic samples. The circles, solid squares and triangles designate experimental data; the lines are fitting to eq. (1) using the parameters listed in the Table 1. Reused from [21] with the Permission from AIP Publishing LLC.

Table 1. Parameter fitting to eq. (1). Reused from [21] with the Permission from AIP Publishing LLC.

Pyrolysis temperature	*A′* (mT)	*B′* (mT/K)	*C′* (mT)	*J* (K)
1000° C	0.02±0.01	-0.0002±0.0002	0.74±0.05	-15.1±0.05
1100° C	-0.06±0.02	-0.0002±0.0002	0.55±0.05	-13.2±0.05
1150° C	-0.01±0.01	-0.0002±0.0002	0.27±0.05	-12.5±0.05

EPR STUDY OF SiCN/Fe AND THE TRANSFORMATIONS AT VARIOUS PYROLYSIS TEMPERATURES

EPR/FMR Spectra of SiCN/Fe Ceramics

An X-band (~9.5 GHz) Bruker ER-200D SRC EPR spectrometer equipped with the nitrogen-flow Bruker variable temperature assembly was utilized to study the SiCN/Fe ceramic samples in the temperature range 77-350 K [49]. Fig. (**4**) represents room-temperature EPR spectra of the different samples synthesized at various temperatures.

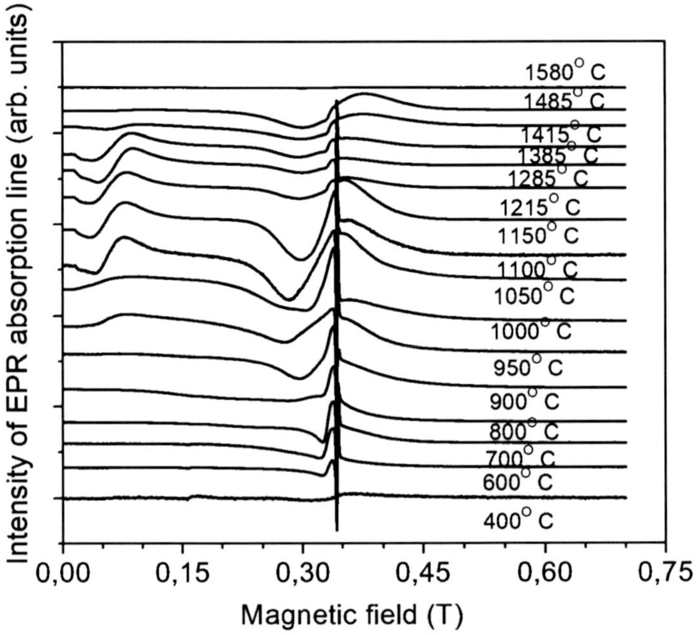

Fig. (4). The plot of Fe FMR/EPR spectra depending on the pyrolysis temperature of the SiCN/Fe samples measured at 295 K. Reproduced from [49] with the Permission from Springer.

The FMR signals, belonging to Fe-containing ferromagnetic crystalline particles, were observed in SiCN/Fe composites. Similarly to SiCN ceramics, narrow EPR line due to dangling bonds was also observed. Numerous transformations of SiCN/Fe ceramics were found from the temperature behavior of these EPR/FMR lines, namely, the dependencies of the linewidth and double-integrated intensity of the first-derivative EPR/FMR signals on the pyrolysis temperature. The explanation of the different EPR/FMR lines can be presented as follows.

(a) *Narrow EPR signal caused by dangling bonds with g = 2.0027*. The double-integrated intensity (per 1 mg) of the first-derivative line owing to sp^2 carbon dangling bonds with *g*-value of 2.0027 is presented in Fig. (**5a**). Fig. (**5b**) shows the dependence of the EPR linewidth on the pyrolysis temperature for the different samples.

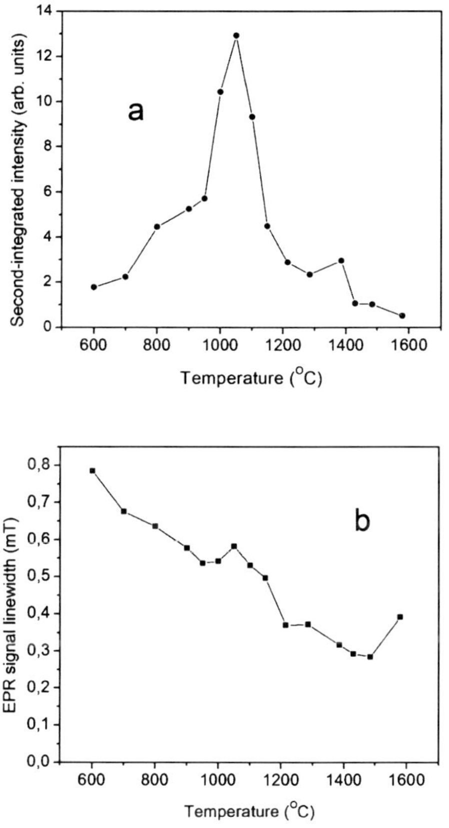

Fig. (5). Dependence of the double-integrated intensity (a) and EPR linewidth (b) of the first-derivative EPR line from dangling bonds on pyrolysis temperature measured in the different SiCN/Fe samples. Reproduced from [49] with the Permission from Springer.

The peak-to-peak width of EPR line from dangling bonds in SiCN/Fe samples pyrolyzed at 1100°C, was found to be larger (0.24 mT and 0.53 mT for 1 and 10 wt. % iron acetylacetonate, respectively) than that in the pure SiCN samples (0.12 mT), pyrolyzed at 1100°C. The dipole-dipole couplings between carbon dangling bonds and Fe^{3+} ions can be the reason of the linewidth increase.

(b) *Fe^{3+} related EPR signal at g ~ 2.0.* The dependence of the double integral of the first-derivative EPR signal (per 1 mg), on the pyrolysis temperature is shown in Fig. (**6**). The peak-to-peak linewidth of the first-derivative EPR signal as a function of the pyrolysis temperature is presented in Fig. (**7**).

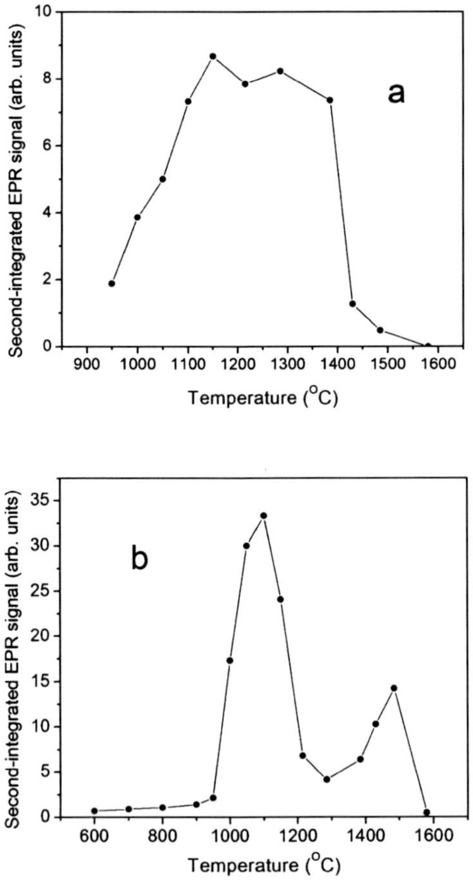

Fig. (6). The plot of the double-integrated intensity of Fe FMR/EPR line depending on pyrolysis temperature of the SiCN/Fe samples: a) for the signal with *g* ~ 10 ; b) for the broad signal with *g*~ 2.0. Reproduced from [49] with the Permission from Springer.

It has been shown in Fig. (**4**) that this line composes of two overlapped broad signals (B1 and B2) with different linewidths for the samples pyrolyzed in the temperature range of 950°C- 150°C.

The linewidth of the EPR signal B1 varies considerably with temperature. The maximum width for EPR signal B1 was observed for the SiCN/Fe sample pyrolyzed at 1000°C (Fig. **7a**). Whereas, the only linewidth of the B2 signal slightly increases (see Fig. **7b**).

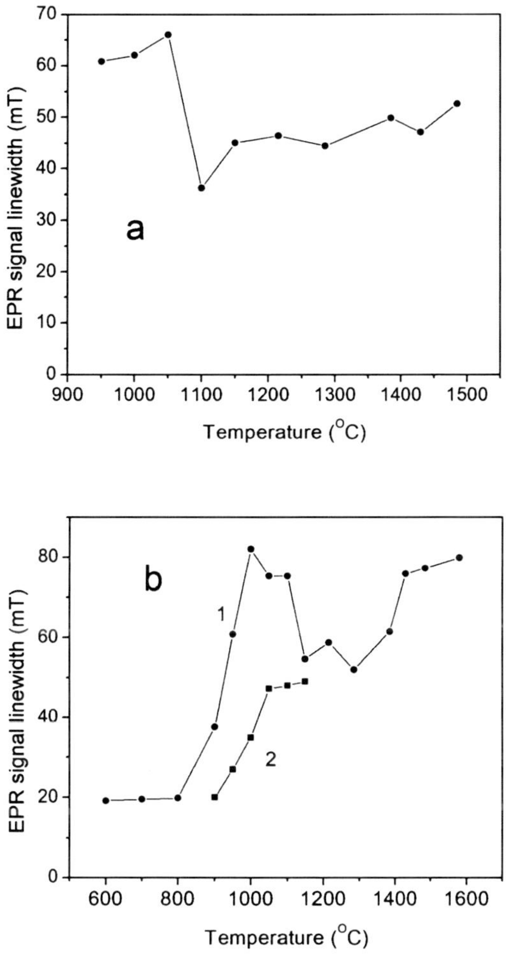

Fig. (7). The plot representing the Fe FMR/EPR linewidth at room temperature depending on the pyrolysis temperature of the SiCN/Fe samples: a) for the signal with g ~ 10; b) for the broad signal with g ~ 2.0 (filled circles and filled squares indicate the two coincided lines). Reproduced from [49] with the Permission from Springer.

The shape of line B1 corresponds to the EPR line of the assembly of superparamagnetic particles diluted in the diamagnetic matrix, as shown by Kliava *et al.* [58].

A correlation between the double-integrated intensity of the EPR signal of dangling bonds depending on the pyrolysis temperature, and the double-integrated intensity of the FMR signal that occurs at $g \sim 2.0$ was observed, as shown in Figs. (**5a** and **6b**).

As the double-integrated intensity of EPR signal related to dangling bonds is proportionate to the amount of free C phase in SiCN ceramics, we could suppose, that these magnetically ordered Fe ions are incorporated into the free-carbon phase. The characteristics of the EPR lines from the different SiCN/Fe samples synthesized at various temperatures can be described as follows.

(i) The Fe ions are in the paramagnetic state in the SiCN samples synthesized at temperatures 600°C-800°C, when SiCN samples are still "preceramized". Their magnetization is rather small, as confirmed by the low magnitude of the double-integrated intensity of EPR signal.

(ii) The double-integrated intensity of the first-derivative Fe-related line increases for the SiCN/Fe samples synthesized at T = 800°C-1150°C, though the total Fe content is unchanged. This assumes that the increase of the Fe magnetic moment is caused by the formation of magnetically ordered Fe-containing nanocrystallites.

(iii) The double-integrated intensity of the wide signal at $g \sim 2.0$ reduced for the SiCN/Fe samples pyrolyzed above 1150°C. However, a third narrow Fe signal, centered at $g \sim 2$, appeared in this sample, assuming that a part of the Fe ions transfers to a new magnetic structural state.

(iv) The double-integrated intensity of the broad EPR signal at $g \sim 2.0$ increased again for the SiCN/Fe samples pyrolyzed at T = 1400°C-1500°C as it follows from Fig. (**6b**). The reason is that EPR signal with $g \sim 10$ vanishes at these temperatures and the Fe ions, preliminarily incorporated in the Fe_5Si_3 crystallites, start to contribute to the EPR signal at $g \sim 2.0$.

(c) *The EPR signal at g ~ 10.0.* This signal is caused by the formation of ferromagnetic crystallites, which arise in the SiCN ceramics at the pyrolysis temperature of 950°C. It totally vanished for the samples pyrolyzed at 1430°C and above. These magnetic crystallites reveal ferromagnetism rather than superparamagnetism, for which a smaller size of the particle by order of SiCN ceramics grain (1.3 nm), is necessary. As shown below, it is probable that these ferromagnetic crystallites are Fe_5Si_3 nanoparticles, which follow from its low value of Curie temperature (T_C = 393 K) [59]. Note that the Curie temperature of Fe_3Si, the other possible nanocrystallite, is much higher, T_C = 800 K [60], therefore, probably only the EPR line near g = 2.0 can be related to such kinds of compounds.

(d) <u>*The weak EPR signal at g ~ 2.0*</u>. This line emerges only in SiCN/Fe samples pyrolyzed at T > 1215°C when the crystallization of SiCN network takes place. The linewidth of this signal is rather narrow, being about 10 mT. It vanishes totally in the samples pyrolyzed at T > 1485°C. It can be assigned to the Fe^{3+} ions incorporated in Si_3N_4 nanocrystallites, which begin to crystallize at 1200°C, as revealed by XRD analysis.

(e) <u>*The samples pyrolyzed above 1500° C.*</u> No EPR signals from the Fe^{3+} ions were observed in these samples, owing to the evaporation of Fe ions from SiCN ceramics at T > 1530°C, corresponding to the iron melting temperature.

Temperature Dependence of Fe^{3+} EPR Lines

(i) <u>*The EPR line at g ~ 10.*</u> The double-integrated intensity (per 1 mg) of the first-derivative EPR line and the peak-to-peak width of the EPR line from the SiCN ceramic sample pyrolyzed at 1100°C in the temperature interval 77 - 350 K are shown in Figs. (**8a** and **8b**), correspondingly. As it follows from Fig. (**8a**), the double-integrated intensity of this signal vanished at about 380 ± 10 K. The behavior of EPR linewidth is also consistent with the temperature variation of the magnetic moment of the ferromagnetic Fe_5Si_3 particles with the Curie temperature of 393 K. It can be proved by fitting with the following formula obtained for the temperature (T) variation of the FMR linewidth of impurity centers in ferromagnetic sample [61]:

$$\Delta B = \Delta B_0 + P|T_c - T|^{-0.7} \qquad \textbf{(2)}$$

The fit to eq. (1) gave the following values ΔB_0 = (22.0 ± 1.0) mT, P = (0.52 ± 0.05) × 10^3 $K^{0.7}$×mT, and T_C = (393 ± 5) K. The Curie temperature, obtained from this fitting, is the same as T_C = 393 K determined for Fe_5Si_3 [59].

The linewidth expression described by eq. (2) is explained as follows. A significant contribution to FMR linewidth in ferromagnetic samples is caused by the magnetization fluctuations M close to the Curie temperature.

As the magnetization shifts the resonance position, magnetization fluctuations cause the resonance signal broadening. Thus, following the theory of second order phase transition [62] and applying the scaling theory [63], the mean-square magnetization fluctuations can be expressed as:

$$\left\langle (\Delta M)^2 \right\rangle = \frac{T_C \chi}{V} \propto \left| T_c - T \right|^{-1.4} \quad \textbf{(3)}$$

here, χ is magnetic susceptibility, V is volume and T_C is Curie temperature.

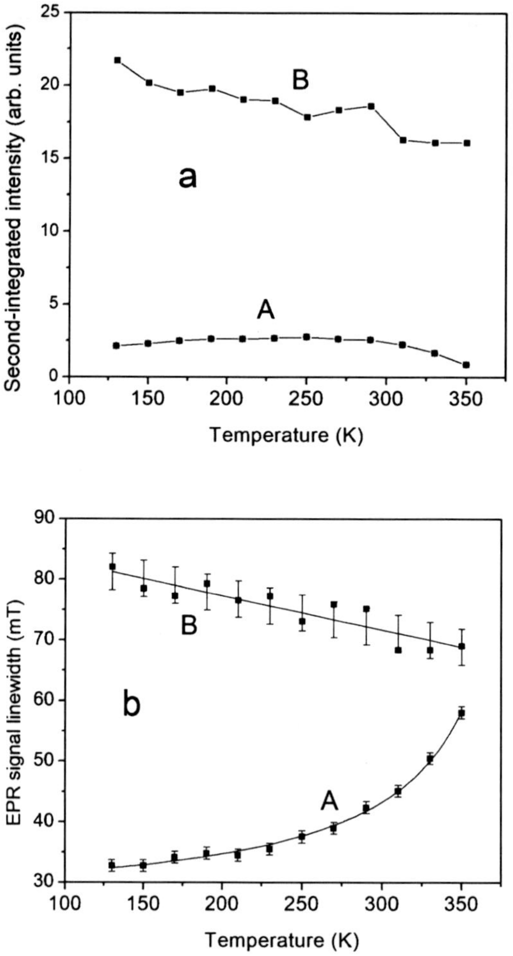

Fig. (8). Temperature variation of the double-integrated intensity (a) and the FMR/EPR linewidth of the first-derivative FMR/EPR line (b) (Reproduced from [49] with the Permission from Springer): a) for the signal at $g \sim 10.$; b) for the signal at $g \sim 2.0$. The solid lines correspond to the fitting results with the parameters reported in the text.

Hence, the fluctuation M is inversely proportionate to the square root of the temperature deviation from the Curie point (T_c) [62]. The EPR linewidth, ΔB, is proportionate to the mean square of magnetization fluctuations M resulting in eq. (2).

(ii) The intense EPR line at g=2.065±0.001. This signal does not reveal considerable temperature dependence. Its linewidth can be described by the linear expression: $\Delta B(T) = \Delta B_0 + P1 \cdot T$, where ΔB = (89.0 ± 2.0) mT and $P1$ = (–0.056 ± 0.008) mT/deg. It is supposed here that the Fe ions are in the superparamagnetic state, confirmed by the following facts.

(a) The *g*-factor is 2.065.

(b) The temperature variation of double-integrated intensity follows, within the range of measurement error, the Curie law in this temperature range of 77-350 K.

(c) The double-integrated intensity, being proportional to its magnetic moment, rises significantly with the enhancement of the pyrolysis temperature of the SiCN/Fe samples (T > 1000°C).

(d) The Fe ions content in the SiCN/Fe samples remain unchanged; therefore, this increase in the double-integrated intensity can only be caused by the formation of the magnetically ordered Fe-containing nanocrystallites.

The investigation of the bulk magnetization of SiCN/Fe ceramics using a vibration sample magnetometer is very helpful along with the study of collective magnetic moments using the EPR/FMR method. The magnetization versus the magnetic field of some SiCN ceramic samples was studied at 295 K in a magnetic field of up to 2.2 T. Almost no hysteresis or remanent magnetization was found in the investigated samples [49]. The coercive field decreases for the SiCN/Fe samples synthesized at 700°C from about 15 mT to zero for the samples synthesized at 1200°C-1500°C. The SiCN/Fe samples definitely exhibit superparamagnetic behavior. Large coercive fields for SiCN/Fe samples prepared at 700°C can be due to ferromagnetic α-Fe particles. The temperature variation of the magnetization at the magnetic field of 0.1 T was studied in the interval of 150 - 700 K as represented in Fig. (**9**).

It was clearly observed from this dependence that there were two various contributions to the magnetization of SiCN/Fe sample synthesized at 1215°C: the ferromagnetic contribution having the phase-transition temperature of T_C = 390 K, and the superparamagnetic contribution, obeying the Curie law.

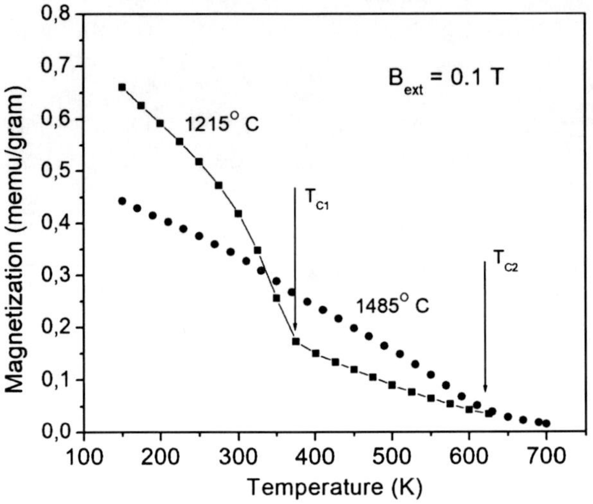

Fig. (9). Temperature variation of the magnetization measured in SiCN/Fe samples prepared at various temperatures at 0.1 T. Reproduced from [49] with the Permission from Springer.

The obtained Curie temperature (390 K) is due to Fe_5Si_3, as confirmed by the EPR/FMR experiments. The low-field FMR line was observed only in the SiCN/Fe sample pyrolyzed above 950°C, that corresponds well to Si-Fe phase diagram [64] for Fe_5Si_3 formation.

The magnetic transition at 620 K was revealed for the SiCN/Fe ceramics synthesized at 1485°C, which is related to a new ferromagnetic phase, most probably, $Fe_{70}Si_xC_{30-x}$ phase with T_C = 620 K [65]. No peculiarities in the dependence of magnetization on temperature near 490 K, the Curie temperature for Fe_3C nanoparticles, were observed [66]. It implies that no such Fe-containing particles were formed in SiCN/Fe. The variation of saturation magnetization with the pyrolysis temperature is analogous to that of the dependence of the double-integrated intensity of the low-field EPR signal. Hence, the main part of the magnetization in SiCN/Fe ceramics is caused by the nanoparticles of Fe-containing crystallites, very likely to those of Fe_5Si_3.

EPR/FMR STUDY OF SiCN/Mn CERAMICS

An X-band EPR Bruker ER-200D SRC spectrometer equipped with nitrogen-flow Bruker variable temperature assembly was utilized to study the SiCN/Mn ceramic samples [50]. The SiCN/Mn samples pyrolyzed at 1100°C were investigated. The EPR/FMR signals measured at various temperatures from 77 to 350 K are represented in Fig. (**10**).

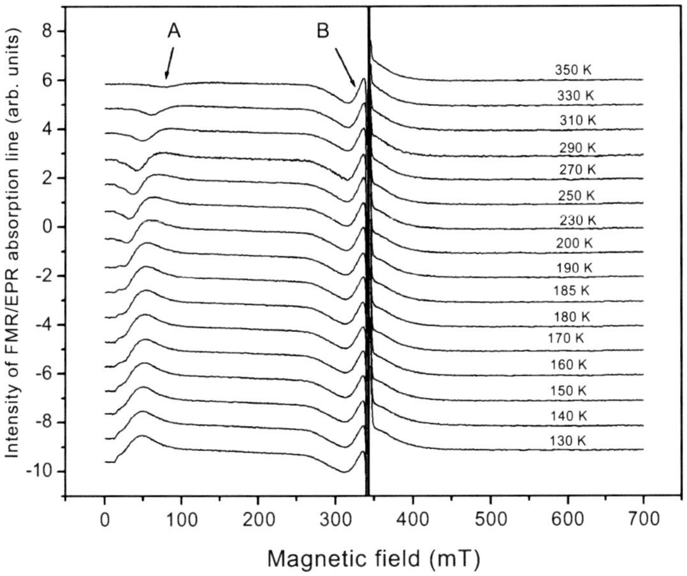

Fig. (10). Temperature dependence of the first-derivative signals. Here A and B designate the FMR and EPR lines, correspondingly. Reproduced from [50] with Permission from Springer.

FMR lines

The FMR signal (A) centered at ~ 65.0 mT with a linewidth of ~ 80.0 mT was observed in this SiCN/Mn sample at $T = 295$ K. The intensity of this signal rises with a decrease in temperature. It represents a structure on the low-field side with a decrease in temperature. This signal is caused by the ferromagnetic particles of the $Mn_5Si_3C_x$ distributed over the SiCN/Mn sample. Additionally, the EPR line (B) was observed, situated at ~340.0 mT with almost constant EPR linewidth (~35.0 mT), which is due to Fe-containing superparamagnetic nanoparticles. The variation of the double-integrated intensity of the absorption line (per 1 mg) and the peak-to-peak linewidth of the first-derivative EPR/FMR signal are presented in the temperature interval 77 - 350 K in Fig. (**11a**) and Fig. (**11b**), respectively.

Critical temperature as determined from the FMR line A. The area under this absorption line, which was obtained by double integration of the first-derivative absorption curve, decreased monotonically from 290 K to the highest temperature of observation (350 K) as seen from Fig. (**11a**). This integral intensity is proportionate to the magnetization of the ferromagnetic particles of $Mn_5Si_3C_x$ in SiCN/Mn. The FMR linewidth of signal A was described by Eq. (2) [61] exhibiting the temperature (T) dependence. The fit yields the values: $\Delta B_0 = (23.0 \pm 1.0)$ mT, $P = (0.26 \pm 0.03) \times 10^3$ $K^{0.7}\times$mT, and $T_C = 363 \pm 2$ K. A significant contribution to the FMR linewidth in ferromagnetic samples is caused by the magnetization fluctuations, *M*, close to the Curie temperature, as explained for the FMR line in SiCN/Fe. The behavior of the FMR line in SiCN/Mn agrees with the temperature variation of the magnetization of the ferromagnetic $Mn_5Si_3C_x$ particles. The Curie temperature of 360 K corresponds to this chemical composition, as derived from the Mn-Si-C phase diagram [67, 68].

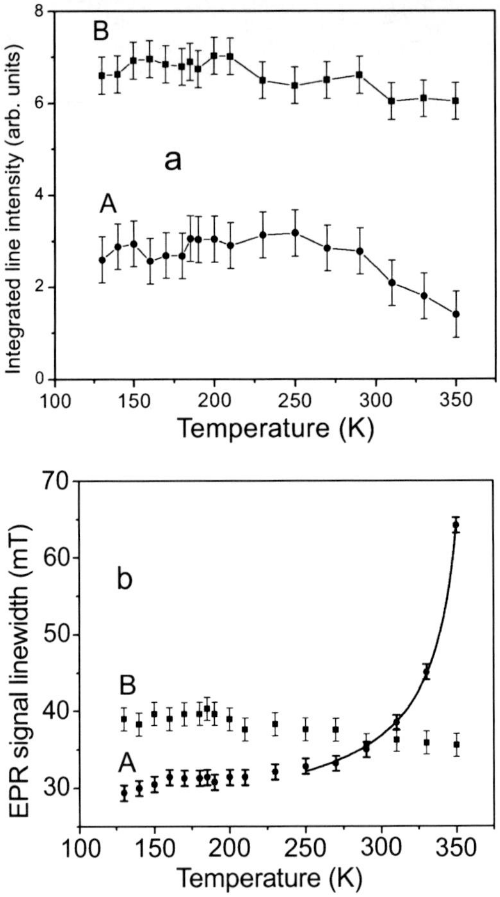

Fig. (11). Temperature dependence of the double-integrated intensity of first-derivative EPR/FMR line (a) and peak-to-peak first-derivative EPR/FMR linewidth (b) for the signals A and B. The solid line is fitting using the parameters, listed in the text. Reproduced from [50] with Permission from Springer.

EPR Lines

EPR lines were observed due to the two species as described below.

(i) EPR signal from the dangling bonds. The sharp, narrow signal with $g = 2.0026$ is caused by the carbon-related dangling bonds. Its EPR linewidth, *e.g.* 0.14 mT

for the sample with 1 wt % of $Mn(C_5H_7O_2)_2$ in the SiCN/Mn sample, was turned out to be slightly larger than those in the pure SiCN samples, *e.g.* 0.12 mT in pure SiCN samples synthesized at 1100°C [21]. The dipole-dipole couplings between C-related dangling bonds and the Mn^{2+} ions in the SiCN/Mn ceramics can be the reason for the additional contribution to the linewidth.

(ii) The EPR signal owing to Mn superparamagnetic species (B). This quite broad signal is centered at ~340.0 mT (g = 2.05 ± 0.01). The position of this line and linewidth do not change in the temperature range 77 – 350 K. The EPR linewidth is ~ 35.0 mT. This EPR signal is caused by the superparamagnetic particles of the Mn-containing compound and it corresponds to the EPR line, calculated by Kliava *et al.* [58] for an assembly of superparamagnetic particles diluted in the diamagnetic matrix.

However, A and B lines correspond to different Mn-containing compounds, because their temperature dependencies are quite different from each other.

Only SiCN/Fe samples pyrolyzed at 1100°C were investigated here and therefore additional investigations are highly desirable for deeper understanding of magnetic transformations of SiCN/Mn ceramics depending on pyrolysis temperature.

CONCLUDING REMARKS AND DISCUSSION

A. Undoped SiCN. The most significant characteristics of the EPR spectra in undoped SiCN samples are as follows:

(i) The two observed EPR signals are caused by C-related sp^2-dangling bonds:

(a) more intense EPR signal at g = 2.0027 relates to the free C phase at the surface;

(b) less intense EPR signal at g = 2.0032 which was well resolved at 170 GHz relates to the bulk of SiCN ceramic network

(ii) The EPR linewidth declines with a rise in pyrolysis temperature of a sample. This is due to exchange narrowing, which enhanced with the increase of a number of carbon-related dangling bonds. This number, in its turn, increases with the increasing pyrolysis temperature due to the removing of hydrogen ions.

(iii) The EPR linewidth increases strongly with decreasing temperature in all samples. The antiferromagnetic interaction among dangling bonds in these samples as estimated from the temperature variation of the EPR linewidth is found to be rather strong, between –12 and –15 K.

(iv) The X-band EPR line, detected in pure SiCN samples pyrolyzed at 1150°C and 1200°C, is stable and very narrow. Therefore, they can be a prospective spin-probe at room temperature.

B. SiCN/Fe. The most important features of the SiCN/Fe samples are as follows.

(i) The SiCN/Fe ceramic samples, which consist of nanoparticles, pyrolyzed in the temperature range 600°C- 1580°C, were investigated by EPR/FMR.

(ii) Three different states of Fe^{3+} ions were revealed in SiCN/Fe samples:

(a) The ordered and amorphous free C phases exist in the SiCN/Fe samples, as well as the transformation from the paramagnetic to superparamagnetic state was observed in the sample pyrolyzed at temperatures 600°C-800°C and 950°C-1150°C, correspondingly.

(b) Fe_5Si_3 particles are characterized by ferromagnetic behavior. The EPR line corresponding to this phase emerges in the SiCN/Fe samples synthesized at 950°C and vanishes in the samples synthesized at 1400°C.

(c) The third type of Fe^{3+} ions exhibits the narrow signaling at $g \sim 2.0$, revealed only in the samples pyrolyzed at $T > 1200$°C. They are possibly incorporated into the Si-C-N nanocrystallites. These lines vanish in the SiCN/Fe samples pyrolyzed at $T > 1530$°C.

(iii). The techniques of EPR/FMR and magnetic measurements provide sufficient data on the characteristics of the SiCN/Fe nanoparticles and show that this material can be exploited for MEMS applications.

C. SiCN/Mn. The most significant features of the SiCN/Mn samples are as follows.

(i) The FMR and magnetization data show that SiCN/Mn samples exhibit ferromagnetic or superparamagnetic behavior and they do not show paramagnetic behavior, which is due to the individual Mn ions in spite of the rather small concentration of Mn (~0.2%).

(ii) Ferromagnetic particles of $Mn_5Si_3C_x$, or other composites, are formed upon pyrolysis. At the same time, superparamagnetic clusters can be formed if the particle size is less than 5 nm. The two various species of Mn ions are:

(a) nanoparticles associated probably with SiCN network, which manifests themselves by the broad EPR signal at $g \sim 2.0$. The magnetization of these particles is almost temperature-independent in the investigated temperature range;

(b) the nanoparticles, which relate to the formation of SiCN network (Mn-Si-N-C) mainly due to the formation of $Mn_5Si_3C_x$ nanoparticles, as derived here from the temperature variation of the low-field FMR line. These particles reveal a transition from ferromagnetic to paramagnetic phase at 363 K, corresponding to the $Mn_5Si_3C_x$ compound.

CONSENT FOR PUBLICATION

Not applicable.

CONFLICT OF INTEREST

The author (editor) declares no conflict of interest, financial or otherwise.

ACKNOWLEDGEMENTS

This research was supported by the Natural Sciences and Engineering Research Council of Canada (NSERC) (SKM); SIA is grateful to Ministry of Education of Russian Federation for partial support within the framework of research project (#3.2166.2017/4.6), allocated to Kazan Federal University.

REFERENCES

[1] Wu, C.H.; Zorman, C.A.; Mehregany, M. Grown of polycrystalline SiC films on SiO_2 and Si_3N_4 on by APCVD. *Thin Solid Films,* **1999**, *355-356*, 179-183. [http://dx.doi.org/10.1016/S0040-6090(99)00494-0]

[2] Liew, L.A.; Zhang, W.; Bright, V. M.; An, L.; Dunn, M. L.; Raj, R. Fabrication of SiCN MEMS by photopolymerization of pre-ceramic polymer. *Sensors and Actuators,* **2001**, *A 89*, 64-70. [http://dx.doi.org/10.1016/S0924-4247(01)00723-3]

[3] Liew, L.-A.; Liu, Y.; Luo, R.; Cross, T.; An, L.; Bright, V.M.; Dunn, M. L.; Daily, J. W.; Raj, R. Fabrication of SiCN MEMS by photopolymerization of pre-ceramic polymer. *Sensors and Actuators,* **2002**, *A 95*, 120-134.

[4] Liew, L-A.; Saravanan, R.A.; Bright, V.M.; Dunn, M.L.; Daily, J.W.; Raj, R. Processing and characterization of silicon carbon-nitride ceramics: application of electrical properties toward MEMS thermal actuators. *Sensors and Actuators. A,* **2003**, *103*, 171-181. [http://dx.doi.org/10.1016/S0924-4247(02)00330-8]

[5] Hermann, A.M.; Wang, Y.T.; Ramakrishnan, P.A.; Balzar, D.; An, L.; Haluschka, C.; Riedel, R. Structure and electronic properties of Si-(B)- C-N ceramics. *J. Am. Ceram. Soc.,* **2001**, *84*, 2260-2264. [http://dx.doi.org/10.1111/j.1151-2916.2001.tb00999.x]

[6] MacLachlan, M.J.; Ginzburg, M.; Coombs, N.; Coyle, T.W.; Raju, N.P.; Greedan, J.E.; Ozin, G.A.; Manners, I. Shaped ceramics with tunable magnetic properties from metal-containing polymers. *Science,* **2000**, *287*(5457), 1460-1463. [http://dx.doi.org/10.1126/science.287.5457.1460] [PMID: 10688788]

[7] Zhang, L.; Wang, Y.; Wei, Y.; Xu, W.; Fang, D.; Zhai, L.; Lin, K.C.; An, L. A silicon carbonitride ceramic with anomalously high piezoresistivity. *J. Am. Ceram. Soc.,* **2008**, *91*, 1346-1349. [http://dx.doi.org/10.1111/j.1551-2916.2008.02275.x]

[8] Andronenko, S.; Stiharu, I.; Packirisamy, M.; Moustapha, H.; Dionne, P. The use of micro

electromechanical systems for surge detection in gas turbine engines. *Proceedings of International Conference on MEMS, NANO, and Smart Systems,* Banff, Alberta, CanadaJuly 24–27, 2005Banff**2005**, pp. 355-358.
[http://dx.doi.org/10.1109/ICMENS.2005.122]

[9] Leo, A.; Andronenko, S.; Stiharu, I.; Bhat, R.B. Characterization of thick and thin film SiCN for pressure sensing at high temperatures. *Sensors (Basel),* **2010**, *10*(2), 1338-1354.
[http://dx.doi.org/10.3390/s100201338] [PMID: 22205871]

[10] Saha, A.; Raj, R.; Williamson, D.V.; Kleebe, H-J. Characterization of nanodomains in polymer-derived SiCN ceramics employing multiple techniques. *J. Am. Ceram. Soc.,* **2005**, *88*, 232-234.
[http://dx.doi.org/10.1111/j.1551-2916.2004.00034.x]

[11] Trassl, S.; Motz, G.; Rossler, E.; Ziegler, G. Characterization of the Free-Carbon Phase in Precursor-Derived Si-C-N Ceramics: I, Spectroscopic Methods. *J. Am. Ceram. Soc.,* **2002**, *85*, 239-244.
[http://dx.doi.org/10.1111/j.1151-2916.2002.tb00072.x]

[12] Trassl, S.; Kleebe, H-J.; Stormer, H.; Motz, G.; Rossler, E.; Ziegler, G. Characterization of the Free Carbon Phase in Precursor-Derived SiCN Ceramics Part II: Comparison of Different Precursors. *J. Am. Ceram. Soc.,* **2002**, *85*, 1268-1274.
[http://dx.doi.org/10.1111/j.1151-2916.2002.tb00256.x]

[13] Trassl, S.; Motz, G.; Rossler, E.; Ziegler, G. Characterization of the free-carbon phase in precursor-derived SiCN ceramics. *J. Non-Cryst. Solids,* **2001**, *293-295*, 261-267.
[http://dx.doi.org/10.1016/S0022-3093(01)00678-0]

[14] Kleebe, H-J.; Störmer, H.; Trassl, S.; Ziegler, G. Thermal stability of SiCN ceramics studied by spectroscopy and electron microscopy. *Appl. Organomet. Chem.,* **2001**, *15*, 858-866.
[http://dx.doi.org/10.1002/aoc.243]

[15] Li, Y-L. Kroke, E.; Riedel, R.; Fasel, C.; Gervais, C.; Babonneau F. Thermal cross-linking and pyrolytic conversion of poly(ureamethylvinil)silazanes to silicon-based ceramics. *Appl. Organomet. Chem.,* **2001**, *15*, 820-832.
[http://dx.doi.org/10.1002/aoc.236]

[16] Resta, N.; Kohler, C.; Treblin, H-R. Molecular dynamics simulations of amorphous Si-C-N ceramics: composition dependence of atomic structure. *J. Am. Ceram. Soc.,* **2003**, *86*, 1409-1414.
[http://dx.doi.org/10.1111/j.1151-2916.2003.tb03484.x]

[17] Berger, F.; Muller, A.; Albinger, F.; Muller, K. Solid state NMR investigation on Si-B-C-N ceramics derived from boron-modified poly(allylmethylsilazane). *Z. Anorg. Allg. Chem.,* **2005**, *631*, 355-363.
[http://dx.doi.org/10.1002/zaac.200400259]

[18] von Bardeleben, H.J.; Cantin, J.L.; Zeinert, A.; Racine, B.; Zellama, K.; Hai, P.N. Spins and microstructure of hydroganated amorphous carbon: a multiple frequency electron paramagnetic resonance study. *Appl. Phys. Lett.,* **2001**, *78*, 2843-2845.
[http://dx.doi.org/10.1063/1.1370980]

[19] Coleman, J.N.; Dalton, A.B.; Curran, S.; Rubio, A.; Davey, A.P.; Drury, A.; McCarthy, B.; Lahr, B.; Ajayan, P.M.; Roth, S.; Barklie, R.C.; Blau, W.J. Phase separation of carbon nanotubes and turbostratic graphite using a functional organic polymer. *Adv. Mater.,* **2000**, *12*, 213-216.
[http://dx.doi.org/10.1002/(SICI)1521-4095(200002)12:3<213::AID-ADMA213>3.0.CO;2-D]

[20] Beuneu, F.; l'Huillier, C.; Salvetat, J-P.; Bonard, J-M.; Forro, L. Modification of multiwall carbon nanotubes by electron irradiation: An ESR study. *Phys. Rev.,* **1999**, *B59*, 5945-5949.
[http://dx.doi.org/10.1103/PhysRevB.59.5945]

[21] Andronenko, S. I.; Stiharu, I.; Misra, S.K. Synthesis and characterization of polyureasilazane derived SiCN ceramics. *J. Appl. Phys.,* **2006**, *99*(113907), 1-5.
[http://dx.doi.org/10.1063/1.2202291]

[22] Tomasella, E.; Spinelle, L.; Bousquet, A.; Rebib, F.; Dubois, M.; Eypert, C.; Gaston, J.P.; Cellier, J.

Structural and Optical Investigations of Silicon Carbon Nitride Thin Films Deposited by Magnetron. *Plasma Process. Polym.*, **2009**, *6*, S11-S16. [http://dx.doi.org/10.1002/ppap.200930103]

[23] Tomasella, E.; Rebib, F.; Dubois, M.; Cellier, J.; Jacquet, M. Structural and optical properties studies of sputtered a-SiCN thin films. *J. Phys. Conf. Ser.*, **2008**, *100*, 082045. [http://dx.doi.org/10.1088/1742-6596/100/8/082045]

[24] Kobayashi, K.; Yokoyama, H.; Endoh, M. Leakage current and paramagnetic defects in SiCN dielectrics for copper diffusion barriers. *Appl. Surf. Sci.*, **2008**, *254*, 6222-6225. [http://dx.doi.org/10.1016/j.apsusc.2008.02.144]

[25] Savchenko, D.; Kulikovsky, V.; Vorliček, V.; Lanček, J.; Kiselov, V.; Kalabukhova, E. Optical and magnetic resonance study of a-SiC_xN_y films obtained by magnetron sputtering. *Phys. Status Solidi, B Basic Res.*, **2014**, *251*, 1178-1185. [http://dx.doi.org/10.1002/pssb.201451041]

[26] Fujimori, H.; Ohnuma, S.; Kobayashi, N.; Masumoto, T. Spintronics in metal-insulator nanogranular magnetic thin films. *J. Magn. Magn. Mater.*, **2006**, *304*, 32-35. [http://dx.doi.org/10.1016/j.jmmm.2006.02.005]

[27] Wolf, S.A.; Awschalom, D.D.; Buhrman, R.A.; Daughton, J.M.; von Molnár, S.; Roukes, M.L.; Chtchelkanova, A.Y.; Treger, D.M. Spintronics: a spin-based electronics vision for the future. *Science*, **2001**, *294*(5546), 1488-1495. [http://dx.doi.org/10.1126/science.1065389] [PMID: 11711666]

[28] Bill, J.; Kamphowe, T.W.; Muller, A.; Wichmann, T.; Zern, A.; Jalowieki, A.; Mayer, J.; Weinmann, M.; Schuhmacher, J.; Muller, K.; Peng, J.Q.; Seifert, H.J.; Aldinger, F. Precursorderived Si–(B–)C–N Ceramics: Thermolysis, Amorphous State and Crystallization. *Appl. Organomet. Chem.*, **2001**, *15*, 777-793. [http://dx.doi.org/10.1002/aoc.242]

[29] Saha, A.; Raj, R.; Williamson, D.L. A model of the nanodomains in polymer-derived SiCO. *J. Am. Ceram. Soc.*, **2006**, *89*, 2188-2195. [http://dx.doi.org/10.1111/j.1551-2916.2006.00920.x]

[30] Saha, A.; Raj, R. Crystallization map for SiCO amorphous ceramics. *J. Am. Ceram. Soc.*, **2007**, *90*, 578-583. [http://dx.doi.org/10.1111/j.1551-2916.2006.01423.x]

[31] Saha, A.; Shah, S.R.; Raj, R.; Russek, S.E. Polymer derived SiCN composite with magnetic properties. *J. Mater. Res.*, **2003**, *18*, 2549-2551. [http://dx.doi.org/10.1557/JMR.2003.0356]

[32] Dumitru, A.; Ciupina, V.; Stamatin, I.; Prodan, G.; Morozan, A.; Mirea, C. Plasma polymerization of ferrocene with silane and silazane monomers for design of nanostructured magnetic ceramics. *J. Optoelectron. Adv. Mater.*, **2006**, *8*, 50-54.

[33] Dumitru, A.; Stamatin, I.; Morozan, A.; Mirea, C.; Ciupina, V. Si–C–N–Fe nanostuctured ceramics from inorganic polymer precursors obtained by plasma polymerization. *Mater. Sci. Eng. C*, **2007**, *27*, 1331-1337. [http://dx.doi.org/10.1016/j.msec.2006.09.004]

[34] Li, Y.; Zheng, Z.; Reng, C.; Zhang, Z.; Gao, W.; Yang, S.; Xie, Z. Preparation of Si-C-N-Fe magnetic ceramics from iron-containing polysilazane. *Appl. Organomet. Chem.*, **2003**, *17*, 120-126. [http://dx.doi.org/10.1002/aoc.400]

[35] Li, J.; Zhang, Z.; Zheng, Z.; Guo, L.; Xu, G.; Xie, Z. Preparation and Magnetic Properties of Fe/Si/C/N Ceramics Derived from a Polymer Precursor. *J. Appl. Polym. Sci.*, **2007**, *105*, 1786-1792. [http://dx.doi.org/10.1002/app.26161]

[36] Mishra, R.; Tiwari, R.K.; Saxena, A.K. Synthesis of Fe–SiC nanowires via precursor route. *J. Inorg.*

Organomet. Polym., **2009**, *19*, 223-227.
[http://dx.doi.org/10.1007/s10904-009-9259-7]

[37] Yan, X.H.; Cheng, X.N.; Han, G.C.; Hauser, R.; Riedel, R. Synthesis and Magnetic Properties of Polymer Derived Metal/SiCN ceramic composites. *Key Eng. Mater.*, **2007**, 353-358, 1485-1488. [http://dx.doi.org/10.4028/www.scientific.net/KEM.353-358.1485]

[38] Park, J.H.; Park, K.H.; Kim, D.P. Superparamagnetic Si_3N_4-Fe-Containing Ceramics Prepared from a Polymer-metal Complex. *J. Ind. Eng. Chem.*, **2007**, *13*, 27-32.

[39] Yan, X.; Cheng, X.N.; Li, C.S.; Hauser, R.; Riedel, R. Synthesis and Low Temperature Magnetic Properties of Metal Elements Filled Polymer-Derived SiCN Ceramic Composites. *Mater. Sci. Forum*, **2007**, *546-549*, 2269-2272.
[http://dx.doi.org/10.4028/www.scientific.net/MSF.546-549.2269]

[40] Hauser, R.; Francis, A.; Theismann, R.; Riedel, R. Processing and magnetic properties of metal-containing SiCN ceramic micro- and nano-composites. *J. Mater. Sci.*, **2008**, *43*, 4042-4049. [http://dx.doi.org/10.1007/s10853-007-2143-3]

[41] Singh, G.; Priya, S.; Hossu, M.R.; Shah, S.R.; Grover, S.; Koymen, A.R.; Mahajan, R.L. Synthesis, electrical and magnetic characterization of core–shell silicon carbo-nitride coated carbon nanotubes. *Mater. Lett.*, **2009**, *63*, 2435-2438.
[http://dx.doi.org/10.1016/j.matlet.2009.08.025]

[42] Francis, A.; Ionescu, E.; Fasel, C.; Riedel, R. Crystallization behavior and controlling mechanism of iron-containing Si-C-N ceramics. *Inorg. Chem.*, **2009**, *48*(21), 10078-10083.
[http://dx.doi.org/10.1021/ic900934u] [PMID: 19788256]

[43] Vakifahmetoglu, C.; Pippel, E.; Woltersdorf, J.; Colombo, P. Growth of one-dimensional nanostructures in porous polymer-derived ceramics by catalyst-assisted pyrolysis. Part I. Iron catalyst. *J. Am. Ceram. Soc.*, **2010**, *93*, 959-968.
[http://dx.doi.org/10.1111/j.1551-2916.2009.03448.x]

[44] Ionescu, E.; Terzioglu, C.; Linck, C.; Kaspar, J.; Navrotsky, A.; Riedel, R. Thermodynamic control of phase composition and crystallization of metal-modified silicon oxycarbides. *J. Am. Ceram. Soc.*, **2013**, *96*, 1899-1903.
[http://dx.doi.org/10.1111/jace.12327]

[45] Zhou, C.; Yang, L.; Geng, H.; Zheng, Q.; Min, H.; Yu, Zh.; Xia, H. Preparation of Si–C–N–Fe magnetic ceramic derived from iron-modified polysilazane. *Ceram. Int.*, **2012**, *38*, 6815-6822. [http://dx.doi.org/10.1016/j.ceramint.2012.05.080]

[46] Zaheer, M.; Schmalz, T.; Motz, G.; Kempe, R. Polymer derived non-oxide ceramics modified with late transition metals. *Chem. Soc. Rev.*, **2012**, *41*(15), 5102-5116.
[http://dx.doi.org/10.1039/c2cs15326b] [PMID: 22337594]

[47] Hojamberdiev, M. Prasad, Ravi Mohan; Fasel, C.; Riedel, R.; Ionescu, E. Single-source-precursor synthesis of soft magnetic Fe_3Si- and Fe_5Si_3-containing SiOC ceramic nanocomposites. *Eur. J. Ceram. Soc.*, **2013**, *33*, 2465-2472.
[http://dx.doi.org/10.1016/j.jeurceramsoc.2013.04.005]

[48] Mera, G.; Gallei, M.; Bernard, S.; Ionescu, E. Ceramic nanocomposites from tailor-made preceramics polymers. *Nanomaterials (Basel)*, **2015**, *5*(2), 468-540.
[http://dx.doi.org/10.3390/nano5020468] [PMID: 28347023]

[49] Andronenko, S.I.; Stiharu, I.; Menard, D.; Lacroix, C.; Misra, S.K. EPR/FMR, FTIR, X-ray, and Raman investigations of Fe-doped SiCN ceramics. *Appl. Magn. Reson.*, **2010**, *38*, 385-402. [http://dx.doi.org/10.1007/s00723-010-0120-5]

[50] Andronenko, S.I. Leo, Alfin; Stiharu, I.; Misra, S.K. EPR/FMR investigation of Mn doped SiCN ceramics. *Appl. Magn. Reson.*, **2010**, *39*, 347 356.
[http://dx.doi.org/10.1007/s00723-010-0163-7]

[51] Bahloul, D.; Pereira, M.; Gerardin, C. Pyrolysis chemistry of polysilazane precursors to silicon carbonitride Part 1. Thermal degradation of the polymers. *J. Mater. Chem.*, **1997**, *7*, 109-116. [http://dx.doi.org/10.1039/a603165j]

[52] Gerardin, C.; Taulelle, F.; Bahloul, D. Pyrolysis chemistry of polysilazane precursors to silicon carbonitride Part 2.- Solid-state NMR of the pyrolytic residues. *J. Mater. Chem.*, **1997**, *7*, 117-126. [http://dx.doi.org/10.1039/a603181a]

[53] Mukaida, M.; Hiyama, I.; Tsunoda, T.; Imai, Y. Preparation of Fe-Si thin films by chemical vapor deposition *Proc. of 17th International Conference on Thermoelectrics,* Nagoya, Japan 24-28 May **1998**, pp. 237-240. [http://dx.doi.org/10.1109/ICT.1998.740360]

[54] Shimizu, T.; Kumeda, M.; Kiriyama, Y. ESR study on sputtered amorphous Si-C, Si-Ge and Ge-C films. *Solid State Commun.*, **1980**, *37*, 699-703. [http://dx.doi.org/10.1016/0038-1098(81)91081-4]

[55] Ishii, N.; Kumeda, M.; Shimizu, T. The g-values of defects in amorphous C,Si and Ge. *Jpn. J. Appl. Phys.*, **1981**, *20*, L673-L676. [http://dx.doi.org/10.1143/JJAP.20.L673]

[56] Dyson, J.F. Electron spin resonance absorption in metals. II. Theory of electron diffusion and the skin effect. *Phys. Rev.*, **1955**, *98*, 349-359. [http://dx.doi.org/10.1103/PhysRev.98.349]

[57] Misra, S.K. Role of exchange interaction in effecting spin-lattice relaxation: new interactions of data on Cr^{3+} in $Cu_{2+x}Cr_{2x}Sn_{2-2x}$ spinel and dangling bonds in amorphous silicon. *Phys. Rev. B,* **1998**, *58*, 14971-14977. [http://dx.doi.org/10.1103/PhysRevB.58.14971]

[58] Kliava, J.; Berger, R. Size and shape distribution of magnetic nanoparticles in disordered systems: computer simulations of superparamagnetic resonance spectra. *J. Magn. Magn. Mater.*, **1999**, *205*, 328-342. [http://dx.doi.org/10.1016/S0304-8853(99)00510-7]

[59] Sawatzky, E. Magnetic and magnetooptical properties of sputtered Fe_5Si_3 films. *IEEE Trans. Magn.*, **1971**, *7*, 374-376. [http://dx.doi.org/10.1109/TMAG.1971.1067099]

[60] Hines, W.A.; Menotti, A.H.; Budnick, J.I.; Burch, T.J.; Litrenta, T.; Niculescu, V.; Raj, K. Magnetization studies of binary and ternary alloys based on Fe_3Si. *Phys. Rev. B,* **1976**, *13*, 4060-4069. [http://dx.doi.org/10.1103/PhysRevB.13.4060]

[61] Altshuler, T.S.; Bresler, M.S.; Goryunov, Yu.V. Ferromagnetic Ordering of Iron Impurities in the Kondo Semiconductor SmB_6. *JETP Lett.*, **2005**, *81*, 475-478. [http://dx.doi.org/10.1134/1.1984033]

[62] *Statistical Physics,* 4th ed; Landau, L.D.; Lifshitz, E.M. Course of Theoretical Physics.Butterworth-Heinemann: Oxford, **1999**, 5, .

[63] Patashinskii, A.; Pokrovskii, V. *Fluctuation theory of phase transitions,* 1st ed; Pergamon: Oxford, **1979**.

[64] Kolel-Veetil, M.K.; Keller, T.M. Organometallic Routes into the Nanorealms of Binary Fe-Si. *Materials (Basel),* **2010**, *3*, 1049-1088. [http://dx.doi.org/10.3390/ma3021049]

[65] Yelsukov, E.P.; Maratkanova, A.N.; Lomayeva, S.F.; Konyagin, G.N.; Nemtsova, O.M.; Ul'yanov, A.I.; Chulkina, A.A. Structure, phase composition and magnetic properties of mechanically alloyed and annealing quasibinary Fe(70)Si(x)C(30-x) alloys. *J. Alloys Compd.*, **2006**, *407*, 98-105. [http://dx.doi.org/10.1016/j.jallcom.2005.05.035]

[66] David, B.; Zboril, R.; Mashlan, M.; Grygar, T.; Dumitrache, F.; Schneeweiss, O. Single ferromagnetic behaviour of nanopowders with Fe_3C. *J. Magn. Magn. Mater.,* **2006**, *304*, e787-e789. [http://dx.doi.org/10.1016/j.jmmm.2006.02.224]

[67] Sürgers, C.; Potzger, K.; Fischer, G. Magnetism of carbon doped Mn_5Si_3 and Mn_5Ge_3. *J. Chem. Sci.,* **2009**, *121*, 173-176. [http://dx.doi.org/10.1007/s12039-009-0019-6]

[68] Gopalakrishnan, B.; Sürgers, C.; Montbrun, R.; Singh, A.; Uhlarz, M. v. Löhneysen, H. Electronic transport in magnetically ordered $Mn_5Si_3C_x$ films. *Phys. Rev. B,* **2008**, *77*, 104414. [http://dx.doi.org/10.1103/PhysRevB.77.104414]

CHAPTER 11

CW and Pulse EPR Study of Paramagnetic Centers in Silicon Carbide Nanomaterials

Dariya Savchenko[1,2,*], **Andreas Pöppl**[3] and **Abdel Hadi Kassiba**[4]

[1] *Department of Analysis of Functional Materials, Division of Optics, Institute of Physics of the Czech Academy of Sciences, Prague, 182 00, Czech Republic*

[2] *Department of Physics and Solid State Physics, National Technical University of Ukraine "Igor Sikorsky Kyiv Polytechnic Institute", Kyiv, 03056, Ukraine*

[3] *Institute of Experimental Physics II, Faculty of Physics and Earth Sciences, Leipzig University, Leipzig, 04103, Germany*

[4] *Institute of Molecules and Materials, UMR-CNRS, Le Mans University, 72085 Le Mans, France*

Abstract: In this chapter, we present the study of SiC nanoparticles obtained by pyrolysis and self-propagating high temperature synthesis (SHS) method using multi-approaches electron paramagnetic resonance (EPR) methods, including continuous wave EPR, field swept electron spin echo (FS ESE), pulsed electron nuclear double resonance (ENDOR) and four-pulse electron spin echo envelope modulation ESEEM (hyperfine sublevel correlation, HYSCORE) spectroscopy. Three paramagnetic defects were observed in SiC nanoparticles. Two of them with g_{iso} = 2.0029(3) and g_{iso} = 2.0043(3) were assigned to carbon vacancy V_C localized in the cubic (*β*) and hexagonal (*α*) phase of the SiC nanoparticles, respectively. The paramagnetic defect with g_{iso} = 2.0031(3) was attributed to the sp^3-coordinated carbon dangling bonds (CDB) located in the carbon excess phase of the SiC nanoparticles. The paramagnetic defect with g_{iso}= 2.0037(3), which was observed only in SiC nanoparticles obtained by SHS method was attributed to the bulk intrinsic defect having a Si-NSi_2 configuration and located in α-Si_3N_4 phase of the SiC nanoparticles. A high delocalization of the electronic wavefunction of the unpaired electron for the carbon vacancy V_C localized in the cubic crystalline phase of the SiC nanoparticles was found from the detail study of the V_C ligand structure by pulse ENDOR and HYSCORE methods.

Keywords: α-Si_3N_4 phase, Carbon dangling bonds, Carbon excess, Carbon vacancy, Davies ENDOR, EPR, Field-swept electron spin echo (FS ESE), Hydrogen, Hyperfine coupling, HYSCORE, Intrinsic defects, Mims ENDOR,

* **Corresponding author Dariya Savchenko:** Department of Physics and Solid State Physics, National Technical University of Ukraine "Igor Sikorsky Kyiv Polytechnic Institute", Kyiv, Ukraine; Tel: +(420) 266 05 2356; Fax: +(420) 286 581 448; E-mail: dariyasavchenko@gmail.com

Dariya Savchenko & Abdel Hadi Kassiba (Eds.)

Nanoparticles, Pulse EPR, Silicon carbide (SiC), Superhyperfine coupling, Spin density, Stoichiometry, α-SiC phase, β-SiC phase.

INTRODUCTION

Many parameters of silicon carbide (SiC) such as good mechanical properties, good oxidation resistance, low thermal expansion coefficient, high electrical and thermal conductivity, high hardness, chemical stability and refractoriness make it important engineering material [1]. On the other hand, these properties become more crucial in nanosized SiC materials. It was found that SiC nanoparticles and their composites revealed size-dependent electrical conductivity [2], photoluminescence efficiency [3], charge carrier specific features [4] and non-linear optical properties [5]. The main parameters governing this behaviour lie in the large specific surface, interfaces and confinement effects as well as the crucial role of the electric active intrinsic defect centers, which are present in the SiC nanoparticles and stabilize them thermodynamically [6].

It is well known that electron paramagnetic resonance (EPR) is a highly relevant and effective method for the detection and identification of electrically active intrinsic defect centers, resulting from defects, impurities, dangling bonds, or doping agents. In this work, multi-approaches EPR methods were applied to the investigation of the SiC nanoparticles. To clarify the origin of the paramagnetic intrinsic defects formed in the SiC nanoparticles, their interaction with the surrounding nuclei, as well as the peculiarities of their electronic structure, the SiC nanoparticles were investigated by continuous wave (CW) EPR, field-swept electron spin echo (FS ESE), pulsed electron nuclear double resonance (ENDOR) and hyperfine sublevel correlation spectroscopy (HYSCORE) methods. As a result, 4 and 3 types of intrinsic paramagnetic defects with isotropic g-values and different spin relaxation times were resolved in the FS ESE spectra of SiC nanoparticles obtained by self-propagating high-temperature synthesis (SHS) and laser pyrolysis methods, respectively.

It is known that pulsed ENDOR and HYSCORE techniques allow measuring weak superhyperfine (shf) interactions between the paramagnetic centers and surrounding ligand nuclei, which are not able to be resolved in the CW EPR spectra and thus, to get detailed information about the local structure of the center. In this work, we use the potential ability of FS ESE, pulsed ENDOR, and HYSCORE methods for the structural analysis of the intrinsic paramagnetic defect in SiC nanoparticles. The investigation of the shf structure of the intrinsic defects by pulse ENDOR and HYSCORE techniques was focused on the defect with $g_{iso} = 2.0029(3)$ that was related to the carbon vacancy (V_C) located in the β-SiC crystalline phase of the SiC nanoparticles.

Taking into account that other defects have insufficient FS ESE signal intensity to be investigated by pulse ENDOR and HYSCORE methods, their identification was carried out by comparison of their experimental EPR data with the microstructure of the SiC nanoparticles. The defect with g_{iso} = 2.0043(3) was assigned to the V_C localized in α-SiC crystalline phase, while the center with g_{iso} = 2.0031(3) was attributed to the carbon dangling bonds (CDB) located in the C-excess phase. The center with g_{iso} = 2.0037(3) observed only in the SiC nanoparticles synthesized by SHS method was assigned to a threefold-coordinated Si atom bonded with one nitrogen atom having Si-Si_2N configuration located in a-Si_3N_4 phase.

SAMPLE CHARACTERIZATION AND EXPERIMENTAL TECHNIQUE

One set of nanosized SiC powders was synthesized by laser pyrolysis of SiH_4 and C_2H_2 gas mixture [7]. The average particle size of the synthesized powder is close to 20-25 nm. Annealing at 1400°C for 1 hour under argon induces a better chemical homogeneity of the powders and their densification without significantly altering the average particle size. The structural investigation by X-ray diffraction and nuclear magnetic resonance (NMR) showed that SiC nanoparticles contain different crystalline polytypes as well as amorphous phases in the following ratios: 40% as cubic β-SiC, 10% as hexagonal (α-SiC) and 49% as amorphous SiC. The presence of the amorphous SiC phase can result either from the disorder at the particle boundaries or to stacking faults in the nanoparticle cores [8].

Another set of SiC nanoparticles was prepared by self-propagating high-temperature synthesis (SHS) method by heating a mixture of a fine powder of semiconductor-pure elemental silicon and thermally exfoliated graphite (TEG) in an argon atmosphere [9 - 11]. The ignition temperature was about 1200-1250°C and process-time of synthesis (heating and cooling) was about 2.5 hours. The obtained β-SiC reaction product in the agglomerated state can be simply reduced to nanoparticles. After synthesis, the unreacted carbon was removed by heating in the air at 700^{0}C. To remove SiO_2 the final cleanup was carried out by etching in hydrofluoric acid, rinsing in distilled water and drying.

As was shown by X-ray diffractometry and high-resolution transmission electron microscopy (HRTEM) measurements the synthesis of the SiC nanoparticles in non-equilibrium conditions leads to the decrease of the β-SiC lattice parameter up to 4.3536 Å. A smaller value of the lattice parameter than that for standard β-SiC (4.3589 Å) can be explained by the partial substitution of Si atoms in the SiC lattice by C, N and O atoms, which have a smaller covalent radius [11]. Reduction of the β-SiC lattice parameter leads to the appearance of the SiC-C solid solution and a small fraction of the nitrogen-containing (Si_3N_4, Si_2N_2O) phases as an impurity, formed as a result of the nitrogen in the air. The complete purification of the synthesized powder showed that the content of these phases does not exceed 3-5% [10]. The HRTEM study of the SiC nanoparticles has shown that the average size of the particles is about 40 nm. Analysis of the Raman modes observed in SiC nanoparticles leads to their assignment to 3C polytype (β-SiC).

The X-band FS ESE, pulsed ENDOR and HYSCORE measurements have been conducted on Bruker ESP 380 and ELEXSYS E580 spectrometers at $T = 6$ K. The CW EPR measurements were performed on RE1308 spectrometer at $T = 4$ K at Q-band frequency. The simulations of the FS ESE spectra were performed using Easyspin toolbox [12]. The schematic representation of the pulse sequences used in various experiments is given in Fig. (**1**).

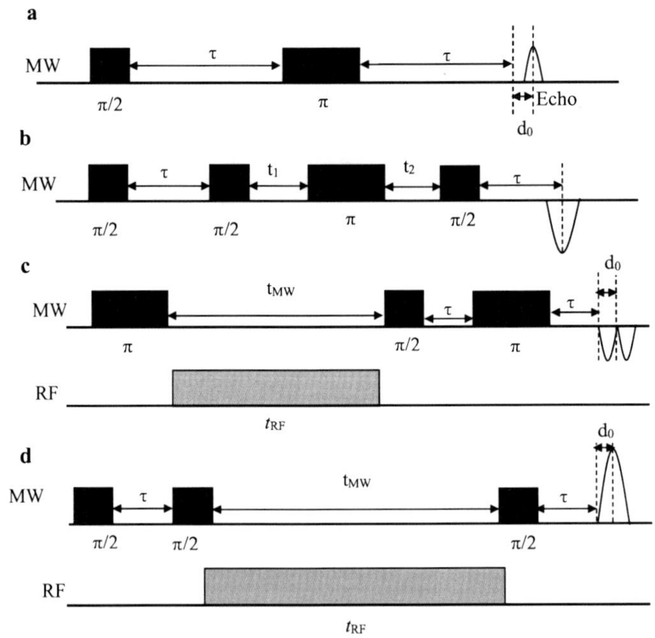

Fig. (1). The schematic representation of the pulse sequences used in various experiments: (a) – two-pulse FS ESE sequence (d0 –detection trigger offset time), (b) – four-pulse (two dimensional) ESEEM (HYSCORE) sequence, (c) – Davies pulse ENDOR sequence, (d) – Mims pulse ENDOR sequence.

The X-band FS ESE spectra were recorded using two-pulse echo sequence [13] (Fig. **1a**) with microwave (MW) pulse durations: $t_{\pi/2} = 16$ ns and $t_{\pi} = 32$ ns and the time between pulses $\tau = 400$ ns (short pulses) as well as $t_{\pi/2} = 96$ and $t_{\pi} = 192$ ns with $\tau = 800$ ns (long pulses). The HYSCORE spectra [14] were obtained using non-selective MW pulses of $t_{\pi/2} = 16$ ns and $t_{\pi} = 32$ ns with the pulse delays of $\tau =$

136 ns and 104 ns between the first two $\pi/2$ MW pulses (see Fig. **1b**). In total, 170 × 170 data points were acquired for t_1 and t_2, and the spectra with the various τ values were added prior two-dimensional (2D) Fourier transformation (FT) and finally, 2D FT magnitude spectra were obtained. The Davies ENDOR spectra [15] were recorded using the following pulse sequence: $t_\pi - t_{MW} - t_{\pi/2} - \tau - t_\pi -$ echo with selective MW pulses of $t_{\pi/2} = 96$ and $t_\pi = 192$ ns and a radio frequency (RF) pulse duration of $t_{RF} = 18$ µs (see Fig. **1c**). For Mims ENDOR experiments [16] the stimulated ESE pulse sequence: $t_{\pi/2} - \tau - t_{\pi/2} - t_{MW} - t_{\pi/2} - \tau -$ echo was used (see Fig. **1d**), with non-selective MW $\pi/2$ pulse durations of $t_{\pi/2} = 48$ ns and the time between pulses was $\tau = 560$ ns. A selective RF pulse of variable frequency and duration of $t_{RF} = 12$ µs was applied between the second and third MW pulses.

EXPERIMENTAL RESULTS

CW and Pulsed EPR Study of the SiC Nanoparticles

Fig. (**2**) shows the CW EPR spectra measured in the SiC nanoparticles grown by SHS method at Q-band frequency and $T = 4.2$ K. The EPR spectrum consists of several unresolved EPR lines centered at the g-factor ~2.0025(2).

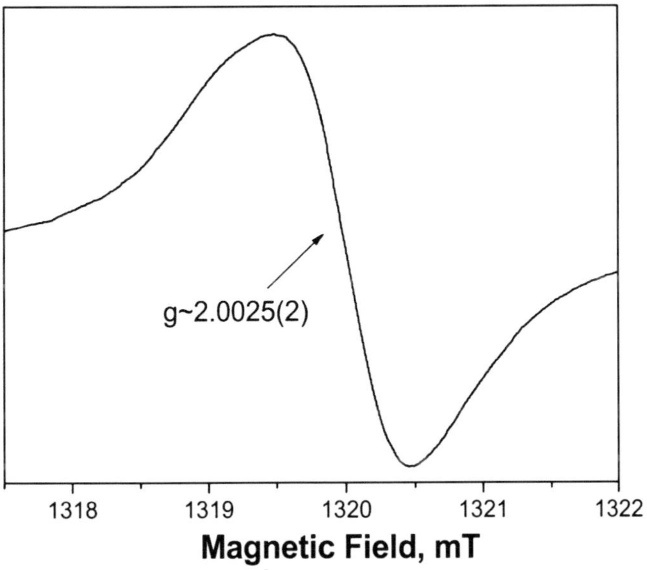

Fig. (2). The Q-band CW EPR spectrum measured in SiC nanoparticles synthesized by SHS at $T = 4.2$ K. Reproduced from Ref [17]. with permission from the Publisher.

The advantage of the pulse EPR technique over the CW EPR is the lack of the magnetic field modulation, which affects the EPR linewidth and the pronounced dependence of the ESE signal on electron spin relaxation times. Considering these advantages FS ESE method was applied to resolve the observed EPR signal.

Fig. (**3**) shows the first derivative of the X-band FS ESE spectra measured in SiC nanoparticles for different pulse length and the time between pulses (delays). A few components in FS ESE spectra with different intensity ratio can be resolved by variation of the pulse length and the time delay between pulses indicating that the electron spin phase memory relaxation times of the intrinsic defects responsible for the components of the FS ESE spectrum are different.

The observed additional shoulders in the wings of the FS ESE spectrum (Fig. **3**) point to the existence of the isotropic shf coupling (A_{iso} = 29.7 MHz) with surrounding ^{29}Si or ^{13}C nuclei. Both isotopes have a nuclear spin I = 1/2 and natural abundances of 4.7% and 1.1%, correspondingly.

The shf signals are symmetrically located regarding the central line at g_{iso} = 2.0029(3) and can be accordingly attributed to this center. The measured intensity ratio between the central signal and shf lines is about 9-10%, which corresponds to the natural abundance of two equivalent Si atoms. The defect with a similar value of the shf splitting was previously assigned to a carbon vacancy (V_C) [2, 7]. Therefore, it can be concluded that the observed center with g_{iso} = 2.0029(3) is also related to the carbon vacancy (V_C).

The identification of other intrinsic defects was carried out by comparing their experimental EPR parameters with the structural characteristic of the SiC nanoparticles. As it was shown previously in Ref [7]. the center with g = 2.0048(6) gradually disappears with decreasing the degree of hexagonality of the SiC nanoparticles. On the other hand, the SiC nanoparticles synthesized both by SHS and CO_2 laser pyrolysis have mostly cubic crystalline structure. Thus, comparing the relative intensity ratio of signals with g_{iso} = 2.0029(3) and g_{iso} = 2.0043(3) it can be concluded that the center with g_{iso} = 2.0029(3) is assigned to V_C localized in the β-SiC crystalline phase, while the center with g_{iso} = 2.0043(3) should be related to the intrinsic defect (possibly V_C) located in hexagonal α-phase of the SiC nanoparticles.

According to Ref [2, 7]. the center with g_{iso} = 2.0031(3) was related to CDBs located in the C-excess phase of the SiC nanoparticles. Considering the presence of the solid solution of carbon (SiC-C) in the crystalline structure of the SiC nanoparticles the center with g_{iso} = 2.0031(3) may attribute to the sp^3-coordinated CDBs located in the C-excess phase of the SiC nanoparticles.

The center with g_{iso}= 2.0037(3) was observed in SiC nanoparticles obtained by SHS method only. The g-factor and a linewidth of 0.60(5) mT of this center are similar to that previously observed for the bulk K-center in amorphous hydrogenated Si_3N_4 layers [18]. Depending on the local bonding configuration Si-X_3 of the K center *g*-factor in different Si_xN_y layers varies from 2.0030 for Si-N_3 to 2.0040 for Si-NSi_2 [18]. In addition, the linewidth of the EPR signal depends on the K center configuration due to the presence of the unresolved shf interaction with the ^{14}N ($I = 1$) nuclei. Thus the center at $g_{iso} = 2.0037(2)$ was assigned with the defect located in *a*-Si_3N_4 phase with Si-NSi_2 configuration. This conclusion was supported by the presence of the peak reflections due to *a*-Si_3N_4 and *a*-Si_2N_2O phases in X-ray diffractometry (XRD) spectrum of SiC nanoparticles.

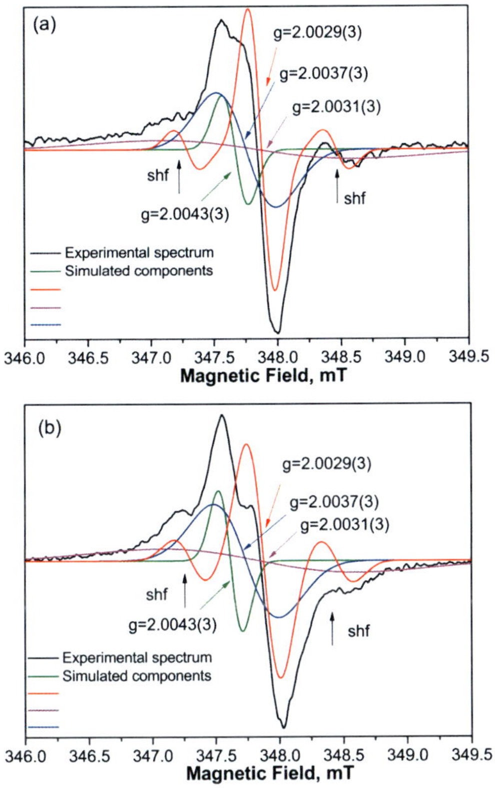

Fig. (3). First derivative of X-band FS ESE spectrum measured on SiC nanoparticles synthesized by SHS at $T = 6$ K with different length of the pulses: short (a) and long (b). Adapted from Ref [17]. with permission from the Publisher.

ENDOR and HYSCORE Study of SHF Structure of V_C in the β-SiC Crystalline Phase

The detailed investigation of the shf structure of the intrinsic defects by pulse ENDOR and HYSCORE techniques was focused on the V_C center located in the *β*-SiC crystalline phase of the SiC nanoparticles synthesized by laser pyrolysis, where only three paramagnetic centers were observed. Other defects have insufficient FS ESE signal intensities to be studied their ligand structure by pulse ENDOR and HYSCORE techniques.

Fig. (**4**) shows the pulse Davies ENDOR spectrum detected at T = 6 K that is more sensitive for detecting large shf interactions in comparison with the Mims pulse sequence.

The Mims ENDOR sequence is usually preferable in the case of small shf interactions when the phase memory time of the sample is enough long to use an optimal τ value or nonselective mw pulses [19].

The ENDOR spectra demonstrate the isotropic character and are consisted of a few doublets positioned at the ^{1}H Larmor frequency ($\nu_n(^1H)$ = 14.80 MHz). Therefore, for the understanding of the ENDOR results, we will take into consideration the isotropic shf coupling (Fermi contact term). The Fermi-contact interaction term a_i reflecting the spin density of the donor electron wavefunction (Ψ) at the nucleus site (r_i) and is described by:

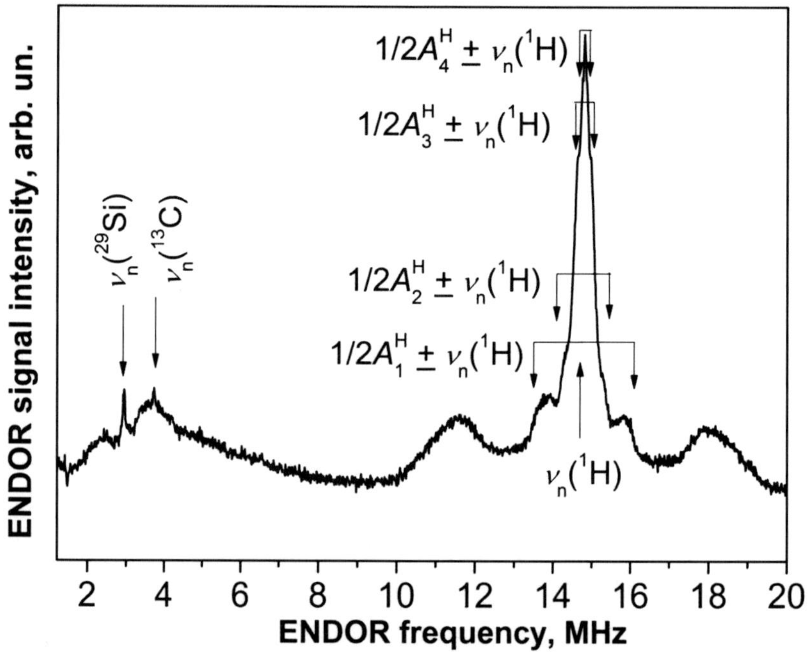

Fig. (4). X-band pulsed ENDOR recorded on the FS ESE position of V_C in β-SiC crystalline phase in SiC nanoparticles obtained by laser pyrolysis at T = 6 K using Davies pulse sequence. Reproduced from Ref [20]. with permission from Elsevier Ltd.

$$a_i = \frac{8\pi}{3}\mu_B\mu_n g_e g_{ni}\left|\Psi(r_i)\right|^2, \qquad \textbf{(1)}$$

where μ_B – is the Bohr magneton of the electron, μ_n – nuclear magneton, g_e – electron g factor, g_{ni} – is the g_n factor of nucleus i, g_n – nuclear g factor.

The ENDOR transitions related to the nuclear spin $I = 1/2$ (*e.g.* ^{1}H, ^{13}C, ^{29}Si) and electron spin $S = 1/2$ have the following frequencies:

$$\nu_{ENDORi} = \frac{1}{h}\left|g_{ni}\mu_n B \pm \frac{a_i}{2}\right|, \qquad \textbf{(2)}$$

Here B is the external magnetic field.

Thus, from the Eq. (2) it follows that each nucleus i results in two ENDOR lines whose frequencies symmetrically located around their Zeeman frequency $g_{ni}\mu_n B_0 / h$ (under conditions of the weak coupling limit: $2g_{ni}\mu_n B_0 / h > a_i$).

As it follows from Fig. (**4**) four doublet ENDOR lines splitting by 2.3 MHz, 1.4 MHz, 0.4 MHz and 0.1 MHz are symmetrically located around the $\nu_n(^1H)$. The observed ENDOR lines are isotropic and belong to the shf interaction with ^{1}H with constants: $A^H{}_1$ = 2.3 MHz, $A^H{}_2$ = 1.4 MHz, $A^H{}_3$ = 0.4 MHz and $A^H{}_4$ = 0.1 MHz. Thus, it could be proposed that the investigated V_C center locates in hydrogenated regions of the SiC nanoparticles. This assumption is suggested by the observation of the ^{1}H ENDOR signal at the proton Zeeman frequency weakly coupled with a unpaired electron by means of electron-nuclear dipolar interaction.

Fig. (**5**) shows the low-frequency part of the Davies and Mims ENDOR spectra recorded with ^{29}Si ($\nu_n(^{29}Si)$ = 2.94 MHz) and ^{13}C ($\nu_n(^{13}C)$ = 3.72 MHz) signals at the same FS ESE magnetic field position. A small shf coupling with ^{29}Si and ^{13}C nuclei that were not resolved in the FS ESE spectra can be measured in ENDOR spectra.

The broad doublet observed in the ENDOR spectrum with the Davies pulse sequence (Fig. **5a**) is symmetrically located around $\nu_n(^{29}Si)$ and corresponds to the minimum and maximum shf interactions with ^{29}Si nuclei (Si(II)) of $A^{Si(II)}{}_{min}$ = 0.5 MHz and $A^{Si(II)}{}_{max}$ = 2.8 MHz. The obtained values can be translated into isotropic and anisotropic shf interaction parameters $A^{Si(II)}{}_{iso}$ = 1.27 MHz and $T^{Si(II)}$ = 0.77 MHz, where $A_{max} = A_{iso} + 2T$ and $A_{min} = A_{iso} - T$ [20].

The obtained data correlate with the total width of the ^{29}Si signal observed in the pulse Mims ENDOR spectrum shown in Fig. (**5b**). Two additional doublets centered at the ^{29}Si and ^{13}C Larmor frequencies, respectively, with the small splitting were observed in the Mims ENDOR spectrum and apparently correspond to the shf couplings with another type of Si nuclei (Si(III) having $A^{Si(III)} = 0.33$ MHz and carbon nuclei (C(I)) with $A^{C(I)} = 0.28$ MHz.

Whereas ENDOR spectroscopy can be effectively applied to the study of the isotropically coupled spin systems in disordered systems, HYSCORE spectroscopy provides additional complementary to ENDOR information about anisotropic shf interactions.

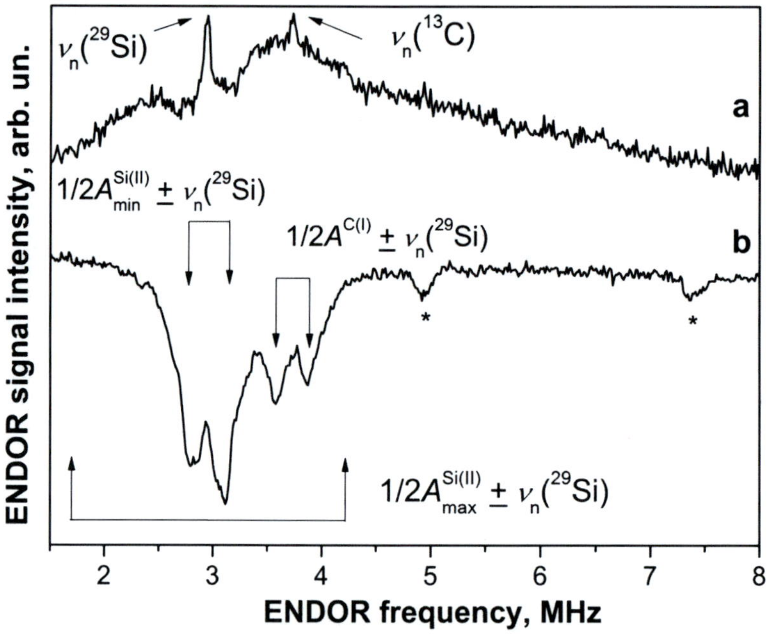

Fig. (5). The Davies (a) and Mims (b) ENDOR spectra of SiC nanoparticles obtained by laser pyrolysis that were recorded at the FS ESE position of V_c in β-SiC crystalline phase at $T = 6$ K. The higher harmonics of the intense distant proton signal were pointed by asterisks (not shown). Reproduced from Ref [20]. with permission from Elsevier Ltd.

Fig. (**6a**) shows the HYSCORE spectrum of SiC nanoparticles recorded at the magnetic field setting that corresponds to the g-value of the V_C located in β-SiC crystalline phase. Besides the intense peaks observed in the low-frequency part of the HYSCORE spectrum belonging to the shf interaction with ^{29}Si, ^{13}C and ^{14}N nuclei, a low-intensity ^{1}H cross peak ridge with a maximum width of about 2.5 MHz centered at $\nu_n(^1H)$ is visible. A similar ^{1}H ENDOR line centered at the ^{1}H

Larmor frequency with $A^H{}_2$ = 2.5 MHz was already deduced from the pulse Davies ENDOR spectrum shown in Fig. (**4**).

Taking into account that no ^{1}H cross peaks at ([11.6, 18.0], [18.0, 11.6]) MHz corresponding to the shf interaction with $A^H{}_1$ = 6.4 MHz were observed in HYSCORE spectrum, the broadening of ENDOR lines can be explained by a distribution in the isotropic shf coupling whereas the contribution from the anisotropic coupling remains negligible. Average isotropic ^{1}H shf interaction $A^H{}_1$ = 6.4 MHz translates into a spin densities in hydrogen 1*s* atomic orbital of about 0.24% - 0.66% for a single interacting proton.

The appearance of the ^{1}H line in the ENDOR spectrum can be explained by the presence of hydrogenated regions in the SiC nanoparticles, which apparently the result of final etching of the samples in hydrofluoric acid.

The intense ^{29}Si cross peak ridge with a total breadth of 4.4 MHz was observed in low-frequency part of the HYSCORE spectrum (Fig. **6b**). The main contribution in the central part of the ridge near v_n(^{29}Si) comes from the silicon nuclei Si(II) and Si(III) while the shf coupling with $A^{Si(II)}{}_{max}$ = 2.8 MHz does not contribute into the ridge width. This means that the paramagnetic center couples to the more distant ^{29}Si nucleus with a maximum shf coupling of $A^{Si(IV)}{}_{max}$ = 4.4 MHz.

Taking into account the shape of the observed ridge breadth the value of the anisotropic shf interaction $T^{Si(IV)}$ should be at least in the order of $1/3(A^{Si(IV)}{}_{max}-A^{Si(II)}{}_{max})$ = 0.5 MHz. Otherwise, the sizable anisotropic shf couplings should lead to a curvature of the cross peak ridges directed to the high frequencies caused by second order effects in the nuclear transition frequencies [19].

According to [21] the anisotropic shf interaction parameter *T* can be calculated from the maximum shift of the cross peak ridges Δv_{max} from the $v_1 = -v_2$ axis:

$$T = \frac{2}{3}\left(\frac{8\Delta v_{max} v_n}{\sqrt{2}}\right)^{1/2} \qquad \textbf{(3)}$$

where v_n is the nuclear Larmor frequency. In our case, the maximum shift of the cross peak ridges is $\Delta v^{Si}_{max} < 0.06$ MHz, which is below the spectral resolution of the experiment and therefore we are able to estimate from eq. (3) only an upper limit for $T^{Si(IV)}$ of 1.4 MHz. Thus, the $T^{Si(IV)}$ was varied in the range between 1.4 MHz $\geq T^{Si(IV)} \geq$ 0.5 MHz.

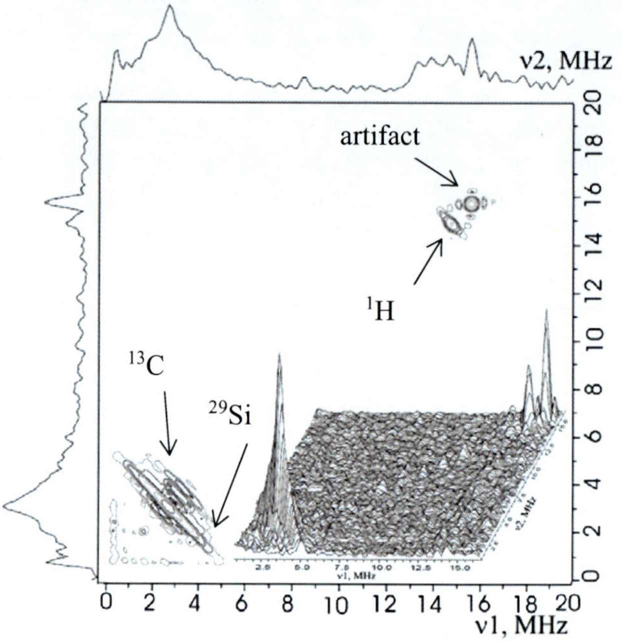

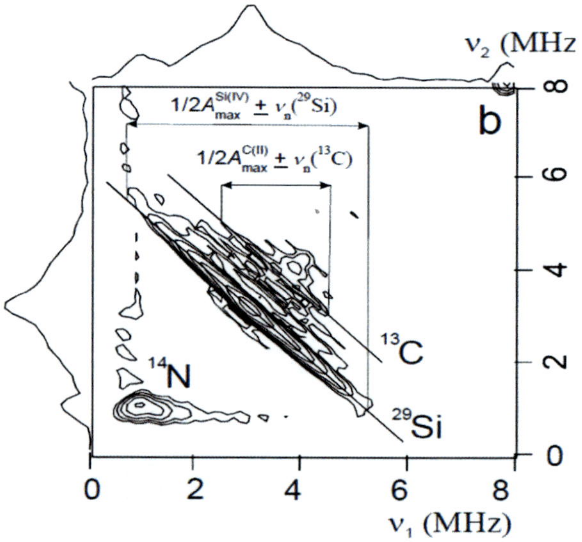

Fig. (6). The contour plot of HYSCORE spectra of SiC nanoparticles recorded at the FS ESE position of V_C in β-SiC crystalline phase. (a) the full spectral range, the inset is the stacked plot of the HYSCORE spectrum, reproduced from Ref [17]. with permission from the Publisher. (b) low-frequency part reproduced from Ref [20]. with permission from Elsevier Ltd. The solid lines in (b) indicate the $\nu_1 = -\nu_2$ axes at the ^{29}Si and ^{13}C Larmor frequencies.

As a result, the isotropic shf coupling for Si(IV) was varied in the range between 1.6 MHz $\leq A_{iso}^{Si(IV)} \leq$ 3.4 MHz. As was seen from Fig. (**6**), a less intense ridge at $\nu_n(^{13}C)$ is also observed in HYSCORE spectra with a breadth of about $A_{max}^{C(II)}$ = 1.6 MHz corresponding to the carbon nuclei C(II). Since $\Delta\nu_{max}^{Si,C}$ < 0.06 MHz defines an upper limit for the anisotropic shf constant $T^{C(II)} \leq$ 0.5 MHz we can evaluate isotropic shf constant for C(II) as ranging between 0.6 MHz $\leq A_{iso}^{C(II)}$ < 1.6 MHz from eq. (3).

From the obtained isotropic and anisotropic ^{29}Si and ^{13}C shf couplings one can estimate the unpaired spin densities ρ_s and ρ_p for 3*s*, 3*p* and 2*s*, 2*p* Si and C atomic orbitals [22] in the vicinity of the V_C (see Table **1**). Assuming that the largest isotropic couplings (Si(I)) belong to the two equivalent ^{29}Si nuclei localized in the nearest environment of the defect site [7], the spin density for the Si atomic 3*s* orbitals amounts of about 1.2% of the total spin density. The isotropic shf couplings found for all other Si and C nuclei and the corresponding their spin densities in *s* atomic orbitals are significantly smaller and therefore correspond to the atoms located in the second or further coordination shells of the defect site [23].

Table 1. Experimentally estimated values of the isotropic (A_{iso}) and anisotropic (T) ^{29}Si and ^{13}C shf interactions and spin densities ρ_s and ρ_p in the Si and C *s* and *p* atomic orbitals obtained for the V_C in β-SiC crystalline phase. (reproduced from Ref [20]. with permission from Elsevier Ltd.).

Nuclei	A_{iso}, MHz	T, MHz	ρ_s, %	ρ_p, %
Si(I)	29.7	-	0.6	-
Si(II)	1.27	0.77	0.028	0.67
Si(III)	0.33		0.007	-
Si(IV)	0.5-1.4	1.6-3.4	0.035-0.075	0.44-1.23
C(I)	0.28		0.007	-
C(II)	0.6-1.6	< 0.5	0.016-0.042	< 0.46

The spin densities in Si and C *p* atomic orbitals for Si(II), Si(IV), and C(II) was found to be by one order of magnitude larger than that in the *s* atomic orbitals. Assuming a comparable ρ_s to ρ_p ratio for Si(I) one can obtain $\rho_p \approx 9\%$, that corresponds to the anisotropic shf coupling of about 10 MHz. And as a result estimation of the upper limit for the overall spin density located at the two Si atoms Si(I) gives the value of about 20%.

Table **2** collects the shf parameters of the V_C in SiC nanoparticles calculated in [6] and hydrogenated carbon vacancy in bulk SiC calculated within the framework of density functional theory (DFT) in [26], V_C in bulk β-SiC [24] along with experimental shf coupling constant obtained for V_C in β-SiC crystalline phase in this work. As it follows from Table **2**, the shf constants of ^{29}Si for V_C are essentially larger in bulk SiC crystals [24] than that in SiC nanoparticles. This means that the wave function of the unpaired electron of the V_C in SiC nanoparticles has higher delocalization than in bulk SiC.

The observation of a ^{1}H ENDOR line at the proton Zeeman frequency weakly coupled with an unpaired electron of the V_C in SiC nanoparticles suggests this conclusion. The remote protons located around paramagnetic center interact with the defect by transfer of the spin density owing to a delocalization of the wave function of the V_C unpaired electron over several bond lengths.

Table 2. The calculated and experimentally obtained parameters of the carbon vacancies V_C in SiC nanoparticles, bulk β-SiC and hydrogenated carbon vacancy (V_C+H) in bulk SiC (S = 1/2) (reproduced from Ref [17]. with permission from the Publisher.)

Ligands	**^{29}Si**		**^{1}H**		**Ref.**
Defect	**$A_{iso,}$ MHz**	**T MHz**	**$A_{iso,}$ MHz**	**T MHz**	
Calcululated V_C in SiC nanoparticles	32.88	-4.35	-	-	[6]
Calcululated V_C in bulk β- SiC	$A_\parallel$ =180 $A_\perp$ = 121				[24]
Calculated V_C+H **H between Si_2, Si_3, Si_4**	-173.98 (Si_{2-4})	-19.7 (Si_{2-4})	5.91	0.66	[26]
	-5.6 Si_1	-5.6 Si_1	-	-	
V_C in β-SiC crystalline phase in SiC nanoparticles	29.7	-	6.4	-	This work

On the other hand, the obtained shf coupling constant of V_C with hydrogen in β-SiC crystalline phase is close to that calculated for the V_C with adjacent hydrogen in SiC bulk crystals in a study [26]. In accordance with the model proposed in [26] the most stable configuration for the hydrogen-related defect (V_C + H) is the H-atom in the center of three ligands Si_3, Si_4 and Si_2. Due to the dynamic Jahn-Teller effect, the (V_C + H) complex exhibits isotropic motional behavior which provides the nearly isotropic shf constant for ^{1}H and only at T = 0 K the anisotropic shf coupling for the hydrogen is expected. Therefore, in accordance with [25, 26] we may suggest that the contamination of the V_C environment with hydrogen may lead to the formation of the hydrogen-related defect (V_C + H) with donor-like behavior in SiC nanoparticles. Further theoretical calculations of the hydrogen-related defect (V_C + H) in SiC nanoparticles are required to confirm this suggestion.

CONCLUSION

The investigation of the intrinsic defects formed in SiC nanoparticles obtained by self-propagating high-temperature synthesis (SHS) and by laser pyrolysis methods was carried out by using EPR, FS ESE, ENDOR, and HYSCORE techniques. Four and three types of the paramagnetic centers with isotropic *g*-values were resolved in the FS ESE spectrum of the SiC nanoparticles owing to their different spin relaxation times. The ligand structure of the paramagnetic center at g_{iso} = 2.0029(3) was studied by pulse ENDOR and HYSCORE techniques. The detail investigation of the ligand structure of the defects was shown that the largest shf splitting (29.7 MHz) was found for the ^{29}Si atoms. The obtained value of the shf interaction with ^{29}Si atoms fits well with that calculated by self-consistent norm-conserving pseudopotential band energy for carbon vacancy V_C in SiC nanoparticles. As a result, the paramagnetic center at g_{iso} = 2.0029(3) was assigned with V_C localized in the β-SiC crystalline phase. On the other hand, the shf interaction of the V_C with ^{29}Si nuclei obtained in SiC nanoparticles significantly smaller than that found for V_C in bulk SiC indicating for the significantly higher delocalization of the V_C unpaired electron in SiC nanoparticles than in bulk SiC. Because the other defects have an insufficient intensity of the FS ESE signals to observe their ligand structure the identification of the defects was performed by comparison their EPR parameters with the microstructure of the SiC nanoparticles. The paramagnetic center at g_{iso} = 2.0043(3) was attributed to the charged carbon vacancy V_C located in hexagonal α-phase of the SiC nanoparticles. The paramagnetic center with g_{iso} = 2.0031(3) was assigned with sp^3-coordinated CDBs located in the C-excess phase of the SiC nanoparticles. The paramagnetic center at g_{iso}= 2.0037(3), which was observed only in SiC nanoparticles synthesized by SHS method was attributed to the K center localized in α-Si_3N_4 phase of the SiC nanoparticles. The bonding configuration Si-X_3 of the K center with g_{iso}= 2.0037(3) was identified as Si-NSi_2. Thus, from the obtained data it follows that every structural phase of the SiC nanoparticles is characterized by the presence of the intrinsic defects which, as predicted in a study [8], are necessary for the thermodynamic stabilization of the particle size and microstructure.

CONSENT FOR PUBLICATION

Not applicable.

CONFLICT OF INTEREST

The author (editor) declares no conflict of interest, financial or otherwise.

ACKNOWLEDGEMENT

The authors are thankful to Prof. G. Gnesin and DrSc. N Gadzyra (Institute for Problems of Materials Science, NAS of Ukraine, Kyiv, Ukraine) for providing the SiC nanoparticles grown by SHS method.

REFERENCES

[1] Niihara, K.; Nakahira, A. Strengthening and Toughening Mechanisms in Nanocomposite Ceramics. *Ann. Chim. Fr.*, **1991**, *16*, 479-486.

[2] Charpentier, S.; Kassiba, A.; Emery, J.; Cauchetier, M. Investigation of the paramagnetic centres and electronic properties of silicon carbide nanomaterials. *J. Phys. Condens. Matter,* **1999**, *11*, 4887-4897. [http://dx.doi.org/10.1088/0953-8984/11/25/308]

[3] Kassiba, A.; Makowska-Janusik, M.; Bouclé, J.; Bardeau, J.; Bulou, A.; Herlin-Boime, N. Photoluminescence Features On The Raman Spectra Of Quasistoichiometric Sic Nanoparticles: Experimental And Numerical Simulations. *Phys. Rev. B,* **2002**, *66*, 155317-15323. [http://dx.doi.org/10.1103/PhysRevB.66.155317]

[4] Kassiba, A.; Bednarski, W.; Pud, A.; Errien, N.; Makowska-Janusik, M.; Laskowski, L.; Tabellout, M.; Kodjikian, S.; Fatyeyeva, K.; Ogurtsov, N.; Noskov, Y. Hybrid Core–Shell Nanocomposites Based On Silicon Carbide Nanoparticles Functionalized By Conducting Polyaniline: Electron Paramagnetic Resonance Investigations. *J. Phys. Chem. C,* **2007**, *111*, 11544-11551. [http://dx.doi.org/10.1021/jp070966y]

[5] Kityk, I.; Makowska-Janusik, M.; Kassiba, A.; Plucinski, K. SiC nanocrystals embedded in oligoetheracrylate photopolymer matrices; new promising nonlinear optical materials. *Opt. Mater.*, **2000**, *13*, 449-453. [http://dx.doi.org/10.1016/S0925-3467(99)00099-3]

[6] Kityk, I.; Kassiba, A.; Tuesu, K.; Charpentier, C.; Ling, Y.; Makowska-Janusik, M. Vacancies in SiC nanopowders. *Mater. Sci. Eng. B,* **2000**, *77*, 147-158. [http://dx.doi.org/10.1016/S0921-5107(00)00466-9]

[7] Kassiba, A.; Makowska-Janusik, M.; Boucle, J.; Bardeau, J.; Bulou, A.; Herlin, N.; Mayne, M.; Armand, X. Stoichiometry and interface effects on the electronic and optical properties of SiC nanoparticles. *Diamond Related Materials,* **2002**, *11*, 1243-1247. [http://dx.doi.org/10.1016/S0925-9635(02)00016-X]

[8] Makowska-Janusik, M.; Kassiba, A.; Bouclé, J.; Bardeau, J.; Kodjikian, S.; Désert, A. Vibrational density of states in silicon carbide nanoparticles: experiments and numerical simulations. *J. Phys. Condens. Matter,* **2005**, *17*, 5101-5110. [http://dx.doi.org/10.1088/0953-8984/17/33/014]

[9] Gadzira, M.; Gnesin, G.; Mykhaylyk, O.; Britun, V.; Andreyev, O. Solid solution of carbon in β-SiC. *Mater. Lett.*, **1998**, *35*, 277-282. [http://dx.doi.org/10.1016/S0167-577X(97)00263-2]

[10] Gadzyra, N. Gnesin, G. Mechanism for the Formation of a Solid Solution of Carbon in Silicon Carbide. *Powder Metall. Met. Ceramics,* **2001**, *40*, 519-525. [http://dx.doi.org/10.1023/A:1014352009750]

[11] Gadzira, M.; Gnesin, G.; Mykhaylyk, O.; Andreyev, O. Synthesis and structural peculiarities of nonstoichiometric β-SiC. *Diamond Related Materials,* **1998**, *7*, 1466-1470. [http://dx.doi.org/10.1016/S0925-9635(98)00201-5]

[12] Stoll, S.; Schweiger, A. EasySpin, a comprehensive software package for spectral simulation and analysis in EPR. *J. Magn. Reson.*, **2006**, *178*(1), 42-55. [http://dx.doi.org/10.1016/j.jmr.2005.08.013] [PMID: 16188474]

[13] Hahn, E. Spin Echocs. *Phys. Rev.*, **1950**, *80*, 580-594. [http://dx.doi.org/10.1103/PhysRev.80.580]

[14] Höfer, P.; Grupp, A.; Nebenführ, H.; Mehring, M. Hyperfine sublevel correlation (hyscore) spectroscopy: a 2D ESR investigation of the squaric acid radical. *Chem. Phys. Lett.*, **1986**, *132*, 279-282. [http://dx.doi.org/10.1016/0009-2614(86)80124-5]

[15] Davies, E. A new pulse endor technique. *Phys. Lett. A*, **1974**, *47*, 1-2. [http://dx.doi.org/10.1016/0375-9601(74)90078-4]

[16] Mims, W. Pulsed Endor Experiments. *Proc. R. Soc. Lond. A Math. Phys. Sci.*, **1965**, *283*, 452-457. [http://dx.doi.org/10.1098/rspa.1965.0034]

[17] Savchenko, D.V.; Pöppl, A.; Kalabukhova, E.N.; Venger, E.F.; Gadzira, M.P.; Gnesin, G.G. Intrinsic Defects in nonstoichiometric ®-SiC nanoparticles studied by pulsed magnetic resonance methods. *SPQEO*, **2010**, *13*, 43-50.

[18] Aubert, P.; von Bardeleben, H.; Delmotte, F.; Cantin, J.; Hugon, M. Electron-paramagnetic-resonance study of the(100)Si/Si3N4interface. *Phys. Rev. B*, **1999**, *59*, 10677-10684. [http://dx.doi.org/10.1103/PhysRevB.59.10677]

[19] Schweiger, A.; Jeschke, G. *Principles of pulse electron paramagnetic resonance spectroscopy*, 1st ed; Oxford University Press: Oxford, **2001**.

[20] Savchenko, D.V.; Poppl, A.; Hoentsch, J.; Kalabukhova, E.N.; Bulois, Y.; Kassiba, A. Intrinsic defects in SiC nanoparticles as studied by pulsed electron paramagnetic resonance. *Sol. St. Comm.*, **2008**, *146*, 83-87. [http://dx.doi.org/10.1016/j.ssc.2008.01.010]

[21] Pöppl, A.; Kevan, L. A Practical Strategy for Determination of Proton Hyperfine Interaction Parameters in Paramagnetic Transition Metal Ion Complexes by Two-Dimensional HYSCORE Electron Spin Resonance Spectroscopy in Disordered Systems. *J. Phys. Chem.*, **1996**, *100*, 3387-3394. [http://dx.doi.org/10.1021/jp9525999]

[22] Morton, J.R.; Preston, K.F. Atomic parameters for paramagnetic resonance data. *J. Magn. Reson.*, **1978**, *30*, 577-582. [http://dx.doi.org/10.1016/0022-2364(78)90284-6]

[23] Lu, H.Y.; Petrich, M.A.; Fan, C.; Hoffman, B.M. Continuous-wave and pulsed electron-nuclear double-resonance study of paramagnetic defects in amorphous hydrogenated silicon-carbon alloy. *Phys. Rev. B Condens. Matter*, **1993**, *48*(3), 1997-2000. [http://dx.doi.org/10.1103/PhysRevB.48.1997] [PMID: 10008583]

[24] Bockstedte, M.; Heid, M.; Pankratov, O. Signature of intrinsic defects in SiC: *Ab initio* calculations of hyperfine tensors. *Phys. Rev. B*, **2003**, *67*, 193102-1-193102-4. [http://dx.doi.org/10.1103/PhysRevB.67.193102]

[25] Gali, A.; Deák, P.; Son, N.; Janzén, E.; von Bardeleben, H.; Monge, J. Calculation of Hyperfine Constants of Defects in 4H-SiC. *Mater. Sci. Forum*, **2003**, *433-436*, 511-514. [http://dx.doi.org/10.4028/www.scientific.net/MSF.433-436.511]

[26] Kalabukhova, E.; Lukin, S.; Savchenko, D.; Mitchel, W.; Greulich-Weber, S.; Rauls, E.; Gerstmann, U. Possible Role of Hydrogen within the So-Called X Center in Semi-Insulating 4H-SiC. *Mater. Sci. Forum*, **2006**, *527-529*, 559-562. [http://dx.doi.org/10.4028/www.scientific.net/MSF.527-529.559]

CHAPTER 12

Size-dependent Effects in Silicon Carbide and Diamond Nanomaterials as Studied by CW and Pulse EPR Methods

Dariya Savchenko*

Department of Analysis of Functional Materials, Division of Optics, Institute of Physics Czech Academy of Sciences, Prague, 182 00, Czech Republic

Department of Physics and Solid State Physics, National Technical University of Ukraine "Igor Sikorsky Kyiv Polytechnic Institute", Kyiv, 03056, Ukraine

Abstract: The great potential of the silicon carbide (SiC) and diamond nanoparticles for future applications in spintronics initiates detailed investigation of the effects of impurities and defects in their electronic characteristics. Among impurities, nitrogen doped SiC nanoparticles are an important item to be studied, because nitrogen donors are common contaminations of an n-type SiC bulk material. The first information about the shallow donor state of nitrogen in SiC nanoparticles and influence of the hydrogen as well as intrinsic defects on electronic properties of nitrogen was presented in this chapter. The delocalization of the nitrogen wave function was observed with the reduction in the nanoparticle size with the onset of about $d < 50$ nm. The delocalization of the nitrogen wave function gives rise to the overlap between wavefunctions of the neighboring donors and transformation of the nitrogen triplet line into one single exchange EPR line. The size-dependent effect was also observed for paramagnetic substitutional nitrogen defects (P1) in nanodiamonds representing free electron interacting with the ^{14}N nuclear spin ($I = 1$). The decrease of size of nanoparticle down to $d < 80$ nm led to a transformation of the hyperfine structure of the P1 defect into a one EPR line caused by dipole-dipole and/or exchange couplings of P1 spins with the rising amount of surface spins, which becomes more effective in nano-sized particles.

Keywords: Carbon vacancy, Compensation, Delocalization, Dipole-dipole interaction, Donor wavefunction, EPR, ENDOR, Exchange interaction, ED EPR, Hydrogen, Hydrogen retention, Hyperfine interaction, Intrinsic defects, Nanodiamonds, Nanoparticles, Nitrogen, Silicon carbide, Size-dependent effect, Superhyperfine interaction, Surface defects.

* **Corresponding author Dariya Savchenko:** Department of Analysis of Functional Materials, Division of Optics, Institute of Physics Czech Academy of Sciences, Prague, 182 00, Czech Republic; Tel: +(420) 266 05 2356; Fax: +(420) 286 581 448; E-mail: dariyasavchenko@gmail.com

Dariya Savchenko & Abdel Hadi Kassiba (Eds.)

INTRODUCTION

Nanoparticles of different materials like metals, semiconductors and oxides were widely investigated owing to their uncommon physical and/or chemical features, which differ from those of bulk materials. Their unusual features are the result of quantum size effect, interface, and/or surface effect. Size-dependent effects (quantum size effects) in nanocrystals are usually investigated by various optical techniques.

In particular, the extensive study of photoluminescence properties of silicon (Si) nanocrystals can open a new opportunity for indirect-gap semiconductors like new materials for photoelectric devices. The correlation between the diameter of Si nanocrystals and the photoluminescence peak energy was shown in a study [1 - 3]. By decreasing the nanocrystals size, the photoluminescence peak shifts to the higher energies and achieves the visible range for Si nanocrystals less than 2 nm. However, the drawback of the optical methods is the contamination of the spectra by spurious emissions or absorptions originating from the effect of the surface, which becomes negligible in magnetic resonance approach.

Electron paramagnetic resonance (EPR) technique has been applied for the first time to the study of the size-dependent effect in Si-doped with phosphorus of different concentrations ranging between $5\cdot10^{15}$ cm^{-3} and $1.5\cdot10^{17}$ cm^{-3} [4]. A gradual reduction of the hyperfine splitting constant A for phosphorus donors in Si nanocrystals versus average Si grain volume was observed. Along with the two hyperfine lines with $A/(g\mu_B) = 4.20$ mT originating from the isolated phosphorus donor ($I = 1/2$), the Lorentzian-shaped line in the center of the EPR spectrum at $g = 1.998$ due to reciprocal overlap between the individual donors appears for the grinded and oxidized Si nanocrystals sample with an average silicon grain volume of about 80-60 nm. Taking into account that the strength of the hyperfine couplings depends on the electron localization, this effect was explained by the delocalization of donor wavefunction and the donor-impurity band formation. However, the evidence for the existence of the quantum confinement effect of phosphorus donors was not obtained due to rather large diameter of the nanocrystals.

The investigation of the electronic states of phosphorus donors in Si nanocrystals with 3-5 nm diameter in a study [5] and in hydrogenated microcrystalline silicon with the mean Si nanocrystals diameter of 20-10 nm in [6] another study showed that the hyperfine splitting was ~10 mT, which significantly exceeded the bulk value of 4.2 mT. The increase of the hyperfine coupling was explained by squeezing of the donor wavefunction by spatial confinement of phosphorus atoms into the regions less than the Bohr radius (1.67 nm) in bulk Si crystals. The defect in the wavefunction became more localized at the phosphorus site as the radius reduced, resulting in the higher amplitude of the wavefunction at the phosphorus core [5].

It should be noted that the size-dependent hyperfine splitting cannot be successfully observed if three requirements are not satisfied [5].

1. The nanocrystal's, size should be close to the effective Bohr radius for shallow donors in bulk material. Besides, there is a "critical size" for nanocrystals, under which, the dopant has a tendency to reside Si atoms close to the surface. For Si nanocrystals having a size less than 2 nm, phosphorus atoms are ejected to the surface [7].
2. The mean number of donors for each nanocrystal must be ~1. If more than 2 donors present in the nanocrystal, the wavefunctions of these donors overlap, leading to the delocalization of electrons and spreading of the hyperfine splitting.
3. To prevent the trapping of the donor carriers at the defects, the number of deep dangling-bond defects should be very small or ideally 0.

Indeed, the EPR investigation of the Si nanocrystals having an average particle size between 4 and 11 nm and a nominal doping concentration of $5 \cdot 10^{20}$ cm^{-3} [8] showed that the intensity of the observed single EPR line at $g = 1.998$ originating from closely located exchange coupled with phosphorous atoms significantly decreased with the decrease in the particle's diameter from 11 nm to 4 nm. This effect is explained by sizable compensation of the phosphorus donors by surface silicon dangling bonds (SiDBs) for smaller nanoparticles, when the ratio of the surface area to the volume of the particle increases.

Compensation of the phosphorous donors by surface defects was also observed in nanocrystals with a diameter of 3-30 nm [9]. In addition, the confinement-induced enhancement of ionization energy of dopants with the reducing nanocrystal size was foreseen through *ab initio* calculations of the dopant formation energy in P doped Si nanocrystals [10] and was suggested experimentally in reference [9].

Thus, the electronic states of shallow impurities in nanometer-size semiconductors and in bulk crystals are different and depend on the nanoparticle sizes.

In this chapter, the investigation of the size-dependent effect on silicon carbide (SiC) nanoparticles doped with nitrogen donors by echo-detected EPR (ED EPR) and pulsed electron nuclear double resonance (ENDOR) is presented. The nitrogen doped SiC nanoparticles are an important subject of investigation, because n-type SiC bulk material usually is contaminated by nitrogen donors.

Though nitrogen has been found to be present in the nanomaterials, there is no experimental confirmation of the shallow nitrogen donor state formation in SiC nanoparticles. The current chapter represents the first data about the size-dependent effect of the shallow nitrogen donor in SiC nanoparticles. The impact of the hydrogen and intrinsic defect formation on the nitrogen donor electronic characteristics was discussed. A similar size-dependent effect was also observed in nanodiamonds for continuous wave (CW) EPR spectra of substitutional nitrogen paramagnetic defect (P1) caused by a single free electron interacting with the ^{14}N nuclear spin [11]. The investigation of the size-dependent physical characteristics of nanodiamonds is of significant practical interest for their potential application in biomedical imaging, drug delivery and other areas of medicine. A gradual decrease of the EPR spectrum intensity from the P1 defect with the reduction of the nanoparticle size was observed. With a reduction of the nanoparticle diameter up to 30 nm, the hyperfine structure of the P1 defect vanished, while two single EPR lines with close values of the *g*-factors remained in the EPR spectrum of nanodiamonds. One of them was attributed to the surface defect caused by grinding of micron-sized synthetic diamonds and the second one to the nitrogen spin exchange broadened EPR line. The influence of defects caused by grinding of micron-sized synthetic diamonds on the hyperfine structure of the paramagnetic defect P1 was also discussed.

SAMPLE CHARACTERIZATION AND EXPERIMENTAL TECHNIQUE

The non-stoichiometric (C/Si > 1) SiC nanoparticles with an average size of $d \leq 50$ nm were prepared by self-propagating synthesis of C and highly active Si at T = 1250°C [12], whereas SiC nanoparticles having 50 nm < d < 100 nm were commercially obtained. The non-stoichiometric (C/Si > 1) green/black SiC nanoparticles having d > 100 nm were produced in an electric resistance furnace using quartz sand and petroleum coke and salt at T = 1750°C. The ED EPR spectra and ENDOR spectra in Davies and Mims pulse sequences were recorded on Bruker ELEXSYS E580 spectrometer at X-band (ν = 9.7 GHz) and T = 6 K.

SIZE EFFECT IN ED EPR SPECTRA OF SiC NANOPARTICLES

Fig. (**1**) represents the ED EPR spectra measured in SiC nanoparticles with various particle sizes at T = 6 K. The triplet lines owing to the hyperfine interaction of nitrogen (^{14}N, I = 1, 99.6%) donors with S = 1/2 and a single line overlapped with the nitrogen central triplet line were revealed in SiC nanoparticles with d > 100 nm. In SiC nanoparticles with 50 nm < d < 100 nm, we have also observed the ^{14}N triplet line of significantly lower intensity than that detected in SiC nanoparticles with d > 100 nm. With the further reduction of the particle size, the intensity of the central line enhanced. Eventually only one central line consisting of two unresolved signals was observed in SiC nanoparticles with $d < 50$ nm. The decrease in the intensity of the ^{14}N hyperfine lines and enhancement of the central EPR line with the reduction of the nanoparticle size was observed to be the result of delocalization and overlap of the donor wavefunction.

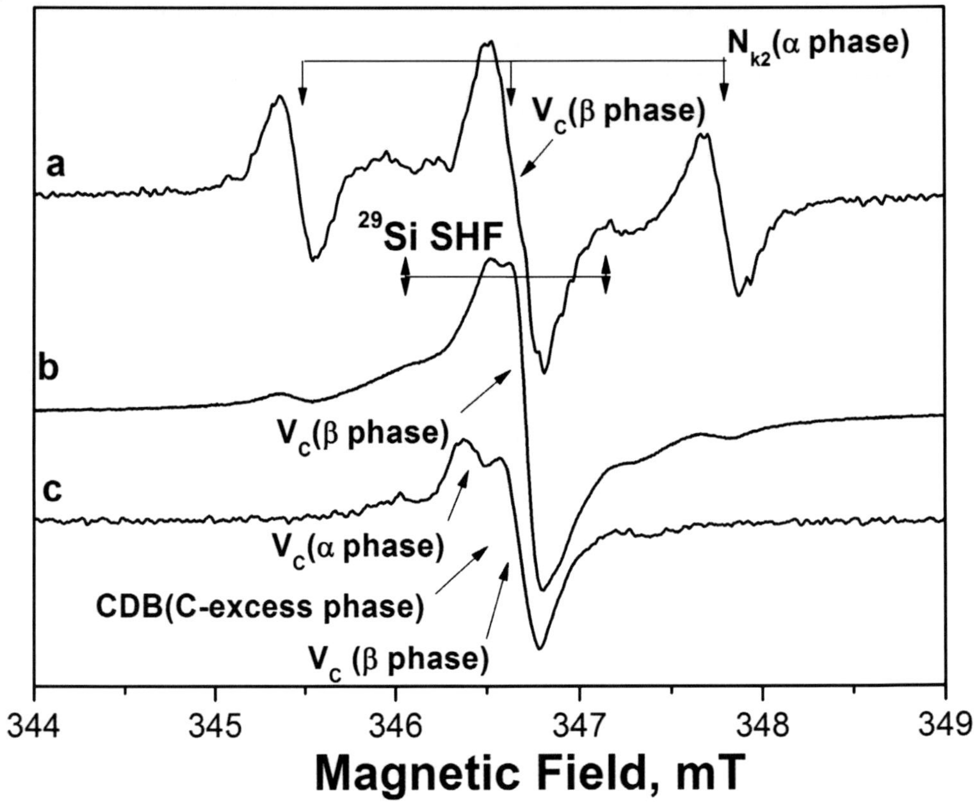

Fig. (1). The X-band ED EPR spectra measured in SiC nanoparticles having $d > 100$ nm (a), 50 nm $< d <$ 100 nm (b), $d < 50$ nm (c) at $T = 6$ K. Reproduced from [13] with the Permission from Institue of Physics PAN.

With the reduction of particle size, ^{14}N line triplet turned into one single line (N_{exch}) overlapped with the central ^{14}N triplet line. However, due to the short spin-lattice relaxation time, N_{exch} single line induced by exchange coupling between nitrogen donors could not be detected in the ED EPR spectra of SiC nanoparticles. At the same time, the observation of exchange EPR line N_{exch} in the CW EPR spectrum of SiC nanoparticles is prevented due to the existence of a few additional unresolved EPR lines overlapped with the central ^{14}N triplet line at the

X-band frequency. Unfortunately, the attempt to resolve overlapped EPR signals at higher microwave frequencies (35 GHz) did not bring success due to their broadening.

Table **1** shows the parameters of the paramagnetic centers taken by fitting and simulating of the ED EPR spectra measured in SiC nanoparticles along with their identification.

Table 1. Spin Hamiltonian parameters of the paramagnetic centers observed in SiC nanoparticles defined by CW and ED EPR techniques.

d, [nm]	Center	$g_{\parallel}$	$g_{\perp}$
$d > 100$	N_{k2} (α-SiC phase)	2.0037(3)	2.0030(3)
	V_C (β-SiC phase)	2.0029(3)	2.0029(3)
$50 < d < 100$	N_{k2} (α-SiC phase)	2.0037(3)	2.0030(3)
	$N_{exch.}$ (α-SiC phase)	2.0037(3)	2.0030(3)
	V_C+H (β-SiC phase)	2.0028(3)	2.0028(3)
$d \leq 50$	V_C+H (β-SiC phase)	2.0028(3)	2.0028(3)
	sp^3-coordinated CDB (C-excess phase)	2.0031(3)	2.0031(3)
	V_C (α-SiC phase)	2.0041(3)	2.0041(3)

The assignment of the additional single lines detected in the ED EPR spectrum of SiC nanoparticles was carried out by comparison of their experimental EPR parameters with those found for the defects preliminary reported in SiC nanoparticles [12, 14].

As it follows from Table **1**, 3 different paramagnetic intrinsic defects were revealed in SiC nanoparticles. The appearance of the superhyperfine lines owing to the coupling with ^{29}Si nuclei (I = 1/2, 4.7%), positioned symmetrically with respect to the central signal at g_{iso} = 2.0029(3), with A_{iso} = 29.7 MHz leading to an argument to attribute it to the C vacancy (V_C) in the cubic (β-SiC) phase [12]. The second signal with g_{iso} = 2.0041(3) was assigned to the V_C in hexagonal (α-SiC) phase [12, 14]. The third signal with g_{iso} = 2.0031(3) was assigned to the sp^3-hybridizied C dangling bonds (CDB) located in the C-excess phase of SiC nanoparticles [12].

SIZE EFFECT IN EPR SPECTRA OF NANODIAMONDS

A similar size-dependent effect was observed in highly purified synthetic micro- and nanodiamonds (for neutrally charged single substitutional nitrogen center N_S^0 (P1) caused by an unpaired electron coupled with the ^{14}N nuclei (see Fig. **3a** in

Chapter 9 and Ref [11].). As can be seen from Fig. (**2**), with the reduction of the nanoparticle's diameter up to $d < 50$ nm, the ^{14}N triplet lines having $g_{iso} = 2.0024$ turned into one single exchange EPR line overlapped with the central ^{14}N triplet line. On the other hand, the gradual reduction of the EPR spectrum intensity from the P1 defect was observed to be accompanied by a gradual increase of the single EPR signal of Lorentzian lineshape with $g = 2.0028(2)$, which was attributed to the surface paramagnetic centers emerging during the diamond grinding process [15]. With the rising number of the surface spins, the dipole-dipole and/or exchange coupling of P1 spins becomes efficient, especially in nano-sized particles. This fact can be considered as a reason for such transformation of the ^{14}N triplet lines.

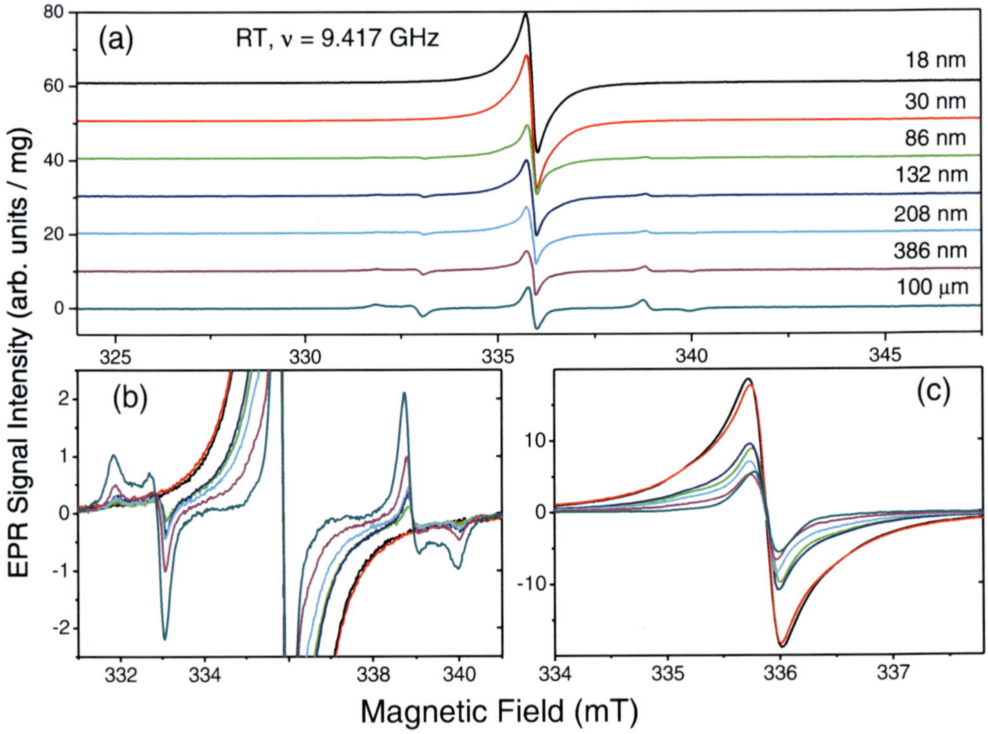

Fig. (2). Taken from [11] with the Permission from IOP Publishing. Room-temperature X-band CW EPR spectra of the series of nanodiamonds as a function of the mean size of particles in the fraction: (a) general view of the $g = 2.00$ region, intensities are comparable; (b) scaling of the hyperfine lines; (c) scaling of the central signal.

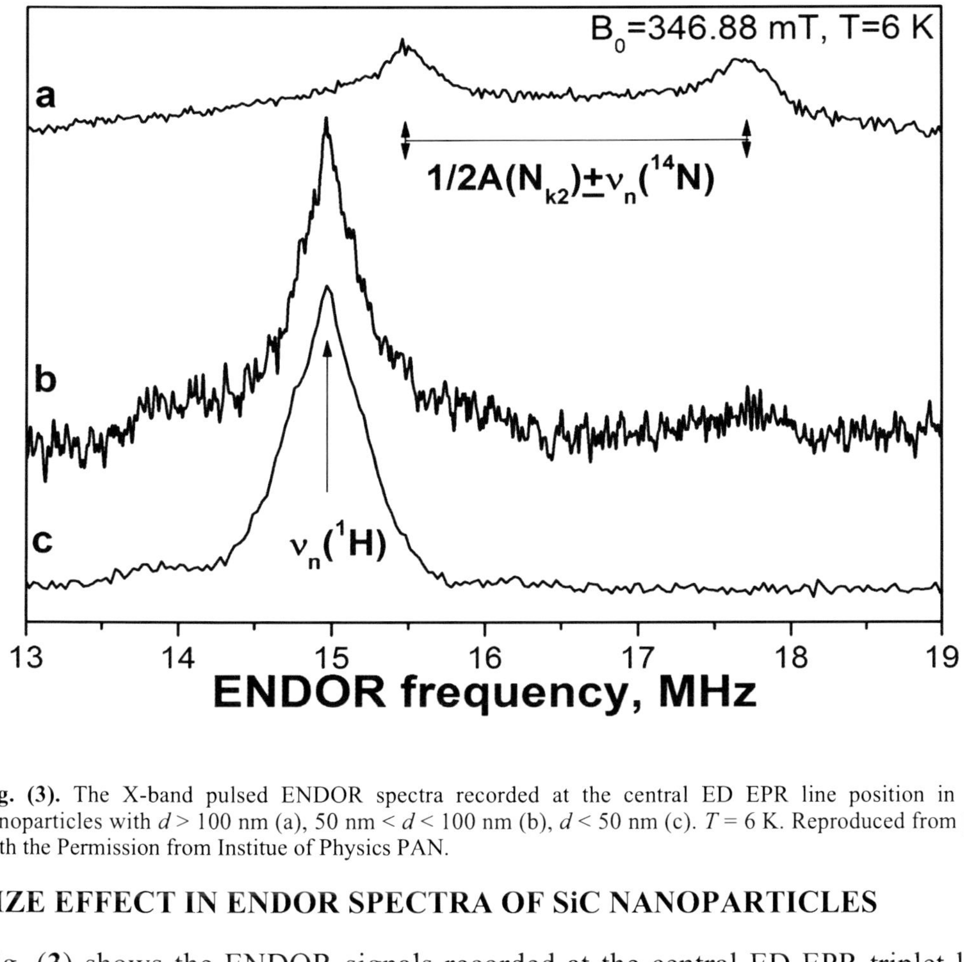

Fig. (3). The X-band pulsed ENDOR spectra recorded at the central ED EPR line position in SiC nanoparticles with $d > 100$ nm (a), 50 nm $< d <$ 100 nm (b), $d < 50$ nm (c). $T = 6$ K. Reproduced from [13] with the Permission from Institue of Physics PAN.

SIZE EFFECT IN ENDOR SPECTRA OF SiC NANOPARTICLES

Fig. (**3**) shows the ENDOR signals recorded at the central ED EPR triplet line position in SiC nanoparticles with $d > 100$ nm and 50 nm $< d <$ 100 nm. Only one doublet line owing to ^{14}N nuclei corresponding to the transitions $1/2A_i \pm \nu_n(^{14}N)$ ($\nu_n(^{14}N)$ – nuclear Larmor frequency of ^{14}N) with the hyperfine coupling constant of about 33.32 MHz was detected in the ENDOR spectrum of SiC nanoparticles having $d > 50$ nm. The observed ENDOR signals with hyperfine interaction constant of 33.32 MHz was observed to be related to the N donor center N_{k2} substituting "k2" quasi-cubic site in 6H SiC crystal lattice [16, 17]. The ENDOR signals from the nitrogen donors residing quasi-cubic "k1" (N_{k1}) and hexagonal "h" (N_h) sites were not detected in the ENDOR spectra of SiC nanoparticles.

The presence of the V_C in SiC nanoparticles ensures the sample compensation. In fact, N_h donor has a significantly shallower energy level (81.04 meV) than that of

N_{k1} and N_{k2} donors and as a consequence, its energy level is not populated in compensated samples. On the other hand, the compensation effect is not responsible for the lack of the signals from the N_{k1} donors in ENDOR spectra because N_{k1} and N_{k2} donors have the close values of the energy levels (137 meV and 142 meV, respectively). Even in highly compensated samples, two triple lines from N_{k1} and N_{k2} donors were observed in the EPR spectrum of 6H SiC monocrystals but with a different intensity ratio [18].

According to the last pulsed ENDOR and general TRIPLE ENDOR literature data, the N_{k2} donor having deeper energy level of 142 meV was assigned to the nitrogen donors, which substitute silicon site [17]. This circumstance gives us the argument to propose that the lack of the N_{k1} residing carbon site is caused by technological preparation of the samples rather than by compensation effect. In fact, according to the site-competition model presented in reference [19], the carbon excess in 6H SiC suppresses the site competition process between the N and C atoms for C sites and thus can cause the N substitution at Si site.

Fig. (**3b,c**), show the ENDOR spectra recorded at the ED EPR line from the V_C center in SiC nanoparticles with d < 100 nm. The observed ENDOR line is centered on the hydrogen (^1H) Larmor frequency. As far as we know, firstly in SiC-based materials, the ^1H ENDOR signals were reported in *a*-SiC:H alloy in [20].

The appearance of the ^1H ENDOR line points out the presence of hydrogenated regions in SiC nanoparticles and probably can be induced by final etching of the SiC nanoparticles in hydrofluoric acid [12]. It is important to mention that the respective ^{29}Si superhyperfine constants for V_C in bulk SiC crystals are substantially larger [21]. This fact indicates that the wavefunction of the unpaired electron of the vacancy in SiC nanoparticles has essentially higher delocalization than in the bulk material. On the other hand, taking into account that the ENDOR spectrum from the V_C center with g_{iso} = 2.0029(3) reveals a weak superhyperfine coupling with hydrogen, we can suppose that the V_C environment can be contaminated with H constituting the hydrogen-related defect (V_C+H) with donor-like characteristics [22, 23].

As seen in Fig. (**3**), in SiC nanoparticles having d > 100 nm, no ^1H ENDOR line was recorded at the central ED EPR triplet line position, which is overlapped with V_C center. This data allows us to assume that hydrogen incorporation in SiC nanoparticles is particle size-dependent. This assumption coincides well with the correlation among diamond grain size and hydrogen storage reported in diamond films [24]. The hydrogen storage in diamond films enhances with reduction in the grain size and suggests that hydrogen is bonded and trapped in the grain boundaries and on the internal grain surfaces of the nanodiamond particles. Therefore, we can consider that the incorporation of hydrogen in SiC nanoparticles depends on the grain size.

CONCLUSION

The nitrogen donor center residing at the quasi-cubic position with deeper energy level in the band gap at 142 meV (N_{k2}) and 3 types of intrinsic defects, including V_C located in the hexagonal and cubic crystalline phase, CDB located in the C-excess phase were observed in the ED EPR spectrum of SiC nanoparticles.

With the reduction of the nanoparticle size, the delocalization of the nitrogen donor wavefunction occurs, and as a consequence, due to the overlap between the wavefunctions of the neighboring donors, the ^{14}N triplet line turns into one single exchange EPR line. It was found that the size-dependent effect became effective for the particle size less than 50 nm (d < 50 nm) both for nitrogen donors in SiC nanoparticles and for neutrally charged single substitutional nitrogen center N_S^0 in synthetic nanodiamonds.

The high compensation degree and technology preparation was considered to be the reason why only one N_{k2} donor center among three donor centers was revealed in the ED EPR and pulsed ENDOR spectra of SiC nanoparticles. The deep intrinsic defects compensate the N_h center, which has substantially more shallow energy level than that of N_{k1} and N_{k2} donors (81.04 meV). On the other hand, taking into account that SiC nanoparticles have a deviation from stoichiometry towards C (C/Si > 1), it was suggested that N tends to substitute Si site and as a result, the N_{k1} donor with energy level becomes close to the N_{k2} donor (137 meV), which substitutes the absence of C site in the SiC nanoparticles. The presence of the ^{1}H ENDOR line in SiC nanoparticles indicates that H incorporation in SiC nanomaterials depends on the grain size. With decreasing of the grain size, the hydrogen retention in SiC nanoparticles enhances. The hydrogen retention in SiC nanoparticle becomes effective for the particle with the size less than 100 nm.

CONSENT FOR PUBLICATION

Not applicable.

CONFLICT OF INTEREST

The author (editor) declares no conflict of interest, financial or otherwise.

ACKNOWLEDGEMENT

The work was supported by GA ČR 13-06697P, MEYS SAFMAT LM2015088 and LO1409 projects.

REFERENCES

[1] Takagi, H.; Ogawa, H.; Yamazaki, Y.; Ishizaki, A.; Nakagiri, T. Quantum size effects on photoluminescence in ultrafine Si particles. *Appl. Phys. Lett.,* **1990**, *56*, 2379-2380. [http://dx.doi.org/10.1063/1.102921]

[2] Schuppler, S.; Friedman, S.L.; Marcus, M.A.; Adler, D.L.; Xie, Y.; Ross, F.M.; Chabal, Y.J.; Harris, T.D.; Brus, L.E.; Brown, W.L.; Chaban, E.E.; Szajowski, P.F.; Christman, S.B.; Citrin, P.H. Size, shape, and composition of luminescent species in oxidized Si nanocrystals and H-passivated porous Si. *Phys. Rev. B Condens. Matter,* **1995**, *52*(7), 4910-4925. [http://dx.doi.org/10.1103/PhysRevB.52.4910] [PMID: 9981675]

[3] Kanzawa, Y.; Kageyama, T.; Takeoka, S.; Fujii, M.; Hayashi, S.; Yamamoto, K. Size-dependent near-infrared photoluminescence spectra of Si nanocrystals embedded in SiO2 matrices. *Solid State Commun.,* **1997**, *102*, 533-537. [http://dx.doi.org/10.1016/S0038-1098(96)00774-0]

[4] Pawlak, B.; Gregorkiewicz, T.; Ammerlaan, C.; Takkenberg, W.; Tichelaar, F.; Alkemade, P. *Experimental investigation of band structure modification in silicon nanocrystals.,* **2001**. [http://dx.doi.org/10.1103/PhysRevB.64.115308]

[5] Fujii, M.; Mimura, A.; Hayashi, S.; Yamamoto, Y.; Murakami, K. *Hyperfine Structure of the Electron Spin Resonance of Phosphorus-Doped Si Nanocrystals.,* **2002**. [http://dx.doi.org/10.1103/PhysRevLett.89.206805]

[6] Müller, J.; Finger, F.; Carius, R.; Wagner, H. Electron spin resonance investigation of electronic states in hydrogenated microcrystalline silicon. *Phys. Rev. B,* **1999**, *60*, 11666-11677. [http://dx.doi.org/10.1103/PhysRevB.60.11666]

[7] Chan, T.L.; Tiago, M.L.; Kaxiras, E.; Chelikowsky, J.R. Size limits on doping phosphorus into silicon nanocrystals. *Nano Lett.,* **2008**, *8*(2), 596-600. [http://dx.doi.org/10.1021/nl072997a] [PMID: 18154366]

[8] Stegner, A.; Pereira, R.; Klein, K.; Wiggers, H.; Brandt, M.; Stutzmann, M. Phosphorus doping of Si nanocrystals: Interface defects and charge compensation. *Physica B,* **2007**, *401-402*, 541-545. [http://dx.doi.org/10.1016/j.physb.2007.09.017]

[9] Almeida, A.; Sugimoto, H.; Fujii, M.; Brandt, M.; Stutzmann, M.; Pereira, R. *Doping efficiency and confinement of donors in embedded and free standing Si nanocrystals.,* **2016**. [http://dx.doi.org/10.1103/PhysRevB.93.115425]

[10] Cantele, G.; Degoli, E.; Luppi, E.; Magri, R.; Ninno, D.; Iadonisi, G.; Ossicini, S. First-principles study of n- and p-doped silicon nanoclusters. Phys. Rev. B 2005, 72, 113303 (1-4). [http://dx.doi.org/10.1103/PhysRevB.72.113303]

[11] Panich, A.; Sergeev, N.; Shames, A.; Osipov, V.; Boudou, J.; Goren, S. Size dependence of ^{13}C nuclear spin-lattice relaxation in micro- and nanodiamonds. J. Phys.: Condens. Matter. 2015, 27, 072203(1-7). [http://dx.doi.org/10.1088/0953-8984/27/7/072203]

[12] Savchenko, D.V.; Pöppl, A.; Kalabukhova, E.N.; Venger, E.F.; Gadzira, M.P.; Gnesin, G.G. Intrinsic Defects in Nonstoichiometric β-SiC Nanoparticles Studied by Pulsed Magnetic Resonance Methods. *SPQEO,* **2010**, *13*, 43-50.

[13] Savchenko, D. Pulsed EPR and ENDOR Study of SiC Nanopowders. *Acta Phys. Pol. A,* **2014**, *126*, 1063-1065. [http://dx.doi.org/10.12693/APhysPolA.126.1063]

[14] Savchenko, D.; Pöppl, A.; Hoentsch, J.; Kalabukhova, E.; Bulois, Y.; Kassiba, A. Intrinsic defects in SiC nanoparticles as studied by pulsed electron paramagnetic resonance. *Solid State Commun.,* **2008**, *146*, 83-87. [http://dx.doi.org/10.1016/j.ssc.2008.01.010]

[15] Samsonenko, N.D.; Sobolev, E.V. Electron paramagnetic resonance produced by surface of diamond. *Pis'ma Z. Eksp. Teor. Fiz.*, **1967**, *5*, 304-307.

[16] Savchenko, D.V.; Kalabukhova, E.N.; Kiselev, V.S.; Hoentsch, J.; Pöppl, A. Spin-coupling and hyperfine interaction of the nitrogen donors in 6H-SiC. *Phys. Status Solidi, B Basic Res.*, **2009**, *246*, 1908-1914.
[http://dx.doi.org/10.1002/pssb.200945082]

[17] Savchenko, D.V.; Kalabukhova, E.N.; Pöppl, A.; Shanina, B.D. Electronic structure of the nitrogen donors in 6H SiC as studied by pulsed ENDOR and TRIPLE ENDOR spectroscopy. *Phys. Stat. Solidi B*, **2012**, *249*, 2167-2178.
[http://dx.doi.org/10.1002/pssb.201248209]

[18] Savchenko, D.; Kalabukhova, E.; Lukin, S.; Mokhov, E.; Hoentsch, J.; Pöppl, A. EPR, ESE and pulsed ENDOR study of the nitrogen donor pairs on quasi-cubic lattice sites in 6H SiC. *Physica B*, **2009**, *404*, 4735-4738.
[http://dx.doi.org/10.1016/j.physb.2009.08.119]

[19] Larkin, D.; Neudeck, P.; Powell, J.; Matus, L. Site-competition epitaxy for superior silicon carbide electronics. *Appl. Phys. Lett.*, **1994**, *65*, 1659-1661.
[http://dx.doi.org/10.1063/1.112947]

[20] Lu, H.Y.; Petrich, M.A.; Fan, C.; Hoffman, B.M. Continuous-wave and pulsed electron-nuclear double-resonance study of paramagnetic defects in amorphous hydrogenated silicon-carbon alloy. *Phys. Rev. B Condens. Matter*, **1993**, *48*(3), 1997-2000.
[http://dx.doi.org/10.1103/PhysRevB.48.1997] [PMID: 10008583]

[21] Bockstedte, M.; Heid, M.; Pankratov, O. 2003Signature of intrinsic defects in SiC:Ab initio calculations of hyperfine tensors. Phys. Rev. B 2003, 67, 193102(1-4).

[22] Gali, A.; Deák, P.; Son, N.; Janzén, E.; von Bardeleben, H.; Monge, J. Calculation of Hyperfine Constants of Defects in 4H-SiC. *Mater. Sci. Forum*, **2003**, *433-436*, 511-514.
[http://dx.doi.org/10.4028/www.scientific.net/MSF.433-436.511]

[23] Kalabukhova, E.; Lukin, S.; Savchenko, D.; Mitchel, W.; Greulich-Weber, S.; Rauls, E.; Gerstmann, U. Possible Role of Hydrogen within the So-Called X Center in Semi-Insulating 4H-SiC. *Mater. Sci. Forum*, **2006**, *527-529*, 559-562.
[http://dx.doi.org/10.4028/www.scientific.net/MSF.527-529.559]

[24] Michaelson, S.; Ternyak, O.; Hoffman, A.; Lifshitz, Y. Correlation between diamond grain size and hydrogen retention in diamond films studied by scanning electron microscopy and secondary ion mass spectroscopy. *Appl. Phys. Lett.*, **2007**, *90*, 031914.
[http://dx.doi.org/10.1063/1.2432996]

CHAPTER 13

Paramagnetic Defects in Amorphous Hydrogenated Silicon Carbide and Silicon Carbonitride Films

Ekaterina Kalabukhova[1,*], **Dariya Savchenko**[2,3] and **Bela Shanina**[4]

[1] *Department of Semiconductor Heterostructures, V.E. Lashkaryov Institute of Semiconductor Physics NAS of Ukraine, Kyiv, 03028, Ukraine*

[2] *Department of Analysis of Functional Materials, Division of Optics, Institute of Physics of the Czech Academy of Sciences, Prague, 182 00, Czech Republic*

[3] *Department of Physics and Solid State Physics, National Technical University of Ukraine "Igor Sikorsky Kyiv Polytechnic Institute", Kyiv, 03056, Ukraine*

[4] *Department of Optics and Spectroscopy, V.E. Lashkaryov Institute of Semiconductor Physics NAS of Ukraine, Kyiv, 03028, Ukraine*

Abstract: In this chapter, the nature of the defects and their relation to the incorporation of carbon, hydrogen, nitrogen and thermal treatment were investigated by electron paramagnetic resonance (EPR) spectroscopy for the fundamental insight of the electronic, optical and magnetic characteristics of the amorphous hydrogenated carbon-rich silicon-carbon (a-$Si_{1-x}C_x$:H) and amorphous silicon carbonitride (a-SiC_xN_y) thin films. The paramagnetic defects due to the silicon dangling bonds (SiDBs), carbon-related defects (CRDs) and K-center with Si-N_2Si configuration were revealed in a-$Si_{1-x}C_x$:H films. The observed strong rise of the CRD spin density in annealed a-$Si_{1-x}C_x$:H films is caused by the hydrogen effusion process that takes place at $T_{ann} >$ 400°C. The rise of the CRD density was occurring with the exchange narrowing of its EPR linewidth owing to the appearance of carbon clusters with ferromagnetic ordering. The temperature variation of g-tensor anisotropy, measured at 37 GHz and 140 GHz frequencies for the CRD EPR line in the a-$Si_{1-x}C_x$:H film annealed at 950°C, was interpreted by the existence of graphite-like sp^2-hybridized carbon clusters and demagnetization field. Examination of the temperature variation of the integrated intensity of the SiDB and CRD EPR lines was demonstrated that their spin systems reveal superparamagnetic and ferromagnetic features, correspondingly. The CDB and Si-related surface defects were observed in a-SiC_xN_y. It was found that the CDB spin concentration significantly increases with the increase of the nitrogen content.

[*] **Corresponding author Ekaterina Kalabukhova:** Department of Semiconductor Heterostructures, V.E. Lashkaryov Institute of Semiconductor Physics NAS of Ukraine, Kyiv, 03028, Ukraine; Tel/Fax: +38-044-525-62-97; E-mail: kalabukhova@yahoo.com

Dariya Savchenko & Abdel Hadi Kassiba (Eds.)

Due to the temperature variation of the linewidth and integrated intensity of the CDB EPR line, it has been supposed that the antiferromagnetic ordering takes place in the spin system.

Keywords: Amorphous hydrogenated carbon-rich silicon-carbon films, Amorphous silicon carbonitride, Annealing, Anisotropy, Carbon clusters, Carbon-related defect, Dangling bonds, Demagnetization, EPR, EPR signal intensity, Exchange interaction, Ferromagnetic ordering, High-frequency EPR, Hydrogen, Multifrequency EPR, Nitrogen, SiC, Stoichiometry, Temperature dependence, Thin films.

INTRODUCTION

Amorphous silicon carbide (*a*-SiC) films attract great attention owing to their tunable electrical, optical, and light-emitting features. The shift of the optical band gap from 1.4 to 3.0 eV as well as the change from photoluminescence (PL) color from red to blue one can be controlled by carbon (C) and hydrogen (H) inclusion in these films [1 - 5]. In particular, it was shown that C-excess in amorphous hydrogenated silicon-carbon films (*a*-$Si_{1-x}C_x$:H) and low-temperature annealing dramatically increases the intensity of the visible PL in these films [6]. Recently it was shown that oxidation during the thermal treatments also enhances the efficiency of the white PL in C-enriched *a*-$Si_{1-x}C_x$:H films [7].

However, the light emission mechanism in *a*-$Si_{1-x}C_x$:H films remains unclear and is still under debate. Simultaneously the carbon enrichment of *a*-$Si_{1-x}C_x$:H films agrees with the rise of the density of paramagnetic centers, amongst which the sp^3-hybridized silicon (Si), C dangling bonds (CDB) and sp^2-hybridized graphite-like C clusters are the most likely candidates. The effect of H inclusion on the paramagnetic defects and features of *a*-$Si_{1-x}C_x$:H thin films has been studied [6, 8]. It was found that thermal treatment of the *a*-$Si_{1-x}C_x$:H films results in the H effusion process and leads to the rise in the spin density of the carbon-related defects (CRDs).

There are several methods to produce *a*-SiC films include microwave (MW) plasma chemical vapor deposition, magnetron sputtering, laser deposition, *etc.* Among them, reactive magnetron sputtering is supposed to be one of the most effective methods to produce the amorphous layers.

In this chapter, we represent the results of the study of the paramagnetic defects in *a*-$Si_{1-x}C_x$:H and their relationship to the incorporation of C, H, nitrogen (N) and thermal annealing by electron paramagnetic resonance (EPR) spectroscopy at high frequencies (37 GHz and 140 GHz) for the fundamental insight of the electronic, optical and magnetic characteristics of this material. The EPR measurements of *a*-$Si_{1-x}C_x$:H samples precipitated by reactive DC-magnetron sputtering of a monocrystalline silicon target on silicon wafer enabled us to distinguish at Q-band different paramagnetic centers, including Si dangling bonds (SiDBs) with g = 2.0059, CRDs with g = 2.0028 and SiDB center bonded with nitrogen atoms in Si-N_2Si configuration having g = 2.0033 [9]. The observed strong rise of the CRDs concentration in annealed *a*-$Si_{1-x}C_x$:H films is caused by H effusion process took place at T_{ann} > 400°C. The rise in the CRD density occurs with the exchange narrowing of its EPR linewidth owing to the appearance of C clusters.

The temperature dependence of *g*-tensor anisotropy for the CRD EPR line at 37 GHz in the films with T_{ann} = 950°C has been interpreted by the existence of graphite-like sp^2-hybridized C clusters. At the same time, the anisotropy of the resonance field position/*g*-tensor observed at 37 GHz and 140 GHz for CRD EPR signal serves as the proof that the demagnetizing field (shape-dependent anisotropy term) exists in the *a*-$Si_{1-x}C_x$:H film. Data on magnetic ordering in the C and Si spin systems was obtained from the study of the temperature variation of their integral intensity and linewidth. It was found that SiDB and CRD spin systems show ferromagnetic and superparamagnetic features, correspondingly. Taking into account that Raman and EPR spectroscopies are complementary techniques for structural investigation of the carbon-based amorphous films the EPR results were compared with those obtained from Raman spectroscopy in Ref [6].

In recent times, amongst the various SiC-based materials, the attention to amorphous silicon carbonitride (*a*-SiC_xN_y) compounds is developed significantly. This attention is mostly focused on the unique features of SiC_xN_y films such as high electrical resistivity that is several orders of magnitude higher than that of SiC, and many physical characteristics similar to those of diamond owing to their short bond length and high bond strength [10]. Additionally, the tunable bandgap over a broader range of 2.2–5.0 eV can be obtained by nitrogen (N) inclusion in *a*-SiC films, enabling to produce on the base of *a*-SiC_xN_y films advanced blue or ultraviolet optoelectronic devices *e.g.* solar cells, flat-panel displays, optical memories, antireflective coatings.

The nitrogen inclusion enables monitoring the optical bandgap and electroconductivity of the films, enabling to produce SiC-based films with semiconductor or insulator features [11]. Consequently, the structural characteristics of *a*-SiC_xN_y films are directly influenced by the precipitation method, the kind of target material and wafer, the dopant concentration, the wafer temperature and are the objects of intense study.

EPR was proven to be one of the best methods to investigate the structural features of the *a*-SiC. The EPR investigation of the *a*-SiC_xN_y films was mostly concentrated on the study of the overall concentration of the paramagnetic centers associated with the DBs and their relation with the nitrogen content in the sample. It turned out that the rise of nitrogen content results in the rise of dangling bonds concentration in *a*-SiC_xN_y samples. Nevertheless, such EPR parameters for DBs as a *g*-factor, linewidth and lineshape were not reported previously [12, 13].

In this chapter, we present the results of the study of the paramagnetic structural defects in *a*-SiC_xN_y thin films precipitated on the SiO_2 wafers by reactive magnetron sputtering using continuous wave (CW) and pulse EPR techniques. The optical and magnetic properties of the carbon dangling bonds (CDBs) and the origin of the interface centers appeared in *a*-SiC_xN_y thin films with various nitrogen content was established in the temperature range T = 4.2-300 K. Raman spectra obtained in the present work revealed the features similar to those observed previously for *a*-SiC_x and *a*-SiC_xN_y films in Ref [2, 13 - 16]. Analysis of Raman spectra points to the existence of C-C, Si-N and C-N bonds in *a*-SiC_xN_y films. It was found that the rise in the nitrogen concentration in the film results in rising of the graphite-like sp^2 fraction and the sp^2 carbon cluster size.

In *a*-SiC_xN_y/SiO_2 films three different EPR signals were found. The most intense one having $g = 2.0033$ has been assigned to the CDB. On the basis of its lineshape and linewidth, we have related this EPR line to the free electron delocalized over sp^2 C cluster. As the N content is increased, the spin concentration of the CDB has considerably risen. The analysis of the temperature variation of the integral intensity and linewidth for the CDB EPR line enables us to conclude that the antiferromagnetic ordering takes place in the spin system. The second EPR line with $g = 2.009$ was ascribed to the interface center corresponding to threefold-coordinated SiDB that could form because of the emergence of the SiO_x on the sample surface. The third EPR line at $g = 2.05$ has been preliminary related to the trapped holes at silicon atom close to Si/SiO_2 interface. It was found that the N incorporation in *a*-SiC_x films has no influence on the spin concentration of both interface centers.

SAMPLE CHARACTERIZATION AND EXPERIMENTAL TECHNIQUE

The *a*-$Si_{1-x}C_x$:H films were precipitated by reactive DC-magnetron sputtering of a monocrystalline Si target on Si substrate. As the working gas a mixture of Ar/CH_4 was used. As the substrates the two-side polished Si (100) substrates (p-type, 40 Ω·cm) were utilized. The wafer temperature upon precipitation was ~200°C. Next to precipitation the films were annealed in a vacuum (10^{-6} Torr) for 15 min. in the interval from 400°C to 950°C. The specification of the precipitation process has been presented elsewhere [8]. The thickness of the *a*-SiC:H films detected by interferometer was estimated to be 500 nm. Different characterization methods such as atomic force microscopy (AFM, Nanoscope), Auger-electron spectroscopy and high resolution scanning transmission electron microscopy (STEM) were employed to get information about morphology, composition and structure of the films. The AFM analysis of the surface morphology of the initial (as deposited) *a*-SiC:H films supports and quantifies its nanostructured columnar morphology, which is characteristic of thin films precipitated at low temperatures [17]. Large columns, with a diameter of 100-150 nm, consist of smaller columns having a diameter of 20-40 nm. The composition of the *a*-$Si_{1-x}C_x$:H films was investigated with Auger-electron spectroscopy JEOL Jump 10s. The Si/C ratio has been evaluated to be ~30/70.

The microscopic structure and the spatial distribution of C bonding states in the film with T_{ann} = 950°C have been studied by the STEM, Jeol-JEM200CX equipped with electron energy loss spectrometer (EELS, Gatan ENFINA 1000). It was turned out that the sample structure is non-homogeneous and involves nano-sized low-density regions (voids) elongated almost normal to the sample surface with a characteristic size of 5 x 50 nm (width x length). The observation of the voids suggests the nanostructured columnar morphology of the initial *a*-SiC:H. The spatial distribution of the graphite-like C agrees well with the spreading of the nano-voids. As a result, it was concluded that graphitic C formed after the high-temperature treatment is mostly located in the nano-voids of the *a*-$Si_{1-x}C_x$:H film and graphite-like clusters mostly condensate on the surface in such a way that they form graphene sheets parallel to the surface.

The magnetic and optical characteristics of the *a*-SiC_xN_y films have been studied on the films precipitated on the polished SiO_2 wafers by reactive DC magnetron sputtering in accordance with [16]. The 3C-SiC polytype powder for the target was obtained by reacting of Si (99.99% pure) and natural graphite (99.5% pure) particles by means of self-propagating high-temperature synthesis. The conductive SiC target (d = 75 mm) with a low C-excess (5-7%) was sputtered at different N_2/Ar gas flow ratios. The elemental composition was examined by the electron probe X-ray microanalysis (EPMA) (JEOL JXA-733 device with Kevex Delta Class V microanalyzer). It was found that the carbon/silicon ratio remained almost unchanged and slightly rises with the rise in nitrogen content, allowing us to study the influence of nitrogen incorporation on *a*-SiC_xN_y films directly.

Analysis of the Raman spectra measured in *a*-SiC_xN_y samples indicates the existence of C(sp^3)-C, C(sp^2)-C, C(sp^3)-N and C(sp^2)-N bonds [16]. It turned out that nitrogen incorporation leads to the existence in Raman spectra of two signals usually designated as: G band due to C(sp^2)-N and the C(sp^2)-C bonds, and D band related to the limitation in the graphite domain size that is caused by grain boundaries or imperfections, such as substitutional nitrogen atoms, C(sp^3)-C, and C(sp^3)-N bonds [18]. It is known that the D-band becomes Raman active owing to the absence of long-range order in *a*-C-based materials. Thus, the observed minor rise in the intensity ratio of D and G bands with the rising of nitrogen content was explained by a higher disorder in the samples caused by grain boundaries, imperfections and also by varying the film composition. Additionally, as the nitrogen content rises, both G and D bands shift towards higher wave numbers that might point to the rise in graphite cluster size and the rise of the sp^2 fraction with the appearance of extra C(sp^2)-N bonds.

EPR measurements of *a*-$Si_{1-x}C_x$:H samples at Q-band (37 GHz) were done using a Bruker EMX spectrometer and a conventional Q-band EPR spectrometer equipped with a liquid helium cryo-system at $T = 4.2$ - 300 K. EPR experiments at 140 GHz were conducted using D-band EPR spectrometer at 4.2 K. The spin concentration was obtained by a double integration of the experimental EPR spectra and comparison with the Si:P reference sample. The evaluated relative accuracy of the N_s estimation is ±20%.

The CW and pulsed EPR methods were utilized for the investigation of the *a*-SiC_xN_y samples. The measurements have been performed on an X-band (9.4-9.7 GHz) Bruker ELEXSYS E580 spectrometer in the temperature interval $T = 4$-300 K. The CW EPR spectra were detected using ER 4123SHQE resonator at the MW power level of 0.004743 mW, a modulation frequency was 100 kHz, the modulation amplitude is 1.0 mT, the conversion time is 60 ms and the spectral resolution was 4096 points. Echo-detected EPR (ED EPR) experiments were performed with EN 4118X-MD4 resonator using two-pulse Hahn echo sequence: $\pi/2 - \tau - \pi - \tau -$ echo with following pulse lengths: $\pi/2 = 16$ ns, $\tau = 140$ ns, $\pi = 32$ ns. The fitting of EPR spectra was performed by means of EasySpin 4.5.1. software package [19].

PARAMAGNETIC DEFECTS IN a-$Si_{1-x}C_x$:H FILMS

EPR Spectra in the Initial *a*-$Si_{1-x}C_x$:H Films

At first, EPR study in the initial *a*-$Si_{1-x}C_x$:H films were conducted at the X-band frequency (9.4 GHz) [6, 8]. The EPR spectrum revealed an isotropic single line with $g = 2.0026$, which was attributed to the CRDs. It was suggested that the observed slight asymmetry of the X-band EPR spectrum is caused by the presence of the second EPR line, possibly caused by the Si dangling bonds (SiDB) with $g = 2.005$ [8]. This suggestion is supported by the fact that optically detected magnetic resonance (ODMR) signal at $g = 2.005$ due to the SiDB was also detected in *a*-$Si_{1-x}C_x$:H films with a low carbon content in Ref [3]. The effect of H inclusion on paramagnetic defects and features of *a*-$Si_{1-x}C_x$:H thin films was reported in [6]. We have found that thermal treatment of the a-$Si_{1-x}C_x$:H films give rise to the H effusion process in a-$Si_{1-x}C_x$:H samples and leads to the rise of the spin density of the CRD. Here the EPR investigation of the initial C-enriched a-$Si_{1-x}C_x$:H film at higher frequencies (35 GHz and 140 GHz) was presented. The measurements at 37 GHz were shown that 3 EPR signals contribute to the isotropic line recorded at X-band [20, 21]. Fig. (**1**) shows experimental and calculated Q-band EPR spectra observed in the initial a-$Si_{1-x}C_x$:H film. The EPR spectrum in the initial a-$Si_{1-x}C_x$:H sample is well simulated with 3 lines of a Gaussian shape having the g-values of 2.0059(2), 20028(2) and 2.0033.

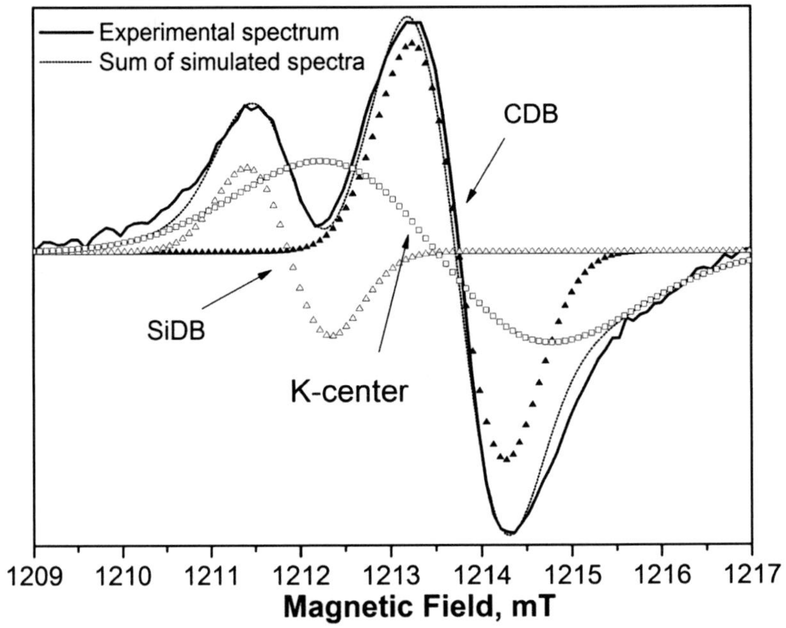

Fig. (1). Experimental (solid lines) and fitted (dashed lines) EPR spectra measured at 37 GHz in the initial a-$Si_{1-x}C_x$:H samples. $T = 4.2$ K.

The EPR signal described by g = 2.0059(2) and linewidth ΔH_{pp} = 1.2 mT was attributed to the SiDBs having g_{iso} = 2.0055(5) and ΔH_{pp} = 0.5-1.5 mT [21]. The slightly higher value of g-factor of the observed EPR signal with respect to that measured for SiDB in a-Si (g = 2.0055) at X-band in [22] can be explained by different structural and chemical environment of the SiDB in a-$Si_{1-x}C_x$:H and a-Si:H films. For comparison g-value of the SiDB in porous Si was found to be g = 2.0059 [23].

The second EPR signal has g_{iso} = 2.0028(2) and ΔH_{pp} = 1 mT that are the characteristic parameters for CRDs observed in C-based materials like *a*-C films [24].

The third EPR line with $g = 2.0033(2)$ and broad linewidth of $\Delta H_{pp} = 2.5(2)$ mT was attributed to the so-called K-center with Si-N_2Si configuration reported earlier in *a*-Si_3N_4:H films [9]. The K centers are assigned to threefold-coordinated SiDBs where a silicon atom can be bonded to 1, 2 or 3 N atoms leading to the appearance of the (Si-NSi_2), (Si-N_2Si), (Si-N_3) complexes, correspondingly. As a consequence, the *g*-value of the K center varies from 2.0030 to 2.0040 and rely on the defect configuration, whereas its linewidth is generally defined by unresolved superhyperfine (SHF) coupling with ^{14}N ($I = 1$) and ^{15}N ($I = 1/2$) nuclei [9]. The low deposition temperature can lead to the occurrence of such defects in *a*-$Si_{1-x}C_x$:H films.

The relative contribution of Si- and C-related paramagnetic centers (N_{Si}/N_C) in the initial samples has been evaluated to be ~1/5 and the overall spin concentration of ~$(5\pm0.1)\cdot10^{19}cm^{-3}$.

The Impact of Thermal Treatment on the EPR Spectra of the *a*-$Si_{1-x}C_x$:H Films

The effect of the thermal annealing on the EPR spectra of the paramagnetic defects was investigated in *a*-$Si_{1-x}C_x$:H films annealed in the interval of $T_{ann} = 400$ - 950°C. From Fig. (**2**) it is seen that the intensity of the EPR line from CRD raised by a factor of 40, whereas the integrated intensity raised no more than by a factor of 8 owing to the EPR line narrowing in *a*-$Si_{1-x}C_x$:H films with $T_{ann} = 950$°C. At the same time, the SiDB EPR signal disappears after the thermal treatment of *a*-$Si_{1-x}C_x$:H annealed at 650°C.

After annealing the sample at 650°C we have observed the strong enhancement in the CRDs concentration and transformation of its lineshape from Gaussian to Lorentzian one. This fact can be interpreted by the H effusion process taken place at the temperatures higher than 400 ^{0}C. The SHF interaction of the unpaired electron with ^1H ($I = 1/2$) cannot be distinguished in the EPR spectra and could only cause the non-homogeneous broadening of Gaussian linewidth likewise as was observed in the initial and *a*-$Si_{1-x}C_x$:H film annealed at low temperatures. As a result, the carbon atom, being free after the hydrogen released from the C-H bonds, contributes to the CRD.

The observed small linewidth narrowing ($\Delta H_{pp} = 0.6 \pm 0.05$ mT) of the Lorentzian CRD EPR line between 4.2 K and 300 K in the *a*-$Si_{1-x}C_x$:H films annealed at high temperatures was explicated by spin-spin exchange coupling occurred among free carriers.

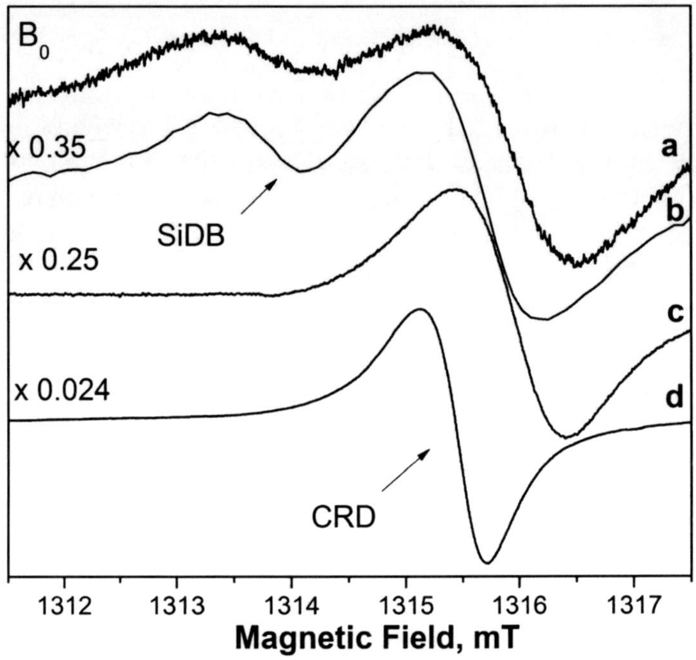

Fig. (2). The Q-band EPR spectra recorded in the initial and annealed a-$Si_{1-x}C_x$:H films at different amplification levels. $T = 4.2$ K. (a) - initial; (b) - $T_{ann} = 450°C$; (c) - $T_{ann} = 650°C$; (d) - $T_{ann} = 950°C$. Adapted from [19] with the Permission from APS.

A similar effect of the thermal annealing on EPR linewidth and lineshape for CRD was observed earlier [24, 25] and related to the appearance of the sp^2 C clusters in *a*-SiC:H films annealed at high temperatures.

The Temperature Dependence of *g*-tensor Anisotropy of the CRD Signal in *a*-$Si_{1-x}C_x$:H Films Annealed at High Temperatures

The investigation of the temperature behavior of the CRD EPR signal was shown that it has the temperature dependent angular dependence of the *g*-factor. The angular dependence of the *g*-tensor of the CRD line detected at Q-band in *a*-$Si_{1-x}C_x$:H film with $T_{ann} = 950°C$ shown in Fig. (**3**) can be described by the following equation [26]:

$$g = g_{\perp} + A^{*} \cdot \cos^2\phi. \quad \textbf{(1)}$$

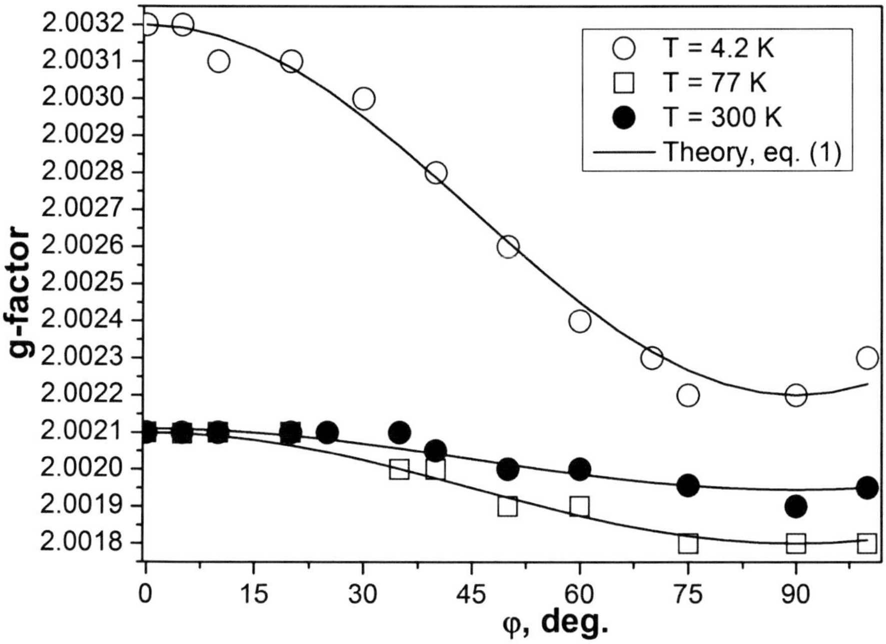

Fig. (3). The angular dependence of the CRD *g*-factor measured at different temperatures in *a*-$Si_{1-x}C_x$:H film with T_{ann} = 950°C. The points represent experimental data; the solid lines are the theoretical fits with eq. (1). ($\vec{B} \perp \vec{c}$ and $\vec{B} \parallel \vec{c}$ in graphite corresponds to the ϕ = 0° and ϕ = 90° in *a*-$Si_{1-x}C_x$:H, correspondingly). Adapted from [21] with the Permission from APS.

Here ϕ denotes the angle between the sample plane and the static magnetic field, ϕ = 0° and ϕ = 90° correspond to the magnetic field directed parallel (B_0) and perpendicular (B_{90}) to the sample plane, respectively.

The $g_\perp$ value is temperature-independent ($g_\perp$ = 2.0023(2)), while the anisotropy valuc A^* highly relies on temperature and rises on lowering of the temperature (A^* = 0.0002 at T = 300 K and A^* = 0.0010 at T = 4.2 K).

The fact that the *g*-factor of the CRD EPR line shows the anisotropy indicates that CRD locates in the ordered phase of the film. The observed temperature dependence of *g*-tensor anisotropy of the CRD line agrees with that reported earlier for the free carriers in monocrystalline graphite, however with the smaller anisotropy factor A^*. As was found in [26] the anisotropy factor A^* for pure graphite is A^* = 0.047 at 300 K and at T = 4.2 K A^* = 0.125.

The *g*-tensor anisotropy for the free carriers was reported in nano- and polycrystalline graphite and in *a*-C:H as well. In these materials the *g*-value in the sheet plane ($\vec{B} \perp \vec{c}$) has the same value of $g_\perp$ = 2.0028 0.0002 and varies for the orientation of the magnetic field perpendicular to the film plane ($\vec{B} \| \vec{c}$) from $g_\|$ = 2.0182 for polycrystalline graphite to $g_\|$ = 2.005 for nanocrystalline graphite and *a*-C at 4.2 K. As was shown in [27], the anisotropy factor A^* strongly depends on the graphite particle size and has the lowest value for the nanoscaled graphitic sp^2 clusters. From the interrelation between the $g_\|$ value and the sp^2 cluster size reported in [26], the size of sp^2 graphite-like clusters appeared in *a*-$Si_{1-x}C_x$:H film was evaluated to be < 5 nm. This conclusion is supported by the presence of the diamond-like sp^3 D-band and graphite-like sp^2 G-bands in Raman spectra measured in *a*-$Si_{1-x}C_x$:H film annealed at 750 and 850 °C associated with the local ordering and the existence of the graphite-like sp^2-hybridized C clusters [28].

Since the *a*-$Si_{1-x}C_x$:H film has the porous columnar morphology of it was concluded that the graphitization takes place at the column surface aligned normal to the sample plane. This finding is confirmed by the fact that the electroconductivity of *a*-$Si_{1-x}C_x$:H parallel to the sample plane is several times lower than the conductivity perpendicular to the sample plane. The anisotropy of the conductivity direction in *a*-$Si_{1-x}C_x$:H coincides well with that reported in graphite. According to [26], the conductivity of graphite along the $\vec{c}$ axis is 100-200 times lower than that in the sample plane. Hence, one can assume, that graphitization in *a*-$Si_{1-x}C_x$:H takes place along the columns, and consequently the $\vec{c}$-axis of the graphite layer is directed roughly along the sample plane.

Evaluation of the sp^2/sp^3 Carbon Ratio in the Initial *a*-$Si_{1-x}C_x$:H Films

Whereas the films contain a combination of the sp^2 and sp^3 carbons, it is desirable to estimate the contribution of the graphite-like sp^2 fraction with regard to the sp^3 C matrix in the initial *a*-$Si_{1-x}C_x$:H samples. The sp^2/sp^3 C ratio in the initial *a*-$Si_{1-x}C_x$:H can be evaluated by using the relation between two paramagnetic subsystems with S = 1/2 coupled by the exchange interaction, which affects the observed magnetic resonance field position [29]:

$$B_{res} = \left(\chi_1 B_{res,1} + \chi_2 B_{res,2}\right) / \left(\chi_1 + \chi_2\right) \quad \textbf{(3)}$$

Here $\chi_{1,2}$ denote the magnetic susceptibilities for the two spin subsystems.

One spin system is related to the CRD with g_{iso} = 2.0028, the second one associates with the free carriers in graphite with the maximum *g*-value g = 2.0496 at T = 300 K. The B_{res} is the observed resonance magnetic field, $B_{res,1}$ and $B_{res,2}$ are the resonance magnetic fields for the individual subsystems in the lack of the exchange coupling. Using eq. (3) one can calculate the χ_1/χ_2 ratio between paramagnetic susceptibilities corresponding to the spin density ratio of the two subsystems. It was found that the concentration of the graphite-like sp^2 dangling bonds reaches 1.1% of the overall concentration of the CDBs in the initial a-$Si_{1-x}C_x$:H.

MAGNETIC PROPERTIES OF THE HIGH TEMPERATURE ANNEALED *A*-$SI_{1-X}C_X$:H FILMS

Magnetization Effect in High Temperature Annealed *a*-$Si_{1-x}C_x$:H Films

For the high-temperature annealed a-$Si_{1-x}C_x$:H sample (950°C) the anisotropy of the CRD resonance field position was determined at 37 GHz and 140 GHz, whereas at lower MW frequencies the anisotropy effect was hidden in the natural linewidth. The orientation effect of the CRD signal resonance field position has been found to be stronger at D-band and 4.2 K with $B_{90} > B_0$, where B_0 and B_{90} are the resonance magnetic field values for external magnetic field $\vec{B}\,||$ and $\vec{B}\perp$ film plane, respectively, which is demonstrated in Fig. (**4**).

The results of fitting of the CRD EPR line at D-band show that it has a Lorentzian lineshape and is independent of the resonance MW frequency, whereas the linewidth changes with the orientation of the magnetic field from 1 mT for $\vec{B}\perp$ to 1.3 mT for $\vec{B}\,||$ film plane. This difference can be explained by the resonance magnetic field scattering effect, which takes place at higher MW frequencies. The linewidth is determined by non-resolved dispersion of g-tensor components which rises proportionally to the MW frequency.

The above anisotropy of the resonance magnetic field/g-factor revealed for the CRD EPR signal indicates to the existence of the demagnetizing field in the a-$Si_{1-x}C_x$:H sample. Unlike the diluted paramagnetic materials, where the demagnetizing fields arc negligible compared to the linewidth, the demagnetizing field turns out to be remarkably for the disordered films where the spins are coupled in clusters.

As the demagnetizing field B_{dem} is proportionate to the magnetization M of the film ($B_{dem} = -4\pi M$), we can estimate the film magnetization M from the following equations obtained in [30] for a thin film:

$$\begin{aligned} (\omega/\gamma)^2 &= B_0\cdot(B_0+4\pi M) \\ (\omega/\gamma) &= B_{90}-4\pi M. \end{aligned} \quad (4)$$

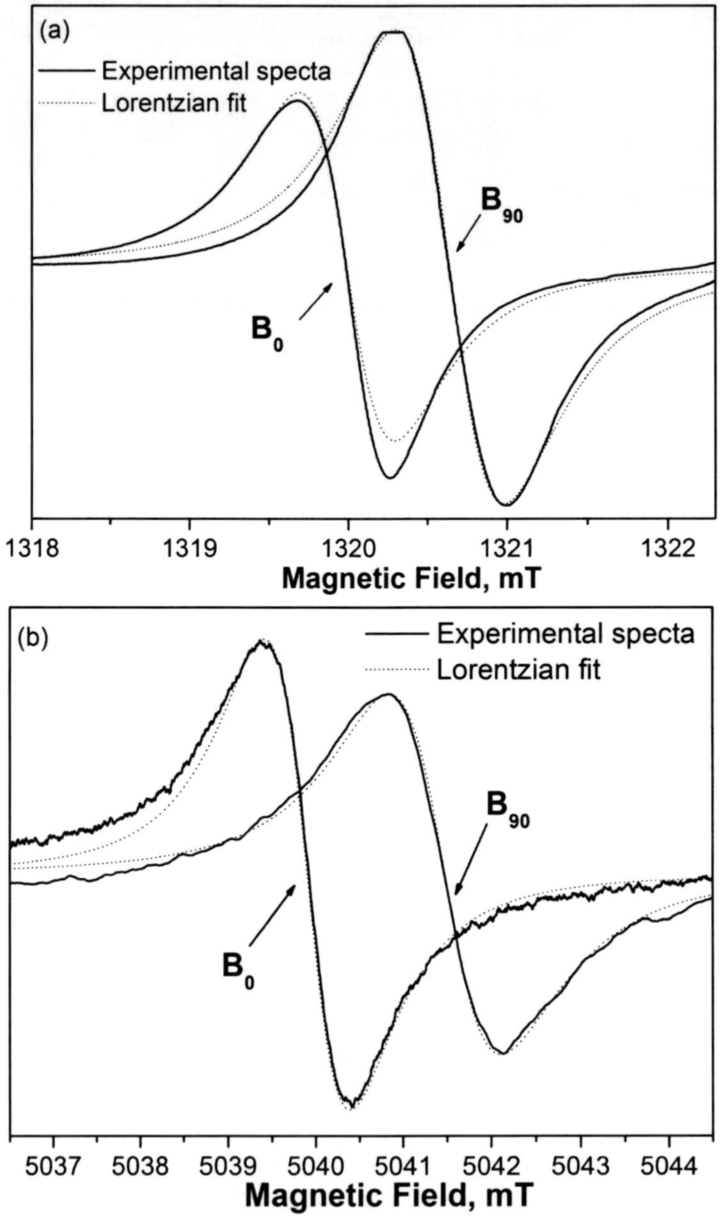

Fig. (4). Anisotropy of the resonance field of CRD EPR line in *a*-$Si_{1-x}C_x$:H film with T_{ann} = 950°C at 37 GHz (a) and 140 GHz (b) for rotation of the magnetic field from B_{90} to B_0 measured at 4.2 K. Dashed lines are the simulated spectra with Lorentzian lineshape. Reproduced from [21] with the Permission from APS.

Here ω denotes the frequency of the MW field; $\gamma = \beta \cdot g/h$ is the gyromagnetic ratio of the spin system in the film; β is the Bohr magneton, g – is the spectroscopic splitting factor, h – is the Plank constant.

From eq. (4) we can obtain two unidentified values: $4\pi M$ and g. By combining these two expressions, we can estimate the film magnetization value at two MW frequencies using the following experimental data: B_{90} = 1320.656 mT and B_0 = 1319.99 mT at Q-band and B_{90} = 5043.14 mT and B_0 = 5041.49 mT at D-band. We have calculated that the demagnetizing field at 4.2 K in a-$Si_{1-x}C_x$:H film with T_{ann} = 950°C is $4\pi M$ = 0.44 mT at Q-band and 1.1 mT at D-band. The gyromagnetic ratio of the spin system obtained from eq. (4) $g = h\omega/(\beta(B_{90} - 4\pi M))$ yields g_0 = 2.0029 at Q-band, g_0 = 2.0028 at the D-band, corresponding to the g-factor of the CRD spin system.

The demagnetization effect in a-C at X-band with B_{dem} of about 0.7 mT at T = 1.75 K [31] and at 285 GHz of about 10 mT at T = 4 K has been reported earlier [32]. This effect becomes considerable at the high spin concentration of the paramagnetic defects, low temperatures ($kT < g\beta B_{res}$, where k – is the Boltzmann constant, B_{res} – denotes the resonance magnetic field position) and higher applied external magnetic field.

Temperature Dependence of Q-band EPR Spectra in *a*-$Si_{1-x}C_x$:H Films

As was shown before the sp^2 C clusters appear in a-SiC:H films annealed at high temperatures. These clusters can exhibit antiferromagnetic, ferromagnetic, or superparamagnetic characteristics, which can be determined by studying the temperature dependence of the magnetization $M(T)$ at a fixed magnetic field. To this end the temperature behavior of the EPR spectrum of the CRD in annealed a-$Si_{1-x}C_x$:H film has been studied. The temperature variation of EPR spectra obtained in a-$Si_{1-x}C_x$:H films with T_{ann} = 400°C for the orientation of magnetic field perpendicular to the film plane is shown in Fig. **5**.

As it is seen from Fig. **5**, the SiDB EPR signal is barely noticeable at T = 50 K because the EPR spectrum in a-$Si_{1-x}C_x$:H film with T_{ann} = 400°C was recorded in the temperature range between 6 K and 50 K at the same amplification level without gain variation. The temperature variation of the integral intensity of SiDB and CRD EPR signals observed in a-$Si_{1-x}C_x$:H with T_{ann} = 400°C for the magnetic field orientated perpendicular and parallel to the sample plane exhibit a different character.

In contradiction to the temperature slope of the SiDB signal, the temperature slope of the EPR line intensity for CRD demonstrates the concave behavior. As it follows from Fig. (**6**), at T < 20 K the CRD line intensity reaches its constant value that is induced by magnetization saturation at low temperatures.

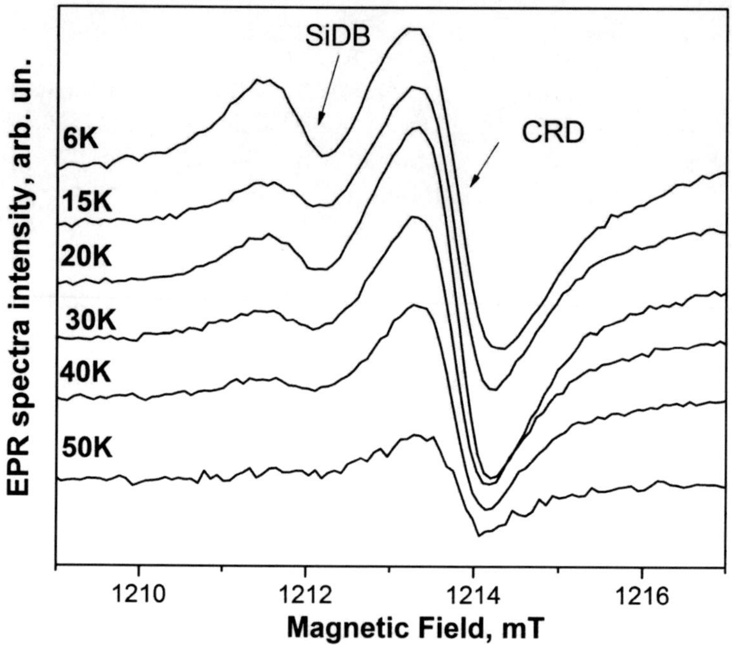

Fig. (5). Temperature behavior of the EPR spectrum measured on the a-$Si_{1-x}C_x$:H film annealed at 400°C with the magnetic field orientation perpendicular to the film plane. Adapted from [21] with the Permission from APS.

Magnetic Ordering in CRD and SiDB Clusters in a-$Si_{1-x}C_x$:H Film

Basically, one can suppose that 3 types of magnetic systems can exist in amorphous material like a-$Si_{1-x}C_x$:H films including paramagnetic, superparamagnetic, ferromagnetic or antiferromagnetic spin systems. The study of the temperature dependence of the integrated intensity of the SiDB and CRD EPR lines has revealed that they do not follow the Curie-Weiss law, which is the characteristic of the paramagnetic system, and their spin systems demonstrate superparamagnetic and ferromagnetic, correspondingly.

Superparamagnetism is usually found for the superspins coupled to the clusters having high magnetic moments. These superspins act like particular spins in paramagnetic materials and accordingly their temperature features are characterized by the Langevin function $L = \coth(\theta/T) - T/\theta$, where $\theta = \beta \cdot H \cdot n/k$, k – is the Boltzmann constant, H – resonance magnetic field; n – the spin number in the cluster; θ (given in the temperature units) – is the energy of the overall magnetic moment ($n \cdot \beta$) in the presence of the magnetic field H. In the real system with the high density of paramagnetic defects, the spin number in the clusters could change in a broad interval. Hence, supposing that the clusters possess a size distribution, the temperature dependence of the integrated intensity of the SiDB signal was described with the integral of the Langevin function [21]:

$$I(T) = A \cdot T \cdot \int_{\frac{\theta_1}{T}}^{\frac{\theta_2}{T}} \left[cth\left(\frac{\theta}{T}\right) - \frac{T}{\theta} \right] d\left(\frac{\theta}{T}\right) = A \cdot T \cdot \ln\left(\frac{sh\left(\frac{\theta_2}{T}\right)}{\left(\frac{\theta_2}{\theta_1}\right) \cdot sh\left(\frac{\theta_1}{T}\right)} \right). \quad \textbf{(5)}$$

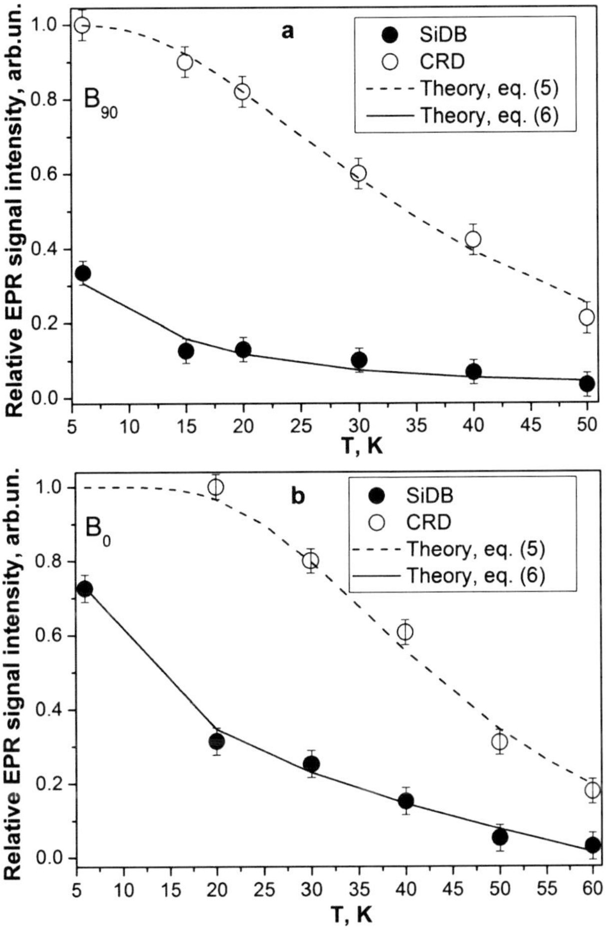

Fig. (6). The temperature dependence of the EPR line intensity of the SiDB and CRD measured in the a-Si$_{1-x}$C$_x$:H film with T_{ann} = 400°C for the magnetic field orientation (a) - perpendicular and (b) - parallel to the sample plane. The points are experimental data, the solid and dash-dot lines are theoretical fits with eq. (5) and (6), correspondingly. Adapted from [21] with the Permission from APS.

The θ_1 and θ_2 are the integration limits corresponding to the minimum and the maximum number of spins in the clusters, respectively.

By calculating the temperature slope in Fig. **6** with eq. (5) we have obtained that the number of spins in Si clusters changes from n_{min} = 16 to n_{max} = 25, for the magnetic field orientation parallel to the sample plane and from n_{min} = 20 to n_{max} = 32 for the magnetic field orientation perpendicular to the sample plane.

The temperature variation of the CRD EPR line intensity agrees with the temperature dependence of magnetization obtained for the clusters having a ferromagnetic ordering in an amorphous system, which is characterized by the percolation mechanism of free charge carriers across the clusters [33]. Owing to the ferromagnetic exchange coupling among spins via the random jumps of free charge carriers, the spin clusters are coupled into the bulk overall magnetic moment. In accordance with the magnetization model in an amorphous system the temperature behavior of the signal intensity of CRD is characterized by the equation [33]:

$$I(T) = I_0 \cdot \left[1 - \exp\left(-v_R \cdot \ln^3\left(J_0 / T\right)\right)\right], \tag{6}$$

where $v_R = 4\pi \cdot n \cdot R^3/3$; n is the number of spins per unit volume, R is an effective radius of the exchange coupling among the nearest spins. J_0 – the energy of the exchange coupling among the nearest spins. The matching of the eq. (6) and the experimental data in Fig. **6** allows one to estimate the exchange coupling energy J_0 among the nearest CRD spins. We have obtained that J_0 = 13.3 meV for the magnetic field orientation in the sample plane and J_0 = 10.7 meV for the magnetic field oriented perpendicular to the sample plane. The v_R value was found to be 0.55 and 0.2 for the magnetic field oriented parallel and perpendicular to the sample plane, correspondingly.

The obtained data are in accordance with the main principle formulated for the C-based materials in [34] stating that "all constituent atoms and molecules should be paramagnetic and interaction should be ferromagnetic". In particular, the two sp^2 C radicals separated by a C atom in sp^3 state are coupled by the ferromagnetic interaction. For a-$Si_{1-x}C_x$:H samples a coexistence of paramagnetic and ferromagnetic spins is supposed and that is confirmed by the fact that 1.1% of the ferromagnetic sp^2-coupled C nanoclusters are dispersed in a sea of the paramagnetic spins having g_{iso} = 2.0028.

PARAMAGNETIC DEFECTS IN *A*-SIC$_X$N$_Y$ FILMS AND THEIR MAGNETIC PROPERTIES

CW and Pulsed EPR Spectra in *a*-SiC$_x$N$_y$ Films

The EPR spectra were studied in the *a*-SiC$_x$N$_y$ films with different nitrogen concentration ranging between 0 at. % and 40.4 at. %. Fig. **7** shows the EPR signals observed at X-band in the *a*-SiC$_x$N$_y$ film with N content of 0 at. % and 40.4 at. %. It consists of three overlapping EPR lines, which are well simulated with 3 lines of Lorentzian shape with $S = 1/2$. The strong narrow EPR signal has $g_{iso} = 2.0033(3)$ and two other weak broad signals are described by a g_{iso} of 2.009 and 2.05, correspondingly. Taking into account the advantage of the ED EPR technique over CW EPR technique in disentanglement of the overlapping EPR lines of paramagnetic centers at the expense of their distinct spin relaxation times the measurements of the ED EPR spectrum in *a*-SiC$_x$N$_y$ films was carried out.

Fig. **8**b shows the ED EPR spectrum measured in *a*-SiC$_x$N$_y$ sample at $T = 10$ K with nitrogen content 40.4 at. %. For comparison, the CW X-band EPR spectrum was presented in Fig. **8**a. As it follows from Fig. **8**, the EPR signal with *g*-factor of 2.0033 was not recorded in the ED EPR spectrum of *a*-SiC$_x$N$_y$ sample owing to its short spin-lattice relaxation time indicating the presence of the spin-spin exchange coupling among free carriers in the spin system.

Other two paramagnetic centers with *g*-factor of 2.009 and 2.05 are well distinguished in the ED EPR spectrum justifying their presence in the X band EPR spectrum of *a*-SiC$_x$N$_y$ film. The same behavior exhibited EPR signals in films with N content of 0 at. %.

Identification of the Paramagnetic Defects in *a*-SiC$_x$N$_y$ Films with Different N Content

As evident from Fig. **9**a, the intensity of paramagnetic centers found in *a*-SiC$_x$N$_y$ films significantly increases with rising of the nitrogen content. Fig. **9**b plots the dependence of the relative intensity of the three EPR lines from the nitrogen content. It can be noted that the intensity of the EPR signal with $g = 2.0033$ considerably rises as the intensity of other two EPR signals remains unaffected.

The obtained *g*-factor ($g = 2.0033$) and linewidth (2.0(2) mT) for the observed EPR line at 4.2 K allowed us to suggest that this line associated with the K center in Si-N$_2$Si configuration with $g = 2.0033$ and linewidth of $\Delta H_{pp} = 2.5$ mT reported in Ref [21]. However, considering that the lineshape of the EPR signal with $g = 2.0033$ is Lorentzian, the contribution to the linewidth from the unresolved ^{14}N SHF coupling is eliminated. In this case, the Gaussian lineshape is assumed.

Hence, the observed EPR line at $g = 2.0033$ was assigned to the CDBs localized within the a-SiC_xN_y film.

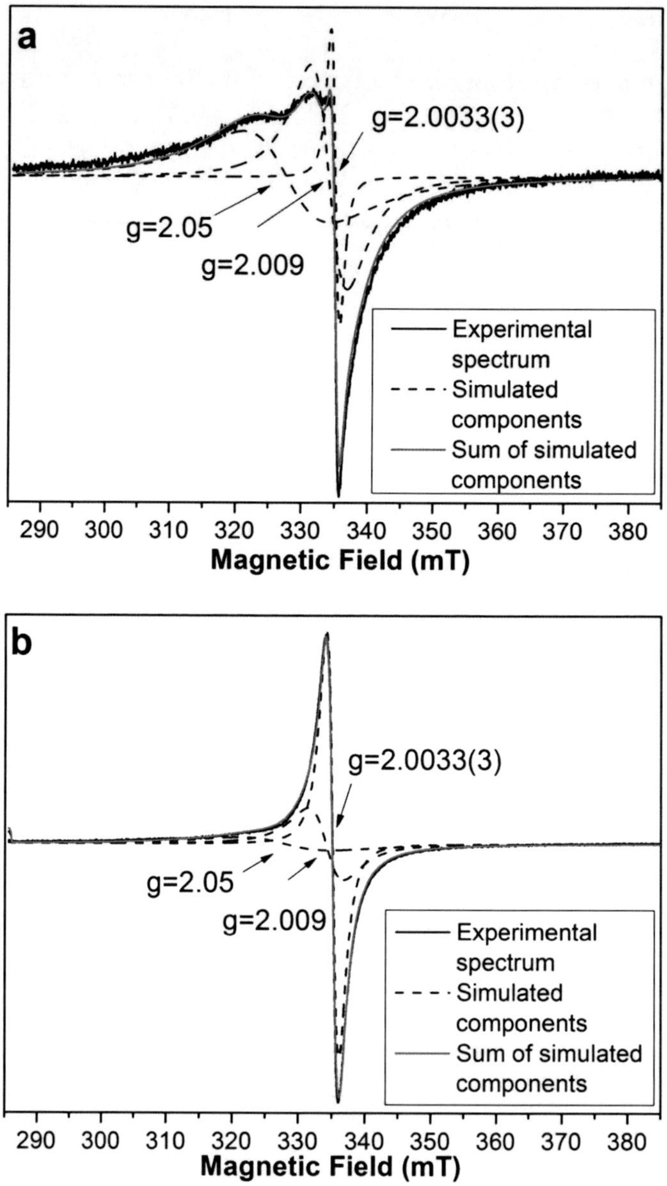

Fig. (7). The experimental (solid lines) and calculated (dashed lines) CW EPR spectra recorded at $T = 10$ K in a-SiC_xN_y samples. (a) –nitrogen content is 0 at.% and (b) – nitrogen content is 40.4 at.%. Adapted from [35] with the Permission from John Wiley and Sons.

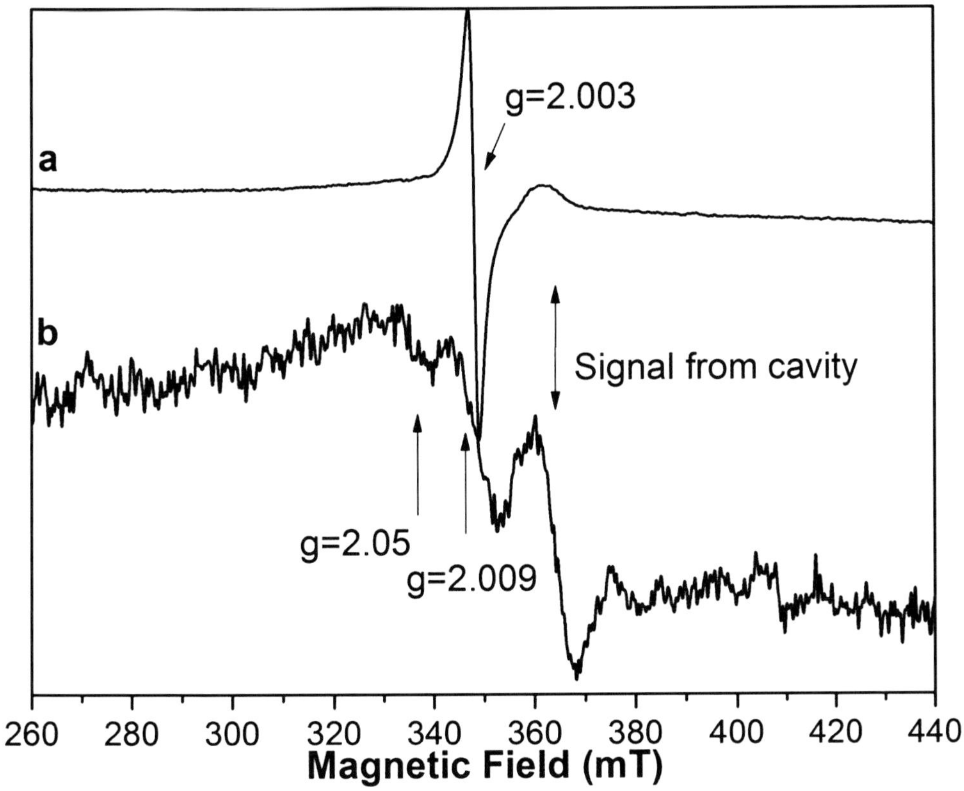

Fig. (8). The X band CW EPR (a) and ED EPR (b) spectra detected in *a*-SiC_xN_y films with nitrogen content of 40.4 at. % at $T = 10$ K. Reproduced from [35] with the Permission from John Wiley and Sons.

The *g*-factor for CDB is typically determined as $g = 2.0024$-2.0029. The shift of *g*-factor to the higher values in the approximate range 2.0032-2.0047 was detected previously in polymer-like, diamond-like films and films where a combination of sp^2- and sp^3-hybridized carbon atoms has been expected [36].

According to [37 - 39] the sp^2- and sp^3-C coordinated state is defined by different EPR linewidth. Owing to the various extents of spin delocalization within the "in-plane" DBs sp^2-C coordinated state is defined by $H_{pp} > 1$ mT, whereas the EPR signal from the sp^3-C coordinated state is defined by linewidth smaller than 1 mT. Considering that the smallest linewidth of the observed CDB EPR line is ~1.2 mT at $T = 292$ K we have assigned it to the CDB in the sp^2-coordinated state. This conclusion agrees with considerable enhancement of the C(sp^2)-C related band in the Raman spectrum with the rise of nitrogen content.

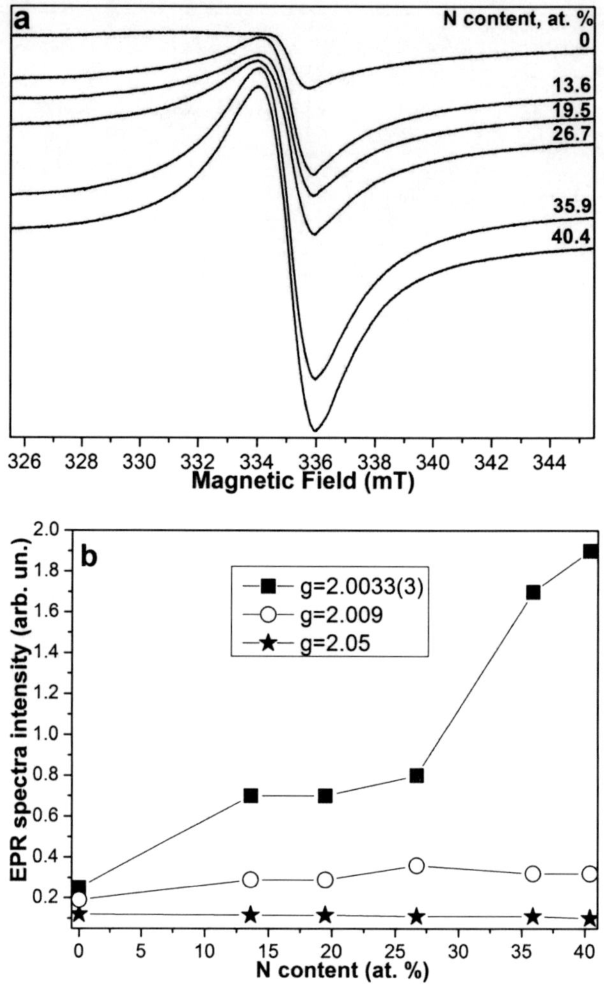

Fig. (9). The dependence of the EPR spectra (a) and relative intensity of observed paramagnetic centers (b) in *a*-SiC_xN_y films from the nitrogen content measured at 10 K. Reproduced from [35] with the Permission from John Wiley and Sons.

The appearance of the second EPR line with the *g*-factor of 2.009 is a sign of the appearance of the interface center in the *a*-SiC_x film similar to P_{b0} defect having $g_{\|}$ ~ 2.0020 and $g\perp$ ~ 2.0090 [40], which represents a SiDB localized at the Si/SiO_2 surface. It is well-known that for chaotically oriented crystallites [41]:

$$g = g_{\perp} + (1/3)\delta g, \tag{7}$$

where δg is dependent on the crystalline lattice deformation, which results in the degeneration of the valence band. Thus, according to eq. (7) the EPR signal at g = 2.009 can be preliminarily assigned to interface center labeled as P_{b0} defect related to threefold-coordinated SiDB, which could form in the *a*-SiC_x film owing to the existence of the O and the appearance of the SiO_x on the sample surface. This consequence is justified by the existence of the Si-Si bonds in Raman spectrum of *a*-SiC_x.

The third EPR line with g = 2.05 was preliminary assigned to the Si interface center with the trapped hole [42].

The essential effect of the N incorporation on the spin concentration of both interface centers was not found. The effect of the N on the P_{b0} defect was reported earlier in Si/Si_3N_4 [9]. The incorporation of nitrogen does cause the structural modifications of the interface P_{b0} center. The isotropic broadening of the EPR linewidth from 0.16 mT observed in the case of SiO_2 interfaces to 0.32 mT in Si/Si_3N_4 was interpreted by the SHF coupling of silicon atoms with the ^{14}N and ^{15}N nuclei. The EPR lines due to the interface centers observed in *a*-SiC_xN_y films are fitted with Lorentzian lineshape and thus the contribution into the linewidth of the unresolved ^{14}N SHF coupling should be excluded. Probably the nitridation of the interface centers was inhibited by the appearance of CN molecules. The effect of the thermal annealing of the *a*-SiC_xN_y thin films up to 1000°C in inert atmosphere on the structural bonding of the paramagnetic centers has not been detected as well.

Magnetic Properties of the Carbon dangling Bonds in *a*-SiC_xN_y Films with High Nitrogen Content

In order to get data about the magnetic characteristics of the CDB, the temperature behavior of the EPR spectrum observed in *a*-SiC_xN_y samples with the nitrogen content of 40.4 at. % was investigated at T = 4.2-300 K. As can be seen from Fig. **10**a, the temperature dependence of integrated intensity obeys a Curie-Weiss law:

$$I(T) = C/(T - \Theta), \qquad \textbf{(8)}$$

where C – Curie-Weiss constant proportionate to the localized spin concentration and Θ – Curie-Weiss temperature corresponding to the strength of exchange coupling among magnetic moments of the localized spins. Its sign depends on the parallel (ferromagnetic) or antiparallel (antiferromagnetic) alignment of the magnetic moments. From the matching of the eq. (8) and the experimental data, it was found that C = 4.6 and Θ = –0.5 K and it demonstrates that the antiferromagnetic coupling takes place in the spin system.

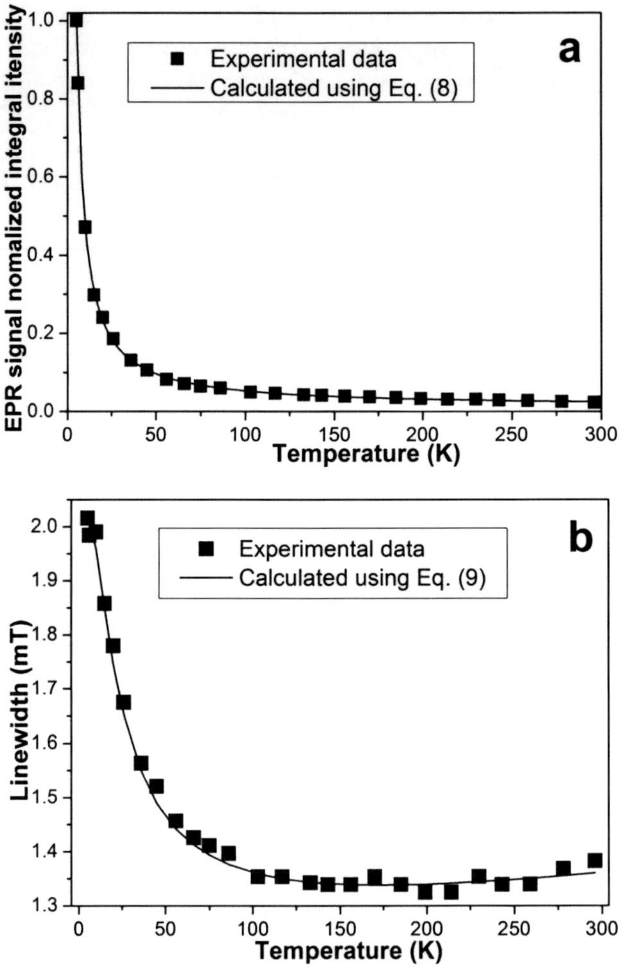

Fig. (10). The temperature variation of the normalized signal integrated intensity (a) and the linewidth (b) of CDB EPR line in the *a*-SiC_xN_y samples with nitrogen content of 40.4 at.%. Adapted from [35] with the Permission from John Wiley and Sons.

The linewidth of EPR line is defined by a combination of the temperature independent dipolar coupling and motional narrowing caused by the hopping motion of CDB electrons at higher temperatures. The observed reduction of the CDB linewidth with the rise of temperature means that narrowing caused by the exchange coupling prevails broadening due to dipolar coupling.

The temperature variation of the EPR linewidth in the low-temperature interval can be well characterized by the exchange couplings in disordered materials among CDBs [43]:

$$\Delta H_{pp}(T) = A' + B'T + C'/[1 + \exp(-|J|/kT)] \quad \textbf{(9)}$$

Here, A', B', C' are constant values, J is the exchange-interaction constant. The 2nd and 3rd terms correspond to the relaxation of spins of an exchange-coupled pair of spins in the disordered solids [43]. From the matching of eq. (9) and experimental data, we have obtained $A' = 0.35$ mT, $B' = 0.00044$ mT/K, $C' = 1.67$ mT.

The negative sign found for J and for Θ in temperature variation of the integrated intensity reveals the antiferromagnetic coupling among CDBs in a-SiC_xN_y films. The constant of antiferromagnetic exchange among CDBs was reported earlier in the SiCN ceramics [43].The antiferromagnetic ordering in the carbon spin system was found in a-C films as well [44].

CONCLUSIONS

Three paramagnetic defects with $S = 1/2$ were found in the EPR spectra of the precipitated hydrogenated C-enriched silicon-carbon (a-$Si_{1-x}C_x$:H) films in the temperature range $T = 4.2$-300 K. Two of them at $g = 2.0028$ and $g = 2.0059$ were attributed to the CDR and SiDBs, respectively. The g-value ($g = 2.0033$), linewidth and lineshape values obtained for the third EPR line are close to the corresponding parameters of the so-called K-center reported in a-Si_3N_4:H films representing threefold-coordinated SiDBs when a silicon atom is bonded with two nitrogen atoms (Si-N_2Si).

The impact of thermal treatment on the EPR spectra of the a-$Si_{1-x}C_x$:H samples was investigated in the temperature interval of $T_{ann} = 400$ - 950°C. It was found that the H effusion process took place in a-$Si_{1-x}C_x$:H at $T_{ann} > 400$°C. Hydrogen loss from C-H bonds starts at $T_{ann} = 450$°C and results in the strong rise in the CRDs concentration. The rise in the CRDs concentration is followed by the exchange narrowing of their EPR linewidth.

The temperature dependent g-factor anisotropy found at 37 GHz for the CRD signal in high-temperature annealed samples (950°C) was interpreted by the existence of graphite-like sp^2-hybridized C clusters. Considering the porous columnar morphology found in the initial a-$Si_{1-x}C_x$:H sample, we have deduced that the graphitization takes place on the column surface and is aligned roughly normal to the sample plane. This assumption agrees with the observed anisotropy of the CRD resonance magnetic field position and the conductivity anisotropy found in a-$Si_{1-x}C_x$:H annealed at high temperatures.

On the other hand, the *g*-value anisotropy found for the CRD EPR line at 37 GHz and 140 GHz is interpreted by the impact of demagnetizing fields, becoming considerable at the high spin concentration of the paramagnetic defects, higher MW frequencies, and low temperatures. We have determined that at 4.2 K the demagnetizing field in a-$Si_{1-x}C_x$:H film with T_{ann} = 950°C is 0.44 mT at Q-band and 1.1 mT at D-band. The study of the temperature variation of the integrated EPR line intensity from SiDB and CRD revealed that they do not follow the Curie-Weiss law, and their spin systems reveal ferromagnetic and superparamagnetic features, correspondingly.

The EPR spectra of a-SiC_xN_y films have shown also 3 paramagnetic defects with $S = 1/2$, however with other properties. One of them with $g = 2.0033$ was assigned to the CDB localized within the a-SiC_xN_y sample. Two others are due to the interface defects, including P_{b0} defect having $g_{\|} \sim 2.0020$ and $g_{\perp} \sim 2.0090$ corresponding to threefold-coordinated SiDB of the form $Si_3{\equiv}Si\cdot$ and Si interface center with the trapped hole at $g = 2.05$. The impact of the nitrogen inclusion on the spin density of both interface defects was not observed as well as the vacuum thermal annealing of the a-SiC_xN_y thin films up to 1000°C has no influence on the structural bonding of the paramagnetic centers.

From the Lorentzian lineshape and linewidth (~ 1.2 mT) the CDB EPR line was related to the free electron delocalized over the sp^2 C cluster. The rise of the CDB spin concentration was found with the rise of nitrogen content in the a-SiC_xN_y film. From the temperature variation of the linewidth and integrated intensity of the CDB EPR line, which follows Curie-Weiss law, we have supposed that the local antiferromagnetic coupling takes place in the spin system. The negative sign found for J supports this conclusion. The observed enhancement of the intensity ratio of D and G peaks (I_D/I_G) in Raman spectra with rising of nitrogen content corresponds to the rise of the disorder degree in the a-SiC_xN_y sample. In contrast to the a-SiC_xN_y thin films the ferromagnetic interaction found in the CRD spin system in a-$Si_{1-x}C_x$:H sample is the consequence of the higher concentration of the CRD (~10^{19} cm^{-3}) and long-range ordering of carbon clusters.

CONSENT FOR PUBLICATION

Not applicable.

CONFLICT OF INTEREST

The author (editor) declares no conflict of interest, financial or otherwise.

ACKNOWLEDGEMENT

The authors are thankful to Dr. V. Kulikovsky (Institute of Physics CAR, Prague, Czech Republic) for providing the *a*-SiCN thin films. The work was partly supported by GA ČR 13-06697P and MEYS SAFMAT LM2015088 and LO1409 projects.

REFERENCES

[1] Bullot, J.; Schmidt, M. Physics of Amorphous Silicon–Carbon Alloys. *Phys. Stat. Sol. (b).*, **1987**, *143*, 345-418.
[http://dx.doi.org/10.1002/pssb.2221430202]

[2] Sussmann, R.; Ogden, R. Photoluminescence and optical properties of plasma-deposited amorphous Si_xC_{1-x}alloys. *Philos. Mag. B Phys. Condens. Matter Electron. Opt. Magn. Prop.*, **1981**, *44*, 137-158.
[http://dx.doi.org/10.1080/01418638108222373]

[3] Liedtke, S.; Lips, K.; Bort, M.; Jahn, K.; Fuhs, W. Photoluminescence and optically detected magnetic resonance of a-$Si_{1-x}C_x$:H films. *J. Non-Cryst. Solids,* **1989**, *114*, 522-524.
[http://dx.doi.org/10.1016/0022-3093(89)90637-6]

[4] Ma, T.; Xu, J.; Chen, K.; Du, J.; Li, W.; Huang, X. Red–green–blue light emission from hydrogenated amorphous silicon carbide films prepared by using organic compound xylene as carbon source. *Appl. Phys. Lett.,* **1998**, *72*, 13-15.
[http://dx.doi.org/10.1063/1.120631]

[5] Niemann, J.; Bauhofer, W. Properties of a-Si_{1-x} C_x:H thin films deposited from the organosilane Triethylsilane. *Thin Solid Films,* **1999**, *352*, 249-258.
[http://dx.doi.org/10.1016/S0040-6090(99)00320-X]

[6] Vasin, A.; Kolesnik, S.; Konchits, A.; Rusavsky, A.; Lysenko, V.; Nazarov, A.; Ishikawa, Y.; Koshka, Y. Structure, paramagnetic defects and light-emission of carbon-rich a-SiC:H films. *J. Appl. Phys.,* **2008**, *103*, 123710.
[http://dx.doi.org/10.1063/1.2946446]

[7] Vasin, A.; Ishikawa, Y.; Rusavsky, A.; Nazarov, A.; Konchitz, A.; Lysenko, V. Photoluminescent properties of oxidized stochiometric and carbon-rich amorphous $Si_{1-x}C_x$:H films. *SPQEO,* **2015**, *18*, 63-70.

[8] Vasin, A.; Kolesnik, S.; Konchits, A.; Kushnirenko, V.; Lysenko, V.; Nazarov, A.; Rusavsky, A.; Ashok, S. Effects of hydrogen bond redistribution on photoluminescence of a-SiC:H films under thermal treatment. *J. Appl. Phys.,* **2006**, *99*, 113520.
[http://dx.doi.org/10.1063/1.2198935]

[9] Aubert, P.; von Bardeleben, H.; Delmotte, F.; Cantin, J.; Hugon, M. Electron-paramagnetic-resonance study of the(100)Si/Si3N4interface. *Phys. Rev. B,* **1999**, *59*, 10677-10684.
[http://dx.doi.org/10.1103/PhysRevB.59.10677]

[10] Chen, C.; Huang, C.; Lin, Y.; Chen, L.; Chen, K. The affinity of Si–N and Si–C bonding in amorphous silicon carbon nitride (a-SiCN) thin film. *Diamond Related Materials,* **2005**, *14*, 1126-1130.
[http://dx.doi.org/10.1016/j.diamond.2004.10.045]

[11] Fraga, M.A.; Maciel, H.S.; Massi, M.; Pessoa, R.S. Applications of SiC-Based Thin Films in Electronic and MEMS Devices. In: *Physics and Technology of Silicon Carbide Devices*; Hijikata Ed.; INTECH Open Access Publisher, **2012**; pp. 313-336.

[12] Tomasella, E.; Spinelle, L.; Bousquet, A.; Rebib, F.; Dubois, M.; Eypert, C.; Gaston, J.; Cellier, J.; Sauvage, T. Structural and Optical Investigations of Silicon Carbon Nitride Thin Films Deposited by Magnetron Sputtering. *Plasma Process. Polym.,* **2009**, *6*, S11-S16.

[http://dx.doi.org/10.1002/ppap.200930103]

[13] Tomasella, E.; Rebib, F.; Dubois, M.; Cellier, J.; Jacquet, M. Structural and optical properties studies of sputtered a-SiCN thin films. *J. Phys. Conf. Ser.,* **2008**, *100*, 082045. [http://dx.doi.org/10.1088/1742-6596/100/8/082045]

[14] Wang, J.; Lu, Y.; Shen, Y. Effect of nitrogen content on phase configuration, nanostructure and mechanical behaviors in magnetron sputtered SiC_xN_y thin films. *Appl. Surf. Sci.,* **2010**, *256*, 1955-1960. [http://dx.doi.org/10.1016/j.apsusc.2009.10.044]

[15] Sundaram, K.; Alizadeh, Z.; Todi, R.; Desai, V. Investigations on hardness of rf sputter deposited SiCN thin films. *Mater. Sci. Eng. A,* **2004**, *368*, 103-108. [http://dx.doi.org/10.1016/j.msea.2003.09.103]

[16] Kulikovsky, V.; Ctvrtlik, R.; Vorlicek, V.; Zelezny, V.; Bohac, P.; Jastrabik, L. Effect of air annealing on mechanical properties and structure of SiC_xN_y magnetron sputtered films. *Surf. Coat. Tech.,* **2014**, *240*, 76-85. [http://dx.doi.org/10.1016/j.surfcoat.2013.12.017]

[17] Messier, R.; Giri, A.; Roy, R. Revised structure zone model for thin film physical structure. *J. Vac. Sci. Technol. A,* **1984**, *2*, 500-503. [http://dx.doi.org/10.1116/1.572604]

[18] Liang, E.; Zhang, J.; Leme, J.; Moura, C.; Cunha, L. Raman analysis of Si–C–N films grown by reactive magnetron sputtering. *Thin Solid Films,* **2004**, *469-470*, 410-415. [http://dx.doi.org/10.1016/j.tsf.2004.09.002]

[19] Stoll, S.; Schweiger, A. EasySpin, a comprehensive software package for spectral simulation and analysis in EPR. *J. Magn. Reson.,* **2006**, *178*(1), 42-55. [http://dx.doi.org/10.1016/j.jmr.2005.08.013] [PMID: 16188474]

[20] Vasin, A.; Kalabukhova, E.; Lukin, S.; Savchenko, D.; Lysenko, V.; Nazarov, A.; Rusavsky, A.; Koshka, Y. High-frequency electron paramagnetic resonance study of the as deposited and annealed carbon-rich hydrogenated amorphous silicon-carbon films. *Proc. MRS,* **2008**, *1069*, 175-180. [http://dx.doi.org/10.1557/PROC-1069-D07-11]

[21] Kalabukhova, E.; Lukin, S.; Savchenko, D.; Shanina, B.; Vasin, A.; Lysenko, V.; Nazarov, A.; Rusavsky, A.; Hoentsch, J.; Koshka, Y. *EPR study of carbon and silicon related defects in carbon-rich hydrogenated amorphous silicon-carbon films.,* **2010**. [http://dx.doi.org/10.1103/PhysRevB.81.155319]

[22] Brodsky, M.; Title, R. Electron Spin Resonance in Amorphous Silicon, Germanium, and Silicon Carbide. *Phys. Rev. Lett.,* **1969**, *23*, 581-585. [http://dx.doi.org/10.1103/PhysRevLett.23.581]

[23] McMahon, T.; Xiao, Y. Electron spin resonance study of the dangling bond in amorphous Si and porous Si. *Appl. Phys. Lett.,* **1993**, *63*, 1657-1659. [http://dx.doi.org/10.1063/1.110727]

[24] Giorgis, F.; Tagliaferro, A.; Fanciulli, M. 'Defects' and their detection in a-C and a-C:H.*Amorphous Carbon: State of the Art*; Silva, S.R.; Robertson, J.; Milne, W.I.; Amaratunga, G.A.J., Eds.; World Scientific: Singapore, **1998**, pp. 143-162.

[25] Christidis, T.; Tabbal, M.; Isber, S.; El Khakani, M.; Chaker, M. Comparative EPR study of hydrogenated and unhydrogenated amorphous silicon carbide thin films. *Appl. Surf. Sci.,* **2001**, *184*, 268-272. [http://dx.doi.org/10.1016/S0169-4332(01)00507-4]

[26] Wagoner, G. Spin Resonance of Charge Carriers in Graphite. *Phys. Rev.,* **1960**, *118*, 647-653. [http://dx.doi.org/10.1103/PhysRev.118.647]

[27] von Bardeleben, H.; Cantin, J.; Zeinert, A.; Racine, B.; Zellama, K.; Hai, P. Spins and microstructure

of hydrogenated amorphous carbon: A multiple frequency electron paramagnetic resonance study. *Appl. Phys. Lett.*, **2001**, *78*, 2843-2845.
[http://dx.doi.org/10.1063/1.1370980]

[28] Vasin, A.; Muto, S.; Ishikawa, Y.; Rusavsky, A.; Kimura, T.; Lysenko, V.; Nazarov, A. Comparative study of annealing and oxidation effects in a-SiC:H and a-SiC thin films deposited by radio-frequency magnetron sputtering. *Thin Solid Films*, **2011**, *519*, 2218-2224.
[http://dx.doi.org/10.1016/j.tsf.2010.11.005]

[29] Hasegawa, H. Dynamical Properties of s-d Interaction. *Progr. Theoret. Phys. (Kyoto)*, **1959**, *21*, 483-500.
[http://dx.doi.org/10.1143/PTP.21.483]

[30] Kittel, C. *Introduction to solid state physics*, 1st ed; Wiley: Hoboken, NJ, **2004**.

[31] Druz, B.; Zaritskiy, I.; Yevtukhov, Y.; Konchits, A.; Valakh, M.; Kolesnik, S.; Shanina, B.; Visotski, V. Filtered Cathodic Arc Deposited Diamond-Like Carbon: Electron Spin Resonance (ESR) and Raman Spectroscopy. *Proc. MRS*, **1999**, *593*, 249-254.
[http://dx.doi.org/10.1557/PROC-593-249]

[32] von Bardeleben, H.; Cantin, J.; Zellama, K.; Zeinert, A. Spins and microstructure of amorphous carbon. *Diamond Related Materials*, **2003**, *12*, 124-129.
[http://dx.doi.org/10.1016/S0925-9635(03)00013-X]

[33] Korenblit, I.; Shender, E. Ferromagnetism of disordered systems. *Sov. Phys. Usp.*, **1978**, *21*, 832-851.
[http://dx.doi.org/10.1070/PU1978v021n10ABEH005686]

[34] Makarova, T. Magnetic properties of carbon structures. *Semiconductors*, **2004**, *38*, 615-638.
[http://dx.doi.org/10.1134/1.1766362]

[35] Savchenko, D.; Kulikovsky, V.; Vorlíček, V.; Lančok, J.; Kiselov, V.; Kalabukhova, E. Optical and magnetic resonance study of a-SiC_xN_y films obtained by magnetron sputtering. *Phys. Stat. Solidi B*, **2014**, *251*, 1178-1185.
[http://dx.doi.org/10.1002/pssb.201451041]

[36] Kausteklis, J.; Cevc, P.; Arčon, D.; Nasi, L.; Pontiroli, D.; Mazzani, M.; Riccò, M. *Electron paramagnetic resonance study of nanostructured graphite.*, **2011**.
[http://dx.doi.org/10.1103/PhysRevB.84.125406]

[37] Trassl, S.; Motz, G.; Rössler, E.; Ziegler, G. Characterization of the Free-Carbon Phase in Precursor-Derived Si-C-N Ceramics: I, Spectroscopic Methods. *J. Am. Ceram. Soc.*, **2004**, *85*, 239-244.
[http://dx.doi.org/10.1111/j.1151-2916.2002.tb00072.x]

[38] Prasad, B.; Sato, H.; Enoki, T.; Hishiyama, Y.; Kaburagi, Y.; Rao, A.; Eklund, P.; Oshida, K.; Endo, M. Heat-treatment effect on the nanosized graphite π-electron system during diamond to graphite conversion. *Phys. Rev. B*, **2000**, *62*, 11209-11218.
[http://dx.doi.org/10.1103/PhysRevB.62.11209]

[39] Erdem, E.; Mass, V.; Gembus, A.; Schulz, A.; Liebau-Kunzmann, V.; Fasel, C.; Riedel, R.; Eichel, R.A. Defect structure in lithium-doped polymer-derived SiCN ceramics characterized by Raman and electron paramagnetic resonance spectroscopy. *Phys. Chem. Chem. Phys.*, **2009**, *11*(27), 5628-5633.
[http://dx.doi.org/10.1039/b822457a] [PMID: 19842480]

[40] Cantin, J.; Schoisswohl, M.; von Bardeleben, H.; Rochet, F.; Dufour, G. Pb1 defect study and chemical characterization of the Si(001)-SiO_2 interface in oxidized porous silicon. *Surf. Sci.*, **1996**, *352-354*, 793-796.
[http://dx.doi.org/10.1016/0039-6028(95)01230-3]

[41] Mrozowski, S. Specific heat anomalies and spin-spin interactions in carbons: A review. *J. Low Temp. Phys.*, **1979**, *35*, 231-298.
[http://dx.doi.org/10.1007/BF00115580]

[42] Nishi, Y. Study of Silicon-Silicon Dioxide Structure by Electron Spin Resonance I. *Jpn. J. Appl.*

Phys., **1971**, *10*, 52-62.
[http://dx.doi.org/10.1143/JJAP.10.52]

[43] Andronenko, S.; Stiharu, I.; Misra, S. *Synthesis and characterization of polyureasilazane derived SiCN ceramics.,* **2006**.
[http://dx.doi.org/10.1063/1.2202291]

[44] Macfarlane, J.; McLintock, I.; Orr, J. Antiferromagnetism in Amorphous Carbon Films. *Phys. Stat. Solidi (b).,* **1973**, *55*, 273-278.
[http://dx.doi.org/10.1002/pssb.2220550128]

SUBJECT INDEX

Dariya Savchenko & Abdel Hadi Kassiba (Eds.)

F

G

H

I

L

M

N

O

P

Q

R

S

T

V

W

X

Z

Made in United States
North Haven, CT
04 November 2026

21727477R00168